AF478898

Hydrogeologic Models of Sedimentary Aquifers

Edited by

Gordon S. Fraser, Indiana Geological Survey, Bloomington, IN
and
J. Matthew Davis, University of New Hampshire, Durham, NH

Copyright 1998 by
SEPM (Society for Sedimentary Geology)

Robert W. Dalrymple, Editor of Special Publications
Concepts in Hydrogeology and Environmental Geology No. 1

Tulsa, Oklahoma, U.S.A   April, 1998

SEPM thanks the following for their generous contributions in support of color figures in
Hydrogeologic Models of Sedimentary Aquifers:

U.S. Department of Energy Savannah River Site
Aiken, South Carolina

and

NIEHS Superfund Grant (ES-04699)

ISBN   1-56576-052-2

PREFACE

Research into the mathematical description of the behavior of groundwater and contaminants in both large- and small-scale problems requires the development of methods to describe more accurately and precisely the spatial variations of aquifer properties. In this volume we intend to show how sedimentological and stratigraphic concepts can be used in hydrologic research to accomplish this task and perhaps to establish a framework of cooperation between the two disciplines. Sedimentologists and petroleum engineers have collaborated for more than 30 years to conceptualize the spatial distribution of petrophysical properties of petroleum reservoirs in order to optimize production. Following the pioneering work of Mary P. Anderson in the early 1980s, many hydrologists and considerably fewer sedimentologists have conducted research aimed at integrating various geometric, geostatistical and process-response models to realistically simulate aquifer properties, groundwater flow and contaminant transport. Although only in the last decade have these models been applied to groundwater supply and protection, they have been particularly useful in visualizing contaminant transport in heterogeneous aquifers, and they have been applied to some of the most critical cleanup sites in the United States.

The emergence of realistic models of aquifer heterogeneity is attributable in equal measure to advances in computing capacity and software design, and the necessity for quantitative predictions. Early models were based on crude two-dimensional representations of aquifers and aquifer systems because insufficient computing power forced hydrologists to grossly simplify assumptions about the geometry, hydraulic properties and boundary conditions of groundwater systems. The real world, however, is complex, three-dimensional, heterogeneous and commonly anisotropic. Therefore, most hydrologists agree that developing a realistic conceptual model of aquifer properties is an important first step toward successful numerical simulation of groundwater flow and contaminant transport.

Toward this end, sedimentologists can work with hydrologists in several ways. First, together they can develop conceptual models that can be used to portray accurately the geometry, boundary characteristics and heterogeneity of aquifers and aquifer systems. Classic facies models developed during the latter half of the 20th century can be used for groundwater exploration, but tapplied hydrologic research has shifted decidedly toward detailed hydrostratigraphic characterization. The need for sophisticated, quantitative descriptions of the three-dimensional distributions of hydraulic parameters, such as porosity and permeability, should drive a renewed interest in the sedimentologic community toward more accurate and detailed delineation of the geometry of internal sedimentary facies and bedding architecture of hydrostratigraphic units, and, second, accurate quantification of sedimentologic parameters that control hydraulic properties. Genetic facies models can be used to define macroscale characteristics, while such tools as ground-penetrating radar and three-dimensional geophysical data can be used to map, interpret and quantify mesoscale heterogeneity within hydrostratigraphic units. Outcrop studies, well-log and core analysis, and laboratory measurements provide detailed petrophysical data for analysis of microscale heterogeneity.

Second, sedimentologists also can work with hydrologists to provide spatially distributed data rather than point data for hydrogeologic analysis. Data from slug tests, pump tests and laboratory conductivity are expensive to obtain and, therefore, are not commonly available in sufficient quantity for accurate hydrologic characterization. Over the past two decades some hydrologists have begun using known or inferred spatial variation of facies parameters as a surrogate for defining spatial distribution of hydrologic parameters. It is within this process that cooperation between hydrologists and sedimentologists can be fruitful. To be useful in the evaluation of water resource potential and movement of contaminants, however, this information must be quantitative and spatially distributed for geohydrologic analysis. Fortunately, a large body of information on modern sand bodies and outcrop analogs is available for use in simulations to predict character, distribution and hierarchy of parameters that control fluid flow and transport in aquifers.

Finally, sedimentologists can work with hydrologists to calibrate the numerical models. Hydraulic parameters (hydraulic conductivity, storage coefficient, recharge rate) within the model units must be calibrated or adjusted until results begin to match field observations (hydraulic heads in wells, flow rates of groundwater discharging to springs streams, or wells). Sedimentologists can use their experience, observations and judgment to make adjustments to the conceptual model and its inputs.

This volume presents reviews and case studies of how sedimentologic information can be used to solve hydrologic problems, including methods of integrating sedimentologic and hydrologic models to diffusive flow, use of sedimentologic information to discretize and parameterize nonhomogeneous aquifers, and examples of how three-dimensional sedimentologic and stratigraphic information and hydrologic data can be integrated and portrayed on maps and in cross sections. We hope that these research results will serve to alert hydrologists to the importance of stratigraphic and sedimentologic controls on the location and characteristics of aquifers; and to inform stratigraphers and sedimentologists about the types and quality of information that hydrologists require in their analyses of the resource potential and sensitivity to contamination of aquifers. This volume is the product of the interaction of many people. The authors, of course, are due special thanks for contributing papers to this experimental venue. Norman Smith, Gail Ashley and Mary Anderson were especially generous with their support and advice during the planning and completion of this volume. Peter Scholle, Dana Ulmer Scholle, Tricia Auberle and Andrea Ash at SEPM showed great patience during the time this volume has been in preparation. Finally, the reviewers listed below generously gave their time and experience in providing exceptionally thorough and helpful reviews of the manuscripts.

Greg Olyphant	John H. Doveton
Steven Brown	Chris Murray
Steven Baedke	Gary A. Smith
Thomas Prickett	Chris Paola
Tim D. Scheibe	Erik K. Webb
Todd Thompson	Gordon M. Sturgeon
Erik Kvale	Graham E. Fogg
Todd Rayne	

Gordon S. Fraser and J. Matthew Davis, Editors
November 1997

TABLE OF CONTENTS

PREFACE..iii

INTRODUCTION
Simulation of the Spatial Heterogeneity of Geologic Properties: An Overview
Erik K. Webb and J. Matthew Davis..1

PAPERS
Using Glacial Terrain Models to Define Hydrogeologic Settings in Heterogeneous Depositional Systems
Anthony H. Fleming..25

Using Glacial Terrain Models to Characterize Aquifer System Structure, Heterogeneity and Boundaries in an Interlobate Basin, Northeastern Indiana
Anthony H. Fleming..47

Using Three-Dimensional Geologic Models to Map Glacial Aquifer Systems: An Example from New Jersey
Scott D. Stanford and Gail M. Ashley..69

Characterizing Aquifer Heterogeneity within Terrigenous Clastic Depositional Systems
William E. Galloway and John M. Sharp, Jr. 85

Hydrogeology and Characterization of Fluvial Aquifer Systems
William E. Galloway and John M. Sharp, Jr. 91

Simulation of Geologic Patterns: A Comparison of Stochastic Simulation Techniques for Groundwater Transport Modeling
Timothy D. Scheibe and Christopher J. Murray..107

A Method for Characterizing Hydrogeologic Heterogeneity Using Lithologic Data
Gregory P. Flach, L. Larry Hamm, Mary K. Harris, Paul A. Thayer, John S. Haselow and Andrew D. Smits..119

Geostatistical Analysis of Facies Distributions: Elements of a Quantitative Hydrofacies Model
David F. Dominic, Robert W. Ritzi, Jr., Edward C. Reboulet and Amber C. Zimmer................................137

Geostatistical Simulation of Hydrofacies Architecture: A Transition Probability/Markov Approach
Steven F. Carle, Eric M. LaBolle, Gary S. Weissmann, David Van Brocklin and Graham E. Fogg..147

Combining Geologic Information and Inverse Parameter Estimation to Improve Groundwater Models
Eileen P. Poeter and Sean A. McKenna..171

SIMULATION OF THE SPATIAL HETEROGENEITY OF GEOLOGIC PROPERTIES: AN OVERVIEW[1]

ERIK K. WEBB[2] AND J. MATTHEW DAVIS[3]

2Power Reactor and Nuclear Fuel Development Corp., Ibaraki-ken, Japan
3Department of Earth Sciences, University of New Hampshire, Durham, New Hampshire 03824

"A major obstacle to performing reliable predictions of plume evolution and to the design of effective detection, monitoring and aquifer remediation strategies is (understanding) the natural heterogeneity of geological materials." The difficulties associated with quantifying both the "nature and degree (of heterogeneity) ... are immense" (Sudicky and Huyakorn, 1991, p. 240).

For those thoroughly immersed in the field of hydrogeology, the purpose of this compilation of papers should be evident. However, for those who are approaching the simulation of spatial heterogeneity of geologic or hydrologic properties for the first time, or who come with more of a sedimentological viewpoint, the overall objectives, successes and remaining challenges may be difficult to place in proper context. This introduction is intended to help the second audience.

For petroleum geologists, the goals and activities here will sound familiar. Hydrogeologists have closely followed the lead of petroleum geologists and engineers in the development of our understanding of fluid flow and solute transport through heterogeneous systems. They differ in several important aspects, however. First, the physical and chemical processes are different; therefore, the important aspects of the heterogeneity differ. Fortunately, groundwater flow and miscible solute transport are much simpler than the multiphase systems in petroleum reservoirs. Spatial and temporal scales also differ. Aquifers typically are smaller, geologically younger (more accessible) and less altered by structural and diagenetic processes. These characteristics make the task of characterizing aquifer heterogeneity much more tenable; however, groundwater problems typically require higher resolution.

The first part of this introduction provides a foundation by discussing the overall objectives of this type of simulation, how this work supports decision making in

hydrogeology, and the current state of the science. This is followed by brief synopses of the individual papers in this volume. We conclude with a summary of some current research directions.

CONTEXT

What is Spatial Geologic Simulation?

In general we can define spatial geologic simulation as a group of conceptual and mathematical approaches used to represent the arrangement of geologic properties in one-, two- or three-dimensional space. The properties of interest range from those familiar to the geologic community, including geologic or sedimentological characteristics, age, consolidation, measurable geophysical characteristics and so forth, to those critical to hydrogeologic analysis, such as porosity, permeability, storativity and geochemical characteristics. Due to the significance and difficulty (both will be discussed later) of accurately estimating these properties, a plethora of approaches has been proposed. The reader is referred to Koltermann and Gorelick (1996) for a comprehensive and well-illustrated review of many spatial simulation techniques. Our purpose here is to provide a foundation for the papers in this volume set for readers who are less familiar with hydrologic literature.

In order to understand these papers, one must understand that the goal is to explicitly include geologic information, over and above direct field measurements of properties, to constrain the results of spatial geologic simulation and thus better constrain the subsequent hydrologic predictions. The importance of including geologic information and potential approaches for doing so were not always evident. However, a handful of individuals have

---

1. This paper is intended to provide the reader with an overview of the subject rather than a comprehensive review. We have tried to include numerous examples and citations, but the list is by no means exhaustive. For a more comprehensive review of simulation approaches, the reader is referred to Koltermann and Gorelick (1996).

championed this concept (Anderson, 1989, 1990; Fogg, 1986, 1989; Phillips et al., 1989), which is now generally accepted. With this characteristic in mind, we can view the spectrum of tools along a continuum between two end members.

Purely Mathematical Approaches.—

One end of the spectrum of approaches is a methodology based entirely on mathematical (and statistical) theory. In the extreme, estimation tools of this type are not constrained by an understanding of geologic processes. Such approaches are, in general, easier to apply, conceptually easier to explain and use direct measurements of the properties of interest. However, the tools provide no direct method for constraining the results by geologic knowledge. This is not to say that the user might not attempt to match reasonable geologic characteristics. This characteristic is both a positive and a negative attribute. On the positive side, an understanding of the underlying geological processes often is lacking or at least difficult to quantify. Thus, a geology-less tool can be applied in the absence of such geologic process constraints. The negative is that geologic knowledge, which is often plentiful, cannot be used directly to constrain the resulting simulations. Thus, a level of detail may be lost, or the simulation results may be geologically unrealistic.

Purely Process Description Approaches.—

The other end member on this spectrum involves understanding in minute detail the physical or chemical characteristics of geologic processes and simulating these complexly interacting processes at fine resolution over large spatial scales. This approach has some scientific appeal in that it can use the vast store of knowledge of geologic processes, and we can investigate the impact of adding or subtracting geologic processes, etc. The negatives are, however, numerous. First, such complex models with high resolution and large space or time scales require immense computational resources. Second, they are difficult to condition in the forward modeling sense to site-specific measured property values. Finally, their results are chaotic in that small changes in the input may propagate through the simulation to produce dramatic changes in the output; thus, they are difficult to calibrate to known site-specific conditions.

Mixed Approaches.—

Both of these rather extreme approaches have been applied over a period of years. However, we are finding value in the middle ground. While it is difficult to use approaches based purely on geologic processes, we know that some of these geologically based patterns are important. For example, when units of high hydraulic conductivity are connected, as we expect in some fluvial deposits, they form preferred flow paths. In such cases, lumped or effective hydraulic parameters are inadequate to describe the pattern of hydrogeologic behavior of the system (Winograd and Pearson, 1976; Fogg, 1986; Anderson, 1990; Desbarats, 1990). The quest then is, first, to distill the essential geological elements (Miall, 1977, 1985; Allen, 1983; Ethridge et al., 1987) from knowledge of geologic processes and the resulting patterns of heterogeneity acquired through many years of careful geologic evaluation (e.g., Leopold and Wolman, 1957; Schumm, 1968; Williams and Rust, 1969; Kessler and Cooper, 1970; Smith, 1970; McDonald and Banerjee, 1971; Wright et al., 1974; Church and Gilbert, 1975; Boothroyd and Ashley, 1975; Miall, 1977; Mosley, 1982; Ashmore, 1991; Brierley, 1991; and many more). Then, we must incorporate those essential elements into quantitative models of aquifer heterogeneity. In many respects the goals are the same for the sedimentologist and hydrogeologist. Both are interested in spatial distribution of geologic materials. The sedimentologist uses the information to infer characteristics of the depositional system, and the hydrogeologist uses the information to predict fluid flow and contaminant migration. The hope is to identify the geologic elements that are most essential for hydrologic processes and develop techniques that enable the sedimentologist's understanding of depositional processes and products to be quantitatively related to the hydrogeologist's need for a suite of three-dimensional images of aquifer properties. The resolution of such images needed for hydrologic forecasting will vary depending on the type of problem and the desired level of confidence in the prediction.

Data Needs of Hydrologic and Contaminant Transport Models

Since the ultimate goal is to solve a hydrogeologic problem, we will briefly review the salient points of hydrogeologic modeling. The reader is referred to the suite of excellent hydrology texts for further explanation and definition of terms (e.g., Fetter, 1994; Anderson and Woessner, 1992; Domenico and Schwartz, 1990; Freeze and Cherry, 1979). Fluid-flow and contaminant transport simulation begins with the formulation of a conceptual model of the system through careful evaluation of all available information, including geologic and hydrologic field data. This conceptual model will include a description of the processes of fluid flow and contaminant migration, location and type of hydrological or chemical boundaries and estimates of the spatial distribution of internal hydrological parameters.

Once the conceptual model has taken form, one must select a numerical representation. We can divide the available approaches into three general classes of models: analytical, numerical and stochastic. In all cases, we represent the hydrologic processes with a governing (differential) equation. The equation describing groundwater flow is often called the hydrodynamic equation, and that describing contaminant transport is called the advection-dispersion equation. The difference between analytical and numerical models is in the method used to solve these governing equations. Analytical models, by assuming regular behavior over finite or infinite areas in space and time, are able to represent either fluid flow or contaminant migration over a given area by a single well-defined equation. Some examples include the Theis solution for flow to a well and the numerous equations for concentration given in many introductory texts. In contrast, numerical models break the area into individual discrete pieces (whose shape depends on the model type) and then represent each small segment by an individual equation. Since these regions are all physically interconnected, the suite of equations that represent them must be solved simultaneously, which usually is done through iterative matrix solution techniques. In numerical models the boundaries and properties must be translated to a spatially arranged grid defined by nodes that represent the spatial arrangement of calculated values within the flow model. This process of transferring a spatial map of some parameter to a grid is known as "discretization" (breaking the continuum of information into several discrete pieces). This discrete form, the required format for numerical flow models, brings with it a number of associated problems that fall under the heading of "upscaling," which will be discussed later. The shape of the grid is dependent on the type of numerical flow code. The most common example of a numerical flow model is MODFLOW (McDonald and Harbaugh, 1988). MT3D (Zheng, 1990) is an example of a numerical model that simulates contaminant migration. Analytical solutions are immensely faster than numerical solutions and provide very accurate results, provided the underlying assumptions are valid. Numerical solutions are plagued with problems of accuracy, large computer power requirements and larger data needs, but they provide a much more detailed representation of the flow or transport system.

The third class of models differs from traditional (deterministic) analytical and numerical models primarily in that they take a probabilistic view of the system (Dagan, 1989; Gelhar, 1986). Instead of predicting a single value of concentration or head, the prediction is one of the "expected concentration" or the "expected head" and the associated uncertainty in the estimate. This class of models also differs from traditional deterministic models in that the hydrologic parameters (e.g., hydraulic conductivity) are assumed to vary in space, and their value at any given location is uncertain. Conceptually, these models consist of a large number of equally possible aquifer configurations, and the prediction of the expected head or concentration represents the average over the ensemble of possibilities. In fact, deterministic numerical models are often used in a "Monte Carlo" analysis to evaluate the probability that a contaminant will reach a receptor within a specified time. The Monte Carlo approach involves compiling a large number of equally probable scenarios and calculating the "expected outcome" from the ensemble. The more we know about the system, the more similar the possible aquifer configurations will be. Thus, the possible outcomes are more similar, and our uncertainty is reduced. One of the most important findings in the development of stochastic models is that we can account for aquifer heterogeneity without knowing the hydraulic conductivity at every location in the aquifer.

If an analytical or numerical solution technique is adopted, the hydrogeologic modeling process usually involves two phases. The first phase is the development of the conceptual model, selection of a solution method and calibration of the model to known hydrologic conditions. The calibration involves the adjustment of parameters until the modeled variables (heads and/or concentrations) provide a reasonable match to those observed in the field. The calibrated model then is used in the second to predict how the system will respond to various natural and imposed conditions. Implementing the stochastic models follows similar steps but uses elaborate techniques that are beyond the scope of the this discussion.

Selecting the most appropriate modeling approach is not trivial, and no single approach has proven superior in all cases. The two main issues are how well the model represents the hydrologic processes of flow and transport and how well the hydrogeologic model represents aquifer heterogeneity. The latter certainly is relevant to this volume in the sense that much of the debate about hydrologic models revolves around the lack of understanding of geologic heterogeneity. Some of the models assume a smooth and continuous spatial variation of hydraulic conductivity with either a single characteristic length scale (e.g.. Gelhar and Axness, 1983; Dagan, 1988; Tompson et al., 1989; Tompson and Gelhar, 1990; Dykaar and Kitanidis, 1992a, b), a few scales of variation (e.g., Rajaram and Gelhar, 1995), or an infinite number of nested scales (e.g., Neuman, 1990). Others feel that discrete internal geological structures produce preferential flow paths that are not directly represented by these analytical continuum models (Winograd and Pearson, 1976; Fogg, 1986; Williams, 1988; Anderson, 1991). One unanswered question is, "Under what geologic conditions are the assumptions of the various models reasonable?"

Regardless of the type of model used, the quality of the predictions relies heavily on the quality of the hydrologic model, which in turn rests upon the accuracy of the conceptual model and the representation of the spatial distribution of hydrogeologic properties. Readers interested in details of modeling approaches may refer to several model indexes (Donley and Deschambault, 1993; Environmental Protection Agency, 1993a, b, 1994; van der Heijde and Kanzer, 1997; International Ground Water Modeling Center, 1994; Peterson, 1994; Moskowitz et al., 1992; and others).

It is important to understand that different physical or chemical transport processes may become important in certain types of contamination problems and thus change completely the suite of important parameters (properties). A good example of this is the migration of a Dense Non-Aqueous Phase Liquids (DNAPL). DNAPLs are a special type of waste composed primarily of chlorinated solvents, coal tar and other dense hydrocarbons. These materials are heavier than water and tend to sink in aquifers. As DNAPLs migrate downward, the path they take is governed by a combination of capillary pressure (a function of grain-size and pore-size distribution), fluids density, viscosity and total volume (saturation). Thus, sedimentary textures or structures defined by differences in grain-size distribution and packing arrangement become a controlling factor in both where and how fast a DNAPL is migrating through a system (Sudicky and Huyakorn, 1991; Kueper, 1989; Cherry, 1987; Webb et al., 1993). In many cases these parameters also can be related to sedimentological or other geological characteristics.

What is the Purpose of Spatial Geologic Simulation?

The array of geologic materials is vast. Their subsurface distribution is complex and ultimately unknowable. However, the quantitative prediction of groundwater flow and contaminant migration at a given site for the purpose of evaluating human safety and environmental impact, or developing engineering designs, usually requires specific descriptions of the spatial arrangement of geologic, hydrologic and/or engineering properties. The uncertainty in spatial distribution of hydrogeologic properties can be reduced with data in the form of specific point measurements or data collected along a continuous line, such as surface seismic or borehole geophysical logs. However, the volume of aquifer for which point data exist is small relative to its entire volume, and indirect geophysical measurements often are difficult to quantify in terms of hydrologic properties. We must then develop means of understanding and constraining our uncertainty. In addition, we must quantify our uncertainty for the decision-making process—when is the remaining uncertainty negligible?

Recognizing the need to represent spatial heterogeneity in greater detail, to better understand our uncertainty and, if possible, to limit it has led to many different approaches for spatial estimation of hydrogeologic properties. First, we look at the components of our uncertainty, then we can discuss methods for reducing it. In a series of papers on hydrogeological decision analysis, Freeze and others (Freeze et al., 1990, 1992; Massmann et al., 1991; Sperling et al., 1992; James and Freeze, 1993) suggested two primary components of uncertainty in a hydrogeologic simulation model. They refer to the first component as geological uncertainty, which encompasses uncertainty in the location of aquifers and aquitards, aquifer boundaries and interaction with bedrock, the sources and sinks of water. The second component, parameter uncertainty, includes uncertainty in the values of the hydrogeologic parameters in space (porosity, hydraulic conductivity, dispersivity and so on). These two types of uncertainty are closely related. One purpose for spatial simulation of geologic materials is to directly reduce geological uncertainty and indirectly reduce parameter uncertainty.

Direct reduction in geological uncertainty may be achieved if the continuity of stratigraphic units can be inferred from information about the depositional system. This inference can be drawn from various sources, each of which differs in the ability to quantify the uncertainty. The most direct means of inferring the preferential orientation of hydrostratigraphic units is through indicator statistics on intensive data sets. Various studies (e.g., Johnson and Dreiss, 1989; Davis et al., 1993; Johnson, 1995; Ritzi et al., 1995; Dominic et al., this volume) have found that preferential orientation of hydrostratigraphic units is consistent with that inferred from local geology. So local geology can be used in conjunction with data to parameterize the uncertainty models. Using indicator statistical techniques enables one to quantify the probability that a specific hydrogeologic unit exists at a specific (unmeasured) location given its presence elsewhere in the system. The studies cited above rely on the assimilation of large data sets to parameterize the statistical models. Another means of inferring spatial continuity of hydrostratigraphic units is the simulation of geologic materials by process simulation techniques. Multiple simulations provide a set of aquifer configurations that are all equally geologically plausible. Conceptually, the more geologic constraints on the system, the more similar are the realizations, and the more the uncertainty is reduced.

Indirect reduction of parameter uncertainty can be achieved by using geologic materials as proxies for hydrogeologic properties. The ability to define such correlation between geologic and hydrologic properties is a central assumption driving the inclusion of geologic data into spatial property simulation. A good example of this is the type of correlation that appears to exist between grain size distributions and estimates of hydraulic conductivity. Several empirical relationships (Hazen, 1893; Masch and Denny, 1966; Bear, 1972) have been derived to describe this relationship. The most important parameter in computing this relationship is some form of representative grain-size diameter (Freeze and Cherry, 1979). The widely known Kozeny-Carman equation (Bear, 1972) for conductivity, takes the form:

$$K = \left(\frac{\rho g}{\mu}\right)\left[\frac{n^3}{(1-n)^2}\right]\left(\frac{d_m^2}{180}\right),$$

where ρ is fluid density, g is gravitational acceleration, μ is viscosity, n is porosity and d_m is a representative grain size. Todd (1980) modified this basic equation to account for various grain shapes and packing. Panda and Lake (1994) re-derived the Kozeny-Carman equation using full grain-size distribution characteristics instead of a single representative grain size. Koltermann and Gorelick (1995) extended the Kozeny-Carman equation to sediments composed of mixtures of gravel, sand and mud. They suggested that the representative diameter, dm, should be either the harmonic or geometric mean of particle sizes, depending on the amount of fines present. Panda and Lake (1995) extended their previous work (Panda and Lake, 1994) to include the effect of cementation on permeability. Their findings suggest that small amounts of cement can significantly reduce permeability. Estimates of hydraulic conductivity require not only understanding of how grain size varies in space but also the spatial distribution of pore-filling cements, when they are present. Spatial geologic simulation techniques can be applied to simulate the spatial distribution of geologic properties, which provides a basis for estimating hydrogeologic properties.

In addition to point values of hydraulic conductivity, some hydrologic models require statistical characteristics of hydraulic conductivity such as its probability density function and correlation structure. Davis (1994) suggested that basic characteristics of the probability density function can be inferred from the types and relative abundance of lithologic facies in a particular setting. Lemons and Chan (pers. commun., 1997) found that the probability distribution of hydraulic conductivity in Pleistocene deltaic deposits is controlled by the types and abundance of lithofacies and is not log-normally distributed, as is often assumed. Davis et al. (1993, 1997) also found that spatial correlation of hydraulic conductivity is closely related to the dimensions of the lithologic facies and facies assemblages. Results of spatial geologic simulations can also help us better understand the spatial statistical characteristics of geologic systems (Scheibe, 1993).

Summary

In summary, detailed and accurate description of groundwater flow or migration of contaminants requires an understanding of spatial distribution of aquifer properties. We rarely have detailed two- or three-dimensional data upon which to base these estimates; instead, we have sparse and widely scattered point-specific data. For numerical simulations, the estimates of properties must be defined for a dense, relatively regular grid of points in one-, two-, or three-dimensional space. The parameters of greatest importance vary with the processes being represented in the contaminant migration simulations. The uncertainty in the aquifer configuration results in uncertainty in our predictions. Following Freeze and others we can classify our predictive uncertainty as a result of uncertainty in the geological characteristics of the aquifer and our uncertainty in the hydrogeologic parameters used in our calculations. The question that remains is, "to what extent can geological and parameter uncertainty be quantitatively reduced by the incorporation of geological information?" There appears to be great promise on both counts. As a supplement to the papers in this volume, we now review some of the significant findings and current activities in this area of research.

WHAT IS THE CURRENT STATE OF THE SCIENCE?

The quest for adequate methods to quantify the spatial arrangement of hydrogeological properties is neither new (Dagan, 1982) nor complete (Yarus and Chambers, 1994; Forster and Merriam, 1996). This discussion focuses on how geological information might be used to reduce the uncertainty in our hydrogeologic models. However, to fully understand the current state of the science, one must be aware of the field-based techniques and experimental activities under way. Thus, we begin with an overview of field investigations, geophysical tools, tracer test experiments and categorization of tools, followed by a description of several major categories of simulation tools.

Field Investigations

Field investigation of geologic properties is as old as the science of geology itself. Most investigations focus on one of two aspects—(1) finding and understanding the arrangement of mineral resources, or (2) interpretation of a time series of processes resulting in the observed distribution of geologic materials. The body of knowledge developed through both types of studies, which are in fact closely related, provides the foundation for current works. If we consider our two sources of uncertainty, two aspects of outcrop studies are particularly useful. First and foremost, outcrop studies further our understanding of the nature of geologic materials as permeable media. The geometry of geologic boundaries, the connectivity of high- and low-permeability units and the hierarchical nature of sedimentary features are all of critical importance in accurately predicting subsurface fluid flow and solute transport. In addition, understanding the quantitative relationships among geological characteristics and hydrogeologic parameters is essential for our parameter uncertainty.

A group at the University of Texas (Austin) conducted one of the most comprehensive studies of outcrop heterogeneity on the Page Sandstone (Jurassic) in northern Arizona. They found notable permeability differences among the three eolian units: wind ripple, grain fall and grain flow (Goggin et al., 1988). Chandler et al. (1989) also noted that the units exhibit different degrees of anisotropy and attributed these differences to bedding characteristics. Kocurek et al. (1991) provided a regional depositional context for the Page Sandstone and application of the permeability data. Goggin et al. (1992) then developed an algorithm for simulating random permeability fields that honors the geological characteristics observed in outcrop.

The study on the Page Sandstone identified, as many have suggested, multiple scales of variability that are associated with the laminae, sets and cosets in the eolian system.

Other studies have found similar hierarchical heterogeneity in fluvial systems. For example, Jordan and Pryor (1992) identified six such levels in their study of a Mississippi River meander belt. Davis et al. (1993, 1997) noted at least three scales of variability associated with sets, cosets and facies assemblages (architectural elements) in the fluvial/interfluvial Sierra Ladrones Formation, New Mexico. While multiple scales of heterogeneity have been identified in various geologic settings, the hierarchical nature has been difficult to quantify. It appears that hierarchical bounding surfaces may provide a link between our geologic and quantitative descriptions (Davis et al., 1997).

Several field studies have also investigated the utility of geologic information for parameter estimation. Dreyer et al. (1990) found good correlation between permeability and lithologic facies in delta-plain distributary sand bodies of the Jurassic Ness Formation (England). However, the fact that Hurst and Rosvoll (1991) found more variation within a specific lithofacies than between suggesting that lithofacies descriptions alone are not good indicators of permeability. Jacobson and Rendall (1990) also found no correlation between lithofacies and permeability in the Jurassic Scalby Formation (England) and attributed the lack of correlation to diagenetic alterations. Delude et al. (1996) found that calcite cementation in a fluvial sandstone has a pronounced affect on heterogeneity.

Geophysical Tools

Numerous geophysical tools are available to help determine hydrological properties and geometries. These are usually categorized as either downhole or surface tools. The form of energy or contrasting property that is used to investigate the subsurface can also subdivide them. These include wave propagation to describe fractures (Hardin et al., 1987); azimuthal resistivity surveys to determine fracture directions of conductivity and porosity (Taylor and Fleming, 1988); borehole temperature and electrical conductivity to estimate hydraulic properties (Michalski, 1989, Silliman and Robinson, 1989; Tsang et al., 1990); surface seismic and self-potential to estimate water table positions (Birkelo et al., 1987; Birch, 1993); map bedrock surfaces (Ayers, 1989) and determine porosity (Sun et al., 1993); electrical resistivity to estimate specific yield (Frohlich and Kelly, 1988); identify waste sources (Ebraheem et al., 1990); delineate groundwater boundaries (Park et al., 1990) and other properties; as well as gamma-gamma and neutron logging to detect the presence of water (Poeter, 1988). Ground-penetrating radar has been used to identify water-table elevations and to delineate near-surface sedimentary unit boundaries (Beres and Haeni, 1991). In addition, cross-hole tomography has been used to produce two- and three-dimensional views of subsurface heterogeneity (Hyndman and Gorelick, 1996).

The sensitivity and the methods of application of these tools continue to expand. Ideally, we would be able to "see" through these remote sensing tools the actual distribution of subsurface materials. However, the interpretation of the signal (e.g., seismic wave energy) recorded by the instruments into a cohesive picture of subsurface materials is dependent on the "model" or concept applied for interpretation and the availability of ground truth or direct measurements of the properties of interest. Thus, the real value here is in combining geophysical data with other direct measurements and using them jointly to condition our interpretive simulation of the subsurface.

Tracer Test Experiments

The great value of tracer experiments lies in our ability to collect information on the integrated response of a hydrologic system, and thus better understand the properties controlling flow and transport over scales of tens to thousands of meters. The two principal issues of interest are how aquifer heterogeneity affects the dispersive process and how aquifer reactivity affects the movement of different chemical species. Gelhar et al. (1992) conducted a comprehensive review of 59 tracer tests conducted in a wide range of geologic materials (mostly sedimentary). Two of the most relevant findings of these experiments are (1) that aquifer dispersivity varies over six orders of magnitude, and (2) a weak correlation exists between the scale of the experiment and the magnitude of the dispersivity. It is generally accepted that the magnitude of dispersivity is proportional to aquifer heterogeneity. These findings suggest that different aquifers of the same scale exhibit significantly different (two to three orders of magnitude) dispersivities and that heterogeneity generally increases with scale. From a geologic perspective, these findings are not particularly new, but they offer challenges in quantifying the differences in geological heterogeneity and in providing as clear an explanation as possible of the hierarchical nature of geologic materials.

Brusseau (1994) reviewed issues that have arisen from the investigation of reactive transport in heterogeneous aquifers. Here the challenge is to understand not only how the spatial variability in physical properties (such as hydraulic conductivity) affects solute transport, but also how geochemical properties of the aquifer affect the transport of reactive solutes. Brusseau points out that much of the "nonideal" behavior observed in the field may be attributed to spatial variation of the aquifer's geochemical properties (i.e., chemical heterogeneity).

Categorization of Tools

Koltermann and Gorelick (1996) conducted the most current and broadest assessment of the array of simulation tools. They divided these tools into three categories: structure-imitating, process-imitating and descriptive. Structure-imitating approaches are not fundamentally based on geologic process knowledge but instead try to match observed patterns of properties. These are like the nongeologic end member described above. Process-imitating approaches attempt to include the ability to replicate how geologic processes work and are closer to the geologic-process end member discussed above. Descriptive approaches are essentially subjective mapping techniques, which are the classical interpretive approach used in geology. We highly recommend Koltermann and Gorelick's description for a detailed review.

Most tools currently being investigated lie somewhere between the two extremes defined by structure-imitating and process-imitating categories. The following discussion does not directly divide individual tools into these two categories, but instead loosely groups them according to the mathematical techniques used in their development. The following list of tools is by no means comprehensive. It does, however, provide a feeling for the variety of approaches and for the depth of detail currently attained.

Simulation or Spatial Map Development Tools

The breadth of spatial mapping and simulation tools is large. This discussion presents in simple terms some examples of the major classes or categories of approaches. Among these, the subjective geologic interpretation approach is most commonly used. It provides a best guess, based on geologic observation and individual judgment. Numerical simulations of hydraulic conductivity fields provide unbiased views of existing data, meaning that the assumptions used to correlate data in space are explicitly stated or are part of the underlying numerical tool. In addition, if the tool allows direct assessment of uncertainty through multiple realizations, the simulator can be used to address lack of knowledge and identify future data collection locations. Finally, these simulators or estimators provide a rapid means of translating discrete geologic measurements into two- and three-dimensional fields of point-based values that can be used in subsequent flow and transport simulation. In other words they provide the connection between field data and spatial estimates of properties used in hydrogeologic simulations.

The critical issue in reviewing the array of possible spatial estimation tools is, "How do we judge which tool is best or at least adequate for a specific geologic setting and hydrogeologic problem?" Answering this question is much more significant than just being able to categorize types of tools. It must be related to the type of output produced by a specific tool. For example, some simulators, by design, produce smooth, continuous fields with few, if any, discrete, drastic changes in properties. Others produce discrete internal structures similar to those observed in geological media.

Based on the criteria of making accurate matches to the property data, we can compare model simulations with field data through techniques like the jackknife (Tukey, 1958) or bootstrap approaches (Efron, 1982; Hall, 1985; Efron and Tibshirani, 1986). However, development of these tools is far from complete, which has prompted Cressie (1993, p. 498) to suggest that one possible research direction is "developing sensible model-selection criteria when the data are spatial." To date, no such criteria have been developed. Judging the value of a specific tool is best done in the context of how the output will be used to address scientific or decision-making issues. Given that there is no "best" approach, the following is a brief description of some of the more common groups of approaches.

Subjective.—
Koltermann and Gorelick referred to these as "Descriptive" tools. Here, we refer to the classical approach of using available direct field information and individual judgment to interpret the depositional system. This interpretation becomes the conceptualization or conceptual model used as the basis for all other spatial estimation activities. Thus, if our goal is to include geologic information as part of the extrapolation process, this is an indispensable step. Good examples of this process are presented in the papers by Fleming, Stanford and Ashley, and Galloway and Sharp in this volume.

One can either move from this conceptual model directly to more numerical estimation techniques described below, or continue with the classical spatial mapping and estimation procedure. Such a process involves both interpolation between data and extrapolation into areas where no data exist (Frind et al., 1987). One clear example is the process of subjective borehole log correlation, which is used to generate cross sections or fence diagrams. Obviously, it also applies to two-dimensional horizontal views of the distribution of geologic and/or hydrologic properties, geologic structure, elevation of the water table, location of geographical features, vegetative cover, soil type, soil chemistry and land use practices. Often this approach is appropriate. All can be judged by one of two criteria. First, if sufficient data exist that the uncertainty in the true map is small, then the mapping process is highly constrained by conditioning data, and all individual interpretations will be similar. This is particularly true for some forms of remote sensing data. Second, if the uncertainty is rather insignificant to the final decision-making process, the deterministic answer is sufficient.

It is important to recognize that the resulting map must be translated into a discrete numerical representation, which can be facilitated by maintaining the map within a numerical spatial data base or geographic information system. If spatial uncertainty is large, capturing this uncertainty usually requires multiple realizations, a process that is difficult to achieve by hand. Thus, we turn to numerical methods.

Regionalization.—
In the coarsest terms, regionalization is much like producing a geological map, except that unit contacts are defined statistically and therefore can represent uncertainty in location boundaries. Within these bounded regions, the value assigned to a given parameter can be either relatively uniform, a statistical distribution or a random field with various structures. Thus, a method must be devised for consistently defining the location of boundaries based on sparse information. The exactness (sharpness) of the estimated region boundaries is a function of the amount of data and their spatial location. The algorithm that describes the location of a boundary can be as simple as a linear interpolation, which would plot the boundary at an average distance between real data points, or the regionalization can assume varying degrees of complexity depending on the types of data available. For example, one might need to estimate the spatial arrangement of hydraulic conductivity based on a few conductivity measurements, a geologic unit map and measurements of water-table elevation. All three types of data should be related. Numerical techniques based on

multivariate statistical analysis can combine these values and help estimate the location of hydraulic conductivity regions (Bohling et al., 1990; Harff and Davis, 1990; Harff et al., 1990).

Geostatistics.—

"Geostatistics offers a way of describing the spatial continuity that is an essential feature of many natural phenomena and provides adaptations of classical regression techniques to take advantage of this continuity," according to Isaaks and Srivastava (1989, p. 3). Extremely good descriptions of the field of geostatistics can be found in Journel and Huijbregts (1978), Isaaks and Srivastava (1989), Deutsch and Journel (1992, 1997), Cressie (1993) and Srivastava (1994). Only a brief introduction is presented here.

Geostatistical techniques can be subdivided into two general classes: parametric and nonparametric. Parametric techniques and models assume a particular probability distribution for the parameter of interest. For example, hydraulic conductivity and transmissivity often are assumed to follow a log-normal distribution. These parametric models, often referred to as Gaussian models, are the basis for many stochastic models of flow and transport (e.g., Gelhar and Axness, 1983; Dagan, 1986; Tompson et al., 1989; Tompson and Gelhar, 1990, and many others). Nonparametric techniques and models make no assumptions regarding the underlying probability distribution. Following Journel (1983), most of the nonparametric methods in hydrology use an indicator random field approach.

Both types of models recognize the fact that hydrogeologic properties are often correlated in space. For example, if a sand unit is observed during drilling, similar material likely exists in the immediate vicinity of the observation. As distance from the observation increases, the likelihood of finding similar material decreases. This concept of spatial correlation is not new, but geostatistics enables us to better quantify our predictions and associated uncertainties. The decrease in strength of the spatial correlation between points as the distance increases is characterized by the experimental variogram. A variogram is a plot of the mean squared difference of observations as a function of the distance between observations. As the strength of the spatial correlation decreases, the magnitude of the mean squared difference increases. The distance at which the variogram levels off is referred to as the *range* and represents the maximum physical distance of spatial correlation. Spatial correlation also can be related to the direction of separation (statistical anisotropy). For example, permeability often is better correlated in the horizontal direction than the vertical. Variogram estimates obtained from data are often fitted with mathematical functions to model the spatial correlation structure.

Two other important concepts in geostatistical models are statistical homogeneity and ergodicity. Statistical homogeneity, a common assumption in applied geostatistics, means that statistical properties are constant in space. For example, if a system consists of three materials (e.g., clay, sand and gravel facies), the chance of encountering any one of the units does not change in space. Statistical homogeneity also applies to the spatial correlation structure. If the statistical properties vary in space, this nonstationary behavior is often referred to as a trend. In a sense, stationarity is a function of scale. If one is interested in the variation of a sand-shale system, the variation within each individual unit may be considered stationary. However, as the size of the domain increases to the point where one of each facies type is encountered, the variation becomes nonstationary. Then, as the size of the domain continues to increase to the point where many sand and shale units are present throughout the domain, the system becomes stationary once again, although with different statistical characteristics than at the scale of the individual facies. These multiple scales of variability are quite common in hydrogeologic problems and result in nonstationary behavior (e.g., Graham and McLaughlin, 1989a, b; Rajaram and Gelhar, 1995).

The ergodicic hypothesis is equally important but more abstract. Geostatistical and stochastic approaches are based on a large number of equally possible aquifer configurations (an ensemble), when in fact, there is only one aquifer, which has a fixed value of permeability at any given location. Use of stochastic theory requires that the statistical properties in any one realization be equal to the statistical properties of the ensemble. This is usually true when the size of the domain is much larger than the range of spatial correlation. The important point is that applicability of the numerous geostatistical and stochastic models relies heavily on an understanding of how geologic materials are correlated over multiple scales in space.

Another important issue in applying Gaussian models to a geologic environment is whether the relatively smooth variability can adequately reproduce discrete geologic units. Fogg (1986) illustrated the importance of discrete features in the conductivity field by showing that for a large-scale, three-dimensional aquifer system, the connection and continuity of high-conductivity paths controlled the circulation pattern and rates of groundwater flow. Scheibe and

Murray (this volume) investigate specifically the limitations of the Gaussian approach in systems that exhibit non-Gaussian characteristics. In regard to simulating aquifer heterogeneity with geostatistical methods, Deutsch and Journel (1992, 1997) provided a comprehensive set of codes that enable variogram analysis, kriging and simulation of both parametric (Gaussian) and nonparametric (indicator) random fields.

Fractal.—

Another mathematical concept used to bridge the gaps between sparse field data and to help estimate the spatial distribution of properties is known as fractal geometry. This fundamental concept assumes that the pattern of heterogeneity looks the same (either visually or mathematically) regardless of the scale of observation. If one viewed a series of maps at different scales (100 km, 1 km, 100 m, 1 m or 1 cm), the pattern of heterogeneity would be fundamentally the same. This set of concepts has been discussed in the hydrologic community (Wheatcraft and Tyler, 1988) and has received some acceptance.

Of the two major groups of conceptual approaches, one involves considering the frequency or density of discrete events or geographic pieces referred to as frequency-size distributions (Turcotte and Huang, 1995). Examples include estimation of grain-size distribution models (Ghilardi et al., 1993; Turcott, 1986), fracture characterization (Hewett, 1994) and determination of petroleum reservoir size (Barton and Scholz, 1995; La Pointe, 1995; Crovelli and Barton, 1995). The other major approach involves continuous distributions, such as landscape patterns (Mandelbrot, 1982), hydrogeologic property patterns (Neuman, 1990; Molz and Boman, 1995; Impey et al., 1996) and stratigraphy (Plotnick and Prestegaard, 1995). Both approaches require the adoption of a "power law" relationship having the basic form:

$$N_i = \frac{C}{r_i^D},$$

which represents the growth in the expected result or the count of the discrete objects (N) with a change in measurement or observation scale (r). The rate of growth is expressed by a factor (D), the fractal dimension. C is a constant of proportionality (Turcotte and Huang, 1995).

Once this relationship, which is based on a fractal dimension, is defined for a specific system, we are able to do spatial estimation for areas between conditioning data using Fourier transform (spectral) methods (Hewett and Behrens, 1990; Impey et al., 1996; and several others). Another approach uses what is called projection onto convex sets (Malinverno and Rossi, 1993). Recently, several commercial codes have been made available to perform such analyses (Impey et al., 1996; Dershowitz et al., 1995).

One of the main implications of these simulation approaches is rather simple. If a scaling law can be determined or assumed for a specific property at one scale, the model can be used over a wide range of scales. For example, data from a few outcrops could be used to model variability at the scale of the entire formation (aquifer). Naturally, there are upper and lower bounds of appropriate scales. That is, any scaling law of hydraulic conductivity can only be applied over scales larger than the representative elementary volume and smaller than the basin.

Geological Process and Geometric Simulation.—

Geological process models attempt to simulate the interaction of physical, chemical and/or biological processes embodied in either theoretical or empirical relationships related to the formation of geological materials. Purely geometrical models are not based on processes, but they attempt to produce spatial patterns similar to those observed in the field using empirically derived geometrical relationships. Either type of model can be used to produce multiple realizations of the spatial distribution of sediment and hydraulic properties for a given type of geologic setting. Simulating processes is more appealing since the equations are based on physical principles, but simulating geometry greatly reduces the computational requirements and allows models to simulate relatively high resolution over large spatial scales. The more powerful geometric models incorporate rules derived from an understanding of the geologic processes.

Early geometrical and process simulation models in geology generally had two goals. First, they were used to check the physical basis for the formation of small-scale sedimentary bedforms (10^{-2} to 10^{-1} m) (Rubin, 1987) and the local forces acting in river channels (10 m) (Bridge, 1977). Second, models assisted in estimating rates of megascale basin fill (10^3 to 10^5 m) and channel avulsion processes (Bridge and Leeder, 1979; Bridge and Diemer, 1983; Allen, 1978). These usually focused either exclusively on surface processes or just the resulting sedimentary deposits. One exception is a model for alluvial fans, developed by Price (1974), that simulated channel migration and used this information to estimate the resulting alluvial sedimentary deposition.

With increased interest in delineating spatial heterogeneity for petroleum and groundwater problems, several groups have tried to use models that are either process based, geometric based or hybrids, to delineate the sedimentary features that control flow at small scales (10^1 to 10^3 m) (Scheibe and Freyberg, 1990) and at regional scales ($>10^3$ m) (Juergens, 1994; Tetzlaff and Harbaugh, 1989; Koltermann and Gorelick, 1992).

Among the more process oriented are a code called SEDSIM (Tetzlaff and Harbaugh, 1989), in its much improved and modified form (Koltermann and Gorelick, 1992), and several two-dimensional basin-filling models. SEDSIM (Tetzlaff and Harbaugh, 1989) models surface water flow, erosion and deposition of sediment. It has been applied at a regional scale to braided stream sediments (Webb and Anderson, 1990a, b) deltaic formations (Tetzlaff and Harbaugh, 1989). The modified form was applied to an alluvial fan environment (Koltermann and Gorelick, 1992). The resulting sediment distributions can be used to infer the spatial distribution of hydrogeologic units and subsequently the hydrogeologic properties within those units. Above, we distinguished between geological and parameter uncertainty. We hope that constraining the spatial distribution of geologic materials with simulators like SEDSIM will reduce our geologic uncertainty. In addition, the grain-size distributions produced in the simulations can be used with empirical models to infer hydrogeologic parameters and likewise reduce our parameter uncertainty. Several limitations restrict SEDSIM use for small-scale or highly variable conditions, including the assumption of a constant vertical velocity profile, use of a spatially averaged erosion equation and a limited number of grid points based on computational power and memory (Webb and Anderson, 1990a). As a result, "SEDSIM cannot simulate small sedimentary features, such as ripples" (Tetzlaff, 1989a). This limitation also may apply to larger features, such as sand dunes, that require vertical velocity variation for their formation.

Cross and Harbaugh (1989) discussed other process simulation models within a hierarchy of quantitative dynamic stratigraphy, including sedimentation models, geodynamic models and stratigraphic models. Sedimentation models, including SEDSIM (Tetzlaff and Harbaugh, 1989), are concerned with time scales up to 20,000 years and spatial scales on the order of tens to thousands of meters. Geodynamic models have time scales of millions of years and spatial scales of tens to hundreds of kilometers. Stratigraphic models are intermediate in both temporal and spatial scales, normally 1 to 10 million years and a few to tens of kilometers in distance (Cross and Harbaugh, 1989). Sedimentation models are of the appropriate scale for comparison with existing tracer studies. These models

simulate "the transport of sedimentary grains (as) an empirical generalization of the ways that turbulent or laminar flow move and deposit particles" (Cross and Harbaugh, 1989). However, the assumptions made to generalize flow conditions and the spatial discretization required for computer simulation limit internal resolution (Webb and Anderson, 1990b). Also, included in Cross (1989) are descriptions of several sedimentary basin-filling models (Syvitski, 1989; Bitzer and Pflug, 1989; Paola, 1989; Ross, 1989; Tetzlaff, 1989b).

The limitations of these geological process models usually stem from one of several points. First, they cannot reproduce the geometry of subsurface deposits but instead focus on surface processes. Second, they operate at scales either larger or smaller than the scale of interest to many contaminant studies (10^1 to 10^3 m). Third, they cannot produce multiple realizations, which are necessary to test the sensitivity of the spatial distribution of hydraulic properties to geological process parameters, because of insufficient computer power and memory requirements. Forth, they cannot be conditioned sufficiently to existing field data. Fifth, they are highly chaotic in that any calibration process is time consuming and sometimes impossible.

Geometric approaches address some but not all of these concerns. They focus primarily on the geometry of subsurface deposits, can operate at scales of interest, are less computer intensive and in some cases can be conditioned to field data. The rules that govern geometric simulation can be highly constrained and therefore less chaotic. The difficulty is that they contain much less direct information on physical and chemical processes and rely heavily on empirically derived geometric relationships.

Geometric models take several forms: those that are purely geometrical, being based on placement of geometric shapes in space, one group of which is known as Boolean models (e.g., Taheri, 1992); those that include both process and geometrical controls (e.g., Dimitrakapoulos and Desbarats, 1992); and those that build on geomorphological observations (Webb, 1995). Models using this general concept also have been developed for fracture networks (Dershowitz and Einstein, 1988; Dershowitz et al., 1991).

Early examples of these geometric methods include Bridge and Leeder (1979) and Allen (1978), who developed models to assess the rate of avulsion and deposition, along with connection of sandy units for homogeneous channel complexes over the range of tens to hundreds of kilometers. The minimum internal resolution of Bridge and Leeder's model was approximately 600 m. Although this resolution is too coarse for application to most reservoir characterization or hydrogeological problems, the approach may still be useful. Bridge (1977) developed a model to simulate the development of point bar deposits in a meander bend at scales of tens and hundreds of meters. This research focused on predicting bedform style and required the initial meander bend location and geometry as input. It did not focus on channel aggradation or long-term meander movement and, consequently, did not simulate the distribution of resulting sedimentary deposits.

Bridge and Diemer (1983) used a code developed by Bridge and Leeder (1979) to assess avulsion and depositional rates for homogenous channel complexes with widths of 600 to 2,000 m. Allen (1978) modified the code to test the effects of channel avulsion but still considered channels about 1,000 m in width. While these models considered sedimentary deposition in two dimensions and allowed the study of sand-unit connectivity, they represent a scale much larger than most contaminant problems in hydrogeology. Price (1974) developed a combined random-walk, stochastic model to simulate alluvial fan deposition at a scale of thousands of meters, but the code was not experimentally validated and produced no estimates of the internal sedimentary structure of the fan complex. At the opposite end of the spectrum, Rubin (1987) developed an algorithm for reproducing the internal bedding structure of individual sedimentary facies that can be applied with resolution of centimeters.

Boolean models and their near cousins, Marked Point Processes, have been discussed by several authors (Koltermann and Gorelick, 1996; Deutsch and Journel, 1992, 1997; Haldorsen et al., 1988). The basic concept is to place geometric objects randomly in space according to fixed probability or other rules. In a Marked Point Process the focus is on the centroid of randomly located objects of various sizes and shapes. The simplest examples are the bombing pattern studies that began during World War II (Cox and Isham, 1980; Ripley, 1987; Stoyan et al., 1987). Among the current applications are a simple code for placing ellipses randomly in a two-dimensional space (Deutsch and Journel, 1992, 1997), more complex channel images in plan view (Tyler et al., 1994) and in cross section (Srivastava, 1994), defining geologic units of contrasting petrologic properties (Haldorsen and Chang, 1986; Haldorsen and MacDonald, 1987; Dubrule, 1989) and more complex codes for three-dimensional reservoir simulation (Bratvold et al., 1994). The obvious and most difficult part of applying such codes is the definition of the appropriate shapes and the rules by which their placement is determined. Examples of these rules, based on classical geo-logical observations, include matching sand-shale or width-depth ratios (Haldorsen and Chang, 1986), object lengths (Haldorsen and Lake, 1984), as well as sand body size and geometry (Lowry and Raheim, 1991).

More recently, a coded called BCS-3D was developed (Webb, 1995; Webb and Anderson, 1996) to simulate the three-dimensional internal geometry of sediment units in braided stream deposits based on the relationship between surface (geomorphological) system and resulting sedimentary deposits. Such a code attempts to identify significant sedimentary architectural units such as the channel geometry described by Miall (1985). However, this and similar codes are difficult to test because of the need to combine surface hydrology/geomorphology data (Mosley, 1982; Brierley, 1989) and resulting subsurface deposition geometry along with the need to define appropriate comparison metrics (Paola, pers. commun., 1992). Attempts to collect such data may soon be possible (Aiken, 1992).

Scheibe and Freyberg (1995) developed a computer code that fills three-dimensional architectural units in a meander-bend sequence with appropriate small-scale sedimentary structures. Hydraulic conductivities were estimated to investigate the effective dispersion properties for this spatial arrangement of sedimentary structures. Refinements that allow the user to specify statistically the distribution of architectural units have made this code relatively powerful. Juergens (1994) simulated channel aggradation of a meandering river system with uniform channel widths at a scale of 0.5 to 2 km, fixed initial channel location (apex) and uniform internal properties. Attached to each channel are three other geometrically defined sedimentary units (channel fills, overbank deposits and splays). This basically represents a three-dimensional expansion of Bridge and Leeder (1979) with the inclusion of geologic units. Again the scale of the model is much larger than scales of interest in most contaminant hydrogeology problems, but a similar approach could be applied at a smaller scale.

While the development of sedimentary models has been under way for 20 years, their application in hydrogeology and petroleum engineering is relatively recent. They require extensive understanding of the anticipated final three-dimensional shapes that are characteristic of the internal geometry of various depositional/hard-rock systems. Little information of this type is available. A great service that sedimentologists could provide is to describe in detail the characteristic three-dimensional shapes and hydrologic properties of the building blocks that comprise the internal geometry of sedimentary deposits, and then correlate them with their genetic processes.

Ultimately, a spatial geological simulator designed for hydrogeological applications should be able to (1) distribute geological features (facies or three-dimensional units) objectively in three-dimensional space at a scale of most hydrologic problems and with high resolution; (2) relate facies deposits to expected/observed hydraulic controls, such as channel location and discharge rates; (3) account for both spatial and temporal variability; (4) allow point sampling at various scales; (5) be conditioned to existing site data; and (6) allow for multiple realizations based on a random component and uncertainty in input parameters. Modifications and new development efforts are needed to solve these problems for various physical systems.

Markov Chain Models.—

Markov chains are functions that list the sequence or order of repetitive items or events. The critical issue is whether one unit follows (lies above, below or is laterally adjacent to) others at a frequency greater than would be expected based on sheer chance. Davis (1986) showed how this has been applied to interpret a portion of depositional sequences using well logs. Numerous sedimentary systems have been investigated using this concept (e.g., Cant and Walker, 1976), and the results used to fashion a conceptual model of order of geologic units. One great difficulty in application of numerical methods of Markov Chain simulation is conditioning internal to the model system. Examples of simulations that have overcome these difficulties include Markov Chain Monte Carlo Methods (Tjelmeland et al., 1994; Hegstad et al., 1993; Oliver et al., 1997) and mixed Markov Indicator Geostatistics (Carle, 1996; Carle and Fogg, 1993, 1996; and Carle et al., this volume).

Others.—

Among the less well known categories or newly appearing application approaches are the following: (1) factor analysis used to map the location of hydrochemical conditions (Voudouris et al., 1997); (2) multifractals to represent one-dimensional well log sequences (Cheng, 1997); and (3) a discrete smooth-interpolation method used to represent spatial objects (Mallet, 1997).

CONTRIBUTIONS

Why This Volume of Papers (To Bring All This Together)

Given the breadth and mass of approaches for estimating the spatial arrangement of geologic properties, it is difficult to recognize the salient issues that help identify those that are most useful. One possible criterion involves the degree to which direct geologic information is included in such modeling approaches. The need for geologically based approaches has been voiced by several authors (Fogg, 1986; Phillips and Wilson, 1989; Phillips et al., 1989; Anderson, 1990), and attempts have been made to develop such methods (Webb and Anderson, 1990a, b; Scheibe and Freyberg, 1992; and the others listed above).

We can view geological data in several ways. For example, we can recognize two categories of information: (1) site-specific measurements of hydrogeologic parameters in space, and (2) information on processes. In most cases information on processes is derived by assimilating large amounts of data from multiple sites and then reducing them to a set of relatively simple physical or chemical statements or geometric rules. The two types of information have different uses. Site-specific data can be used to develop local, site-specific parameter descriptions (e.g., the site-specific variogram), as conditioning data for site specific simulation, or to help define the overall geometry of the system (which in some ways is similar to conditioning). Process information, on the other hand, is used to constrain the set of possible simulation outcomes by placing rules or limits on the mathematics of the simulators.

In reviewing the following papers you will see that explicit inclusion of geologic data, observations and processes is the unifying theme. However, each approach uses these two types of data differently. Thus we can ask two primary questions. First, what are the underlying assumptions of the method presented in each paper, and how are these spatial estimations constrained by geologic processes? Second, how well does the estimation approach condition to site specific information (measured points)?

Synopses of Included Papers

All the papers contained in this volume fall at the "process-oriented" end of the spectrum described above. Thus, the following description do not discriminate between "process-oriented" or "mathematics-oriented" papers. Instead, they will focus on the underlying approach, the data used and the results of each method.

*Using Glacial Terrain Models to Define Hydrogeologic
Settings in Heterogeneous Depositional Systems*
A.H. Fleming

This paper presents a description of "terrain models," which are applied to near-surface, unconsolidated material deposited by continental glaciers in the upper Midwestern United States. This by no means limits the range of application of this concept. Terrain model development is an extension of the classical surface geologic mapping approach that includes details of landscape type (geomorphology), vertical sequences as observed in local boreholes, and interpreted facies associations. This joint interpretation allows a more detailed understanding of the subsurface to be annotated on a spatial map. Consequently, this areal interpretation can be used to infer local geologic and hydrologic conceptual models and/or to form the basis for more detailed local investigations. While any direct or indirect geologic data can be used in this interpretive approach (e.g., water well records, geotechnical borings and so on), the example focuses on drill cuttings, downhole logs and detailed descriptions of large surface exposures. The primary result is a spatial map showing regional scale "terrains" with their associated descriptions, and in some cases, the more local scale "terrain elements," which identify subareas that vary in some significant way from this regional norm. This approach provides a framework—solidly rooted in an understanding of the depositional history, resulting landforms and deposits—that can be used as a conceptual model for the heterogeneity (variation within units) likely to be seen at the local scale in a hydrogeologic problem.

Using Glacial Terrain Models to Characterize Aquifer System Structure, Heterogeneity and Boundaries in an Interlobate Basin, Northeastern Indiana.
A.H. Fleming

This second paper, building on the ideas presented in Fleming's first paper, provides a more detailed description of an application to the Huntertown aquifer system in Allen County, Indiana. Throughout this example, it is clear that the Terrain Models based primarily on geologic descriptions correlate strongly with hydrogeologic properties, such as transmissivity and potential well yield. In addition, the description provides two of the critical elements of any hydrologic study—gross geometry and hydrologic boundaries of the system. Thus, both the boundaries of the system and the internal property distributions are described.

Using Three-Dimensional Geological Models to Map Glacial Aquifer Systems: An Example from New Jersey
S.D. Stanford and G.M. Ashley

The authors make the point that in many geologic environments, mapping should be used to define the gross geo-

logic units, their boundaries and relationships. Then, if necessary, mathematical techniques can be used to fill these larger defined units with smaller scale heterogeneity. In addition, they point out that directing this type of field mapping exercise toward a definition of permeability units, their boundaries and properties, provides the connection between sedimentology and hydrologic analysis. While such an approach provides a conceptual model, as in the work described by Fleming, it also provides the foundation for numerical analysis. Three sections are devoted to different components of this type of field analysis, as applied to a glacial system in New Jersey. The first section describes the four major descriptive elements of this type of sedimentary system—type of contact/geometries that can be expected between the glacial sediments and underlying units, major types of units and their controlling factors and generalized shapes, interunit contacts, and internal variability. The second section summarizes the four-step approach to mapping—(1) geomorphic investigation, (2) field exposure mapping of contacts, (3) gathering of subsurface information, and (4) compilation of results. The first three provide insight about the type of data one must acquire. For example, lithologic data are critical to understanding hydrologic properties, but direction of ice migration and timing may affect the spatial interpretation of the gross unit geometry. The third section describes a site application to a glacial valley-fill system in New Jersey.

Characterizing Aquifer Heterogeneity within Terrigenous Clastic Depositional Systems
W.E. Galloway and J.M. Sharp, Jr.

The authors assert that "deterministic delineation of hydraulic conductivity and other properties is the prerequisite for accurate prediction (or retrodiction) of fluid flow and solute transport." The process for performing this type of delineation is founded in analysis of deposition systems, including their physical characteristics and genetic history. The authors first review three major classes of heterogeneity: (1) layer cake, (2) jigsaw puzzle, and (3) labyrinth. Each is the result of different but understood geologic depositional systems. The level of site-specific data necessary to delineate individual units is least in type 1 and most in type 3. The specific heterogeneity patterns within these systems must be described at a series observational scales different from the entire depositional systems or stratigraphic sequences (gigascopic), through individual hydrostratigraphic units (megascopic), to the scale of individual sedimentary facies (Macroscopic), to internal bedding structure (mesoscopic), and finally to the individual grain/pore scale (microscopic). At each scale the authors provide information on the relative importance or properties of potential sediment structures.

Hydrogeology and Characterization of
Fluvial Aquifer Systems
W.E. Galloway and J.M. Sharp, Jr.

Building on the terminology of the previous article, the authors focus on fluvial systems by first describing three types of fluvial deposition based on sediment load/channel types: (1) trunk-stream bed load, (2) mixed load and (3) suspended-load channel systems. Some background on significant characteristics or values to be measured during their characterization are presented. Then, for three typical alluvial basin fills they describe characteristic groundwater flow patterns, specifically, recharge and discharge characteristics, relative hydraulic conductivity, and subsystem components. Basin types include: (1) closed terrestrial basins with mixture of depositional types, (2) alluvial valleys dominated by fluvial and alluvial fan systems, and (3) coastal plains with alluvial fan, fluvial, deltaic, and shore-zone systems. Both of these papers present a relatively detailed description of the important hydrologic characteristics of several depositional systems, and thus bring together a wide array of geologic information that would be inaccessible to the average hydrologist. Additionally, they provide access to more detailed descriptions of each geologic setting by including system-specific references.

Simulation of Geological Patterns:
A Comparison of Stochastic Simulation Techniques for
Groundwater Transport Modeling
T.D. Scheibe and C.J. Murray

The authors compare how well three different stochastic simulation algorithms represented a heterogeneous flow and transport field. These simulators are compared with a synthetic "fully" described data set. Care was taken to focus on the differences in the underlying model assumptions rather than model parameterization uncertainty. The three simulation tools—Sequential Gaussian Simulation, Sequential Indicator Simulation and 2D Markov Chain Simulation—are all applied in two dimensions to produce a set of 50 unconditioned realizations. The results of the spatial simulations are used for flow and transport simulation. The results were compared with a "perfect" knowledge from the synthetic aquifer based on total water flux,

early arrival time (5%), median arrival time, late arrival time (95%) and apparent longitudinal dispersivity. Early arrival times for all models were overestimated (late). Median arrival time was slightly overestimated, and late arrival was underestimated (passes too quickly). Global flux and bulk conductivity were predicted correctly, but dispersivity was too small. Results indicate that models based on continuum field assumptions tend to undervalue (misrepresent the significant contribution to transport) discrete spatial structure. The authors present useful discussion of maximum entropy and how the concept must be applied to model uncertainty in addition to parameter uncertainty. Finally, they point out the necessity of extending this work to three dimensions but that this will exacerbate current model weaknesses.

A Method for Characterizing Hydrogeologic
Heterogeneity Using Lithologic Data
G.P. Flach, L.L. Hamm, M.K. Harris, P.A. Thayer,
J.S. Haselow and A.D. Smits

The authors present an approach for directly using lithologic borehole data as the basis for estimating a 3D hydraulic conductivity field. The process is relatively straightforward. Field data, in this case lithologic information on the fine-grained fraction from vertical boreholes at 1-ft intervals, are extrapolated over a relatively coarse three-dimensional grid (cell size of 500 ft by 500 ft horizontally and 1 ft vertically). After reviewing several approaches, the authors point to a comparison by Boman et al. (1995) which suggests that a simple approach is best. Thus, they impose a method using a series of areal two-dimensional spline fits to produce a laterally smoothed but vertically discrete heterogeneous field of clay fraction information. Then, the mud fraction data are translated to conductivity for a regional flow model through a two-step process. The first step combines site-specific mud fraction-to-conductivity correlation data and generalized grain-size equations to establish initial estimates of conductivity. The initial conductivity estimates are used in a simple flow model. Results are calibrated to gross measures of recharge and discharge in order to refine the mud fraction-to-conductivity correlation. Then, a translation process was applied to interpolate from the initial sedimentary grid mud-fraction values to a second flow-model-grid conductivity values. The translation used an areal smoothing process (cubic splines) and a vertical averaging process.

This resulting estimate of conductivity formed the basis for a standard, iterative calibration process that matches flow model results with hydraulic head, recharge and discharge data. During this process the estimates of conductivity are adjusted by the model user to best fit the calibration targets. The resulting simulations represent smaller scale preferential flow paths much better than models that do not include spatial geologic data.

This process draws on several classical approaches, including correlation of grain size information (mud %) with hydraulic conductivity and the iterative calibration approach for refinement of a groundwater-flow and particle-tracking model. The choice of a relatively simple spatial correlation model for this regional scale application seems reasonable. The authors point out that such an approach can be applied to any suite of geologic or geophysical information that can be acquired over the spatial system, and due to its relative simplicity it can be applied more rapidly than other, interpretive subjective modeling approaches.

Geostatistical Analysis of Facies Distributions:
Elements of a Quantitative Facies Model
D.F. Dominic, R.W. Ritzi, Jr.,
E.C. Reboulet, and A.C. Zimmer

The authors present a definition, general philosophy behind, and current details of an ongoing implementation of "quantitative facies models." The basic concept is a numerical tool capable of delineating "(1) the geometry of individual facies, (2) the proportions and spatial distribution of facies within facies assemblages, or (3) trends in ways that facies assemblages replace one another vertically and/or laterally." The choice in this case is to follow the pattern described by Carle and Fogg to interpret existing data from outcrop and numerous boring logs. Geologic and hydrologic property interpretation and conceptualization guide the choice of important units, the general field boundaries and the type of observations used to develop a numerical representation. In this case, statistical measures of facies unit abundance along with indicator geostatistics combined with transition probabilities in the vertical direction and standard variography in the horizontal direction are used to interpret the three characteristics listed above. Application of these numerical tools in turn helps identify or refine understanding of the geologic system.

The goal of this approach is to couple this type of detailed numerical description, the quantitative facies model, with information on the genesis and field observable characteristics of specific facies units so that the information from detailed studies can be applied to the interpretation of other sites where less information is available. Under-

standing geologic processes and site genesis helps to define analogous locations. As such, this paper presents the strongest example, of the papers in this collection, of how numerical tools can actually be used to help interpret the geologic system.

The approach is illustrated through application to an aquifer system underlying the Miami River in Ohio. Two major facies types are identified based on gross hydrologic properties. Each hydrofacies or stratigraphic facies consists of several types of sedimentary units that, with current data, are difficult to distinguish in the field and may be insignificant from a hydrologic viewpoint. The abundance of these two facies types is used to define three facies assemblages, which are then mapped areally in sections of the river valley. The process of this site description is still in progress.

Specific approaches for deriving estimates of unit length and thickness are based on indicator variograms and vertical transition probabilities following the work of Carle and Fogg (1996). The scale of these problems is fairly large, corresponding more closely to water supply problem scale, but it provides a framework for smaller scale work and could be applied at smaller scales with appropriate data.

Geostatistical Simulation of Hydrofacies Architecture:
A Transition Probability/Markov Approach
S.F. Carle, E.M. LaBolle, G.S. Weissmann,
D. Van Brocklin and G.E. Fogg

In the most mathematically rigorous of the papers in this volume, Carle and others provide both the conceptual framework and the details of how to implement an integrated Markov Chain/Transition Probability-based interpretation of spatial correlation with Sequential Indicator Simulation (SIS) (refined by Simulated Annealing). In relatively simple terms, this approach takes observations in either the vertical or horizontal direction, with vertical being most common in practice, and develops a covariance relationship based on the frequency of transition from one geologic unit to another. Where horizontal (along strike and dip) data are sparse, the vertical transition matrix is used to develop a horizontal correlation relationship. This covariance relationship can be subjectively refined or at least cross checked against geologic interpretive understanding of unit lengths, frequencies, relative abundance and so forth. The simulation approach uses SIS for a first estimate of the spatial field, which seems to miss certain features of the transition probability description. The SIS image is refined through multiple rounds of annealing. The final image compares well with the initial (guiding) transition probabilities and thus spatial correlation, while maintaining the spatial conditioning. This approach was developed using data from alluvial fan deposits at Lawrence Livermore National Laboratory in California and has been applied to two other sites: the Kings River alluvial fan and mixed alluvial fan and fluvial deposits in the Salinas Valley, California. A great strength of this approach is the ability to relate spatial correlation to geologically observable or interpretable characteristics. Another significant accomplishment is the ability to include locally variable anisotropy, which is almost universally present in such geologic systems as fan-shaped deposits or fining-upward sequences.

Combining Geologic Information and Inverse Parameter Estimation to Improve Groundwater Models
E.P. Poeter and S.A. McKenna

This paper, through application to two synthetic examples and one field example, shows how to include subjective stratigraphic correlation, geophysical interpretation and hydrologic interpretation of sedimentary rules into inverse flow modeling. The goal is to refine both a description of spatially distributed hydrologic properties and the resulting flow and transport estimates. This paper is a significant step in the progress of mixing descriptive/subjective information with complex numerical simulation tools.

The authors initially acknowledge the positive and negative attributes of subjective and inversion model calibration approaches, then suggest that the sensitivity of inverse models to internal unit geometry be used as a method for "checking" conceptual models of spatial structure. Through three examples (two artificial and one field example) they provide insight into the types of geologic information that can be used to constrain or guide selection of potential hydrologic conceptual models. The models then are tested through application of inverse modeling. If one or more of the potential conceptual models fails to fit the known geologic conditions, it can be eliminated. The three major classes of geologic information used to constrain these simulations are (1) location and character of geologic unit boundaries, (2) the degree of continuity and direction of trends of units, and (3) relative values of properties between units. The specific tools used include a relatively powerful, publicly available geostatistical simulation tool, UNCERT, which is capable of Gaussian and Indicator geostatistics. The indicator geostatistical application to the field example explicitly used "soft information," meaning that geophysical data are used to evaluate the probability of occurrence of a given unit at a particular location. The authors evaluate these types of geologic information and then suggest how to proceed. Options include (1) finding more quantitative methods to delineate unit boundaries/-orientations/continuity based on field data; (2) including more "pore" oriented information (permeability, porosity) in sedimentary descriptions; (3) collecting data at multiple scales for better understanding of spatial correlation, and (4) developing a method for identifying correlation scales within units during field work. In an appendix the authors provide a description of geostatistical simulation.

FUTURE

In some ways the future of spatial geologic simulation will mirror the past. Given that it is much simpler to develop numerical methods that do not include the complex relationships found in geologic observations, tools that do not explicitly incorporate this information will advance rapidly in type and in detail. At the same time, field-based observation approaches have begun and will continue to become more numerical, and thus provide better sources of data for both geologic and nongeologic spatial estimation approaches. In evaluating these type of tools, we will face two questions, "Under what circumstances is the added effort to incorporate geologic processes into the simulation approach worthwhile, and under what geologic circumstances are relatively less complex analytical (deterministic and stochastic) techniques adequate?"

In some ways all these tools will be judged on how well they address certain hydrologic issues. The first issue involves the problem of "upscaling," in which measurements of properties made at usually small scales are interpreted to represent much larger areas or volumes. A second issue involves selecting either calibration targets for forward models or conditioning data for those techniques that can be conditioned. We also would like to be able to constrain the simulations by as much data as are available regardless of tool and regardless of the form of the information (soft descriptions, point values, geophysical logs, tracer tests, etc.). The processes for including these types of data will continue to evolve. A third issue involves quantifying uncertainty both in sets of individual model simulations from one tool, and in determining which tool is best for a given geologic setting and problem. Finally, we must integrate these spatial geologic simulation tools with the decision-making process to guide site characterization, cost optimization and communication.

The following are brief discussions of these issues.

Upscaling

Upscaling is the process of interpreting small-scale measurements in order to estimate values at a larger scale. The need arises because measurement of specific properties over large areas at high resolution (field mapping and sampling) is not practical and most likely impossible. Furthermore, current sampling techniques are limited in their scales of observation or have a characteristic scale of observation. To define measured properties at larger scales, we must consider the spatial heterogeneity of the property at that larger scale. For this interpretation, we can use some of the tools described above to estimate the arrangement of spatial properties. Then, by averaging over larger and larger spaces or by applying flow models to sample the effects of the small-scale properties over larger spaces, we can estimate the "equivalent homogeneous" properties.

This approach requires the user to accept the underlying structural model of heterogeneity incorporated into the numerical simulation code. This approach may be used to develop scaled-up, effective parameters (Lasseter et al., 1986; Bachu et al., 1990; Dimitrakopoulos, 1990a, b) using an assumed geometrical distribution of properties (Desbarats, 1987; Dagan, 1986; Gelhar, 1986; Gelhar and Axness, 1983) and spatial correlation structures (Desbarats, 1990). Such an approach works best if no major discrete features are believed to exist below the scale of the spatial simulation; thus, continuum field simulators, such as fractal fields and Gaussian geostatistics, can be applied. Specific techniques for performing these simulation-based scaling approaches are under development (Tidwell and Wilson, 1997). Field studies (Davis et al., 1993; Goggin et al., 1988; Fogg et al., 1993) that do not rely on spatial assumptions of simulation tools are much more believable but require immense effort, and the results probably vary from one geologic setting to another. Thus, using field data from one site to extrapolate to others is somewhat uncertain.

Future work must extend this type of detailed field sampling to a broad range of geologic settings so that we can determine whether general rules about geologic heterogeneity structure can be used to guide upscaling. The numerical techniques for upscaling need to be discussed in light of these types of detailed field studies.

Conditioning Data and/or Calibration Criteria for Forward Simulation Tools

Calibration targets generally consist of accepted, observed system characteristics that any model simulating that system must reproduce to be considered adequate. In addition, the definition of calibration targets should specify how close one must be to the target for calibration to be complete. Such is the case in fluid hydraulics or reservoir engineering, where comparison of field data and model estimates for some measures of potential (such as head) and flux rate are used to test the model results. Some of the geometric and process simulation tools described above use what is called "forward simulation" with no guarantee that a specific model run will adequately match measured data (calibration targets). In this case the models must be run iteratively, where the user, by hand or through an inverse algorithm, guides the process of matching results to calibration targets. Such models have one advantage in that they can be matched to many kinds of data, such as spatial patterns of facies units or size and shape distributions of units, over and above being constrained to match point-specific measurements. Nevertheless, the process is difficult and may be the main reason such tools never reach the same level of application as tools that directly condition to data. In addition, which measures to use and the appropriate techniques for determining accuracy and adequacy are unclear at present. One point in their favor is that geometric and process simulation approaches may produce better spatial representations (average behavior) without matching specific points with extreme accuracy. Knowing how to judge the trade-off between correct process representation and correct data matching is a critical issue.

Among the simulation tools described above, those that directly condition to measured data, and thus exactly match measured data (calibration targets), require no calibration. Instead, the problem is how to condition to multiple, very different data types. These might include both point specific measurements and continuous data in multiple dimensions, such as geophysical surveys. Such a demand requires development of tools that have multivariate conditioning capability. As stated above, numerically based approaches are moving forward at a greater rate, and the ability to incorporate multiple data types is proliferating.

Accommodating Uncertainty

As described above, the uncertainty in spatial geologic systems can be divided into two broad categories—geological uncertainty (also referred to as conceptual model or process uncertainty) and parameter uncertainty. Geologic uncertainty is the uncertainty in the presence and location of geologic units. This in turn results both from our lack of knowledge about genetic processes and from misrepresentation that results from applying numerical simulation models. This division of uncertainty is, in some ways, an oversimplification since some of the parameters guiding a model (such as the correlation length of a variogram) may represent a difference in the underlying model structure. Nevertheless, we can limit the definition of geologic (or process) uncertainty to "lack of understanding of the geologic processes." The group of possible conceptual models that result from this geologic uncertainty correspond to the different equations and assumptions underlying different geological simulations. Consequently, parameter uncertainty is the range of different ways that a single geologic simulator can be implemented.

In considering parameter uncertainty first, if we run a specific model several or many times using a range of reasonable values and then compile the suite of output to evaluate the range of possible simulations that all match our current knowledge and data, then we have evaluated parameter uncertainty. Several simulation tools described in Koltermann and Gorelick (1996) contain functions to perform this multiple-simulation realization approach.

Geologic process uncertainty must be handled differently. We currently have many approaches for representing the spatial arrangement of properties. Each tool, as noted above, possesses different underlying assumptions that we must accept when using any of them. If we do not know that one set of assumptions is correct, we have geological uncertainty. Investigating this type of uncertainty involves several sets of reasonable assumptions (constrained by existing data) and their corresponding models. In other words, using a fractal and a Gaussian and a Markov chain approach and investigating the differences in their results for a specific problem. Several studies of this type have been performed but always within the context of determining the pros and cons of each tool as if one approach was best (D.A. Zimmerman, pers. commun., 1997). Nevertheless, this particular area has not received much attention.

Data Value and Decision Analysis

Understanding the issue of uncertainty relates directly to the issues of data value and decision analysis. Specifically, decisions analysis is really taking a formal approach to evaluating options. The difficulty in choosing an option exists because of various uncertainties in predictions. Once uncertainty is quantified, the risk of an incorrect decision can be weighed against the cost of collecting additional data. This process of specifying additional data based on its expected return is called Data-Value or Data-Worth optimization. To do this assumes that one is following a Bayesian decision process—make an initial analysis, use it to define additional data needs, collect that data, and finally update the analysis in order to make a decision. The process can be used multiple times. In areas such as manufacturing, these types of decision processes have been demonstrated and refined using classical statistical tests and statistical simulation approaches. Equivalent applications to geologic settings are still being developed (e.g., Freeze et al., 1990, 1992; McKenna, 1996; Rosen and Gustafson, 1996; James and Freeze, 1993; Journel, 1988; Englund and Heravi, 1993; Olea, 1984; Conrad et al., 1993).

SUMMARY

The field of spatial geologic simulation is expanding rapidly with new, innovative modeling techniques, many of which are based solely on mathematical approximations and contain little geologic information on processes or structures. Most of these have the advantage of direct conditioning to observed site-specific properties and relatively rapid simulation times. Geologic process and geometry-based approaches also are being developed, but at a slower pace. These tools take advantage of the broad array of geologic process knowledge, but in many ways they require longer simulation times and are difficult to condition. Between these two generalized end members is an array of other options that attempt to take the best of both. The papers in this volume fall into this middle ground. They are neither pure geological process nor purely mathematical. Each has its own way of incorporating geologic information of either the site-specific measurement data type or the generalized geologic-process type. Choosing the best tool for a given geologic setting and hydrogeologic problem continues to be a difficult task. However, we seem to be making significant strides in improving our ability to incorporate geologic information into our predictive models of fluid flow and solute transport.

REFERENCES

AIKEN, J.S., 1992, The three-dimensional arrangement of sedimentary facies in coarse glacial outwash—Influence on contaminant movement: Unpub. M.S. Thesis, University of Wisconsin, Madison.

ALLEN, J.R.L., 1978, Studies in fluviatile sedimentation—An exploratory quantitative model for the architecture of avulsion-controlled alluvial suites: Sedimentary Geology, v. 21, p. 129-147.

ALLEN, J.R.L., 1983, Studies in fluviatile sedimentation—Bars, bar complexes and sandstone sheets (low-sinuosity braided streams) in the Brownstones (L. Devonian), Welsh Borders: Sedimentary Geology, v. 33, p. 237-293.

ANDERSON, M.P., 1989, Hydrogeologic facies models to delineate large-scale spatial trends in glacial and glaciofluvial sediments: Geologic Society of America Bulletin, v. 101, p. 501-511.

ANDERSON, M.P., 1990, Aquifer heterogeneity—A geological perspective, *in* Bachu, S., ed., Parameter Identification and Estimation for Aquifer and Reservoir Characterization: Proceedings of the 5th Annual Canadian/American Conference on Hydrogeology, Canada, National Water Well Association, p. 3-22.

ANDERSON, M.P., 1991, Comment on "Universal scaling of hydraulic conductivities and dispersivities in geologic media," by S.P. Neuman: Water Resources Research, v. 27, p. 1381-1382.

ANDERSON, M.P., AND WOESSNER, W.W., 1992, Applied groundwater modeling—Simulation of flow and advective transport: New York, Academic Press, 381 p.

ASHMORE, P.E., 1991, How do gravel-bed rivers braid?: Canadian Journal of Earth Sciences, v. 28, p. 326-34.

AYERS, J.F., 1989, Application and comparison of shallow seismic methods in the study of an alluvial aquifer: Ground Water, v. 27, p. 550-563.

BACHU, S., CUTHIELL, D., KRAMERS, J., AND YUAN, L.P., 1990, Reservoir characterization case study—Bimodally heterogeneous facies, *in* Bachu, S., ed., Parameter Identification and Estimation for Aquifer and Reservoir Characterization: Proceedings of the 5th Annual Canadian/American Conference on Hydrogeology, p. 23.

BARTON, C.C., AND SCHOLZ, C.H., 1995, The fractal size and spatial distribution of hydrocarbon accumulations—Implications for resource assessment and exploration strategy, *in* Barton, C.C., and La Pointe, P.R., eds., Fractals in Petroleum Geology and Earth Processes: New York, Plenum Press, p. 13-34.

BEAR, J., 1972, Dynamics of Fluids in Porous Media: New York, American Elsevier, 764 p.

BERES, M., JR., AND HAENI, F.P., 1991, Application of ground-penetrating radar methods to hydrogeologic studies: Ground Water, v. 29, p. 375-386.

BIRCH, F.S., 1993, Testing Fournier's method for finding water table from self-potential: Ground Water, v. 31, p. 50-56.

BIRKELO, B.A., STEEPLES, D.W., MILLER, R.D., AND SOPHOCLEOUS, M., 1987, Seismic reflection study of a shallow aquifer during a pumping test: Ground Water, v. 25, p. 703-709.

BITZER, K., AND PFLUG, R., 1989, DEPO3D, A three-dimensional model for simulating clastic sedimentation and isostatic compensation in sedimentary basins, *in* Cross T.A., ed., Quantitative Dynamic Stratigraphy: Englewood Cliffs, New Jersey, Prentice Hall, p. 335-348.

BOHLING, G.C., HARFF, J., AND DAVIS, J.C., 1990, Regionalized classification—Ideas and applications, *in* Bachu, S., ed., Parameter Identification and Estimation for Aquifer and Reservoir Characterization: Proceedings of the 5th Canadian/American Conference on Hydrogeology, Banff, Alberta, Canada, National Water Well Association, p. 229-243.

BOMAN, G.K., MOLZ, F.J., AND GUVEN, O., 1995, An evaluation of interpolation methodologies for generating three-dimensional hydraulic property distributions from measured data: Ground Water, v. 22, p. 247-258.

BOOTHROYD, J.C., AND ASHLEY, G.M., 1975, Processes, bars morphology, and sedimentary structures on braided outwash fans, northeastern Gulf of Alaska, *in* Jopling, A.V., and McDonald, B.C., eds., Glaciofluvial and Glaciolacustrine Sedimentation: Society of Economic Paleontologists and Mineralogists Special Publication 23, p. 193-222.

BRATVOLD, R., HOLDEN, L., SAVANES, T., AND TYLER, K., 1994, STORM—Integrated 3D stochastic reservoir modeling tool for geologists and reservoir engineers: Society of Petroleum Engineers, Paper SPE 27563.

BRIDGE, J.S., 1977, Flow bed topography, grain size and sedimentary structure in open channel bends—A three-dimensional model: Earth Surface Processes, v. 2, p. 401-416.

BRIDGE, J.S., AND DIEMER, J.A., 1983, Quantitative interpretation of an evolving ancient river system: Sedimentology, v. 30, p. 599-623.

BRIDGE, J., AND LEEDER, M., 1979, A simulation model of alluvial stratigraphy: Sedimentology, v. 26, p. 617-644.

BRIERLEY, G.J., 1989, River planform facies models—The sedimentology of braided, wandering and meandering reaches of the Squamish River, British Columbia: Sedimentary Geology, v. 61, p. 17-35.

BRIERLEY, G.J., 1991, Bar sedimentology of the Squamish River, British Columbia—Definition and application of morphostratigraphic units: Journal of Sedimentary Petrology, v. 61, p. 211-225.

BRUSSEAU, M.L., 1994, Transport of reactive contaminants in heterogeneous porous media: Reviews in Geophysics, v. 32, p. 285-313.

CANT, D.J., AND WALKER, R.G., 1976, Development of braided fluvial facies model for Devonian Battery Point sandstone, Quebec: Canadian Journal of Earth Sciences, v. 12, p. 102-119.

CARLE, S.F., 1996, A transition probability-based approach to geostatistical characterization of hydrostratigraphic architecture: Unpub. Ph.D. Dissertation, University of California, Davis, 233 p.

CARLE, S.F. AND FOGG, G.E., 1993, Simulating heterogeneity in an alluvial fan aquifer using a Markov-indicatory approach (abs.): Eos, Transactions of the American Geophysical Union, v. 74, p. 136.

CARLE, S.F., AND FOGG, G.E., 1996, Transition probability-based indicator geostatistics: Mathematical Geology, v. 28, p. 453-476.

CHANDLER, M. A., KOCUREK, G., GOGGIN, D.J., AND LAKE, L.W., 1989, Effects of stratigraphic heterogeneity on permeability in eolian sandstone sequence, Page Sandstone, northern Arizona: American Association of Petroleum Geologists Bulletin, v. 73, p. 658-668.

CHENG, Q., 1997, Discrete multifractals: Mathematical Geology, v. 29, p. 245-266.

CHERRY, J.A., 1987, Groundwater occurrence and contamination in Canada, *in* Healey, M, ed., Canadian Aquatic Resources: Canadian Bulletin of Fisheries and Aquatic Science, No. 215, p. 387-426.

CHURCH, M., AND GILBERT, R., 1975, Proglacial fluvial and lacustrine environments, *in* Jopling, A.V., and McDonald, B.C., eds., Glaciofluvial and Glaciolacustrine Sedimentation: Society of Economic Paleontologists and Mineralogists Special Publication 23, p. 22-100.

CONRAD, S.H., WEBB, E.K., AND GLASS, R.J., 1993, Probabilistic mapping of spatial heterogeneity—An approach for finding DNAPLs (abs.): Eos, Transactions of the American Geophysical Union, v. 74, p. 135.

COX, D., AND ISHAM, V., 1980, Point Processes: New York, Chapman and Hall, 188 p.

CRESSIE, N.A.C., 1993, Statistics for Spatial Data (rev. ed.): New York, John Wiley and Sons, 900 p.

CROSS, T.A., (ed.), 1989, Quantitative Dynamic Stratigraphy: Englewood Cliffs, New Jersey, Prentice Hall, 625 p.

CROSS, T.A., AND HARBAUGH, J.W., 1989, Quantitative dynamic stratigraphy—A workshop, a philosophy, a methodology, *in* Cross, T.A., ed., Quantitative Dynamic Stratigraphy, Englewood Cliffs, New Jersey, Prentice Hall, p. 3-20.

CROVELLI, R.A., AND BARTON, C.C., 1995, Fractals and the Pareto distribution applied to petroleum accumulation size distributions, *in* Barton, C.C., and La Pointe, P.R., eds., Fractals in Petroleum Geology and Earth Processes: New York, Plenum Press, p. 59-72.

DAGAN, G., 1982, Stochastic modeling of groundwater flow by unconditional and conditional probabilities, part 2—The solute transport: Water Resources Research, v. 18, p. 835-848.

DAGAN, G., 1986, Statistical theory of groundwater flow and transport—Pore to laboratory, laboratory to formation and formation to regional scale: Water Resources Research, v. 22, p. 120S-134S.

DAGAN, G., 1989, Theory of Flow and Transport in Porous Formations: New York, Springer-Verlag.

DAVIS, J.C., 1986, Statistics and Data Analysis in Geology: San Diego, Academic Press, 550 p.

DAVIS, J.M., 1994, A conceptual sedimentological-geostatistical model of aquifer heterogeneity based on outcrop studies: Unpub. Ph.D. Dissertation, New Mexico Institute of Mining and Technology, Socorro, 234 p.

DAVIS, J.M., LOHMANN, R.C., PHILLIPS, F.M., AND WILSON, J.L., 1993, Architecture of the Sierra Ladrones Formation, central New Mexico—Depositional controls on the permeability correlation structure: Geological Society of America Bulletin, v. 105, p. 988-1007.

DAVIS, J.M., WILSON, J.L., PHILLIPS, F.M., AND GOTKOWITZ, M.B., 1997, Relationship between fluvial bounding surfaces and the permeability correlation structure: Water Resources Research, v. 33, p. 1843-1854.

DELUDE, N.A., DAVIS, J.M., MOZLEY, P.S., HALL, J.S., 1996, Effect of cementation on permeability heterogeneity in the Sierra Ladrones Formation, NM (abs.): Geological Society of America Annual Meeting, Denver, Colo., Abstracts with Programs, v. 28, no. 7, Abstract A-349.

DERSHOWITZ, W.S. AND EINSTEIN, H.H., 1988, Characterizing rock joint geometry with joint system models: Rock Mechanics and Rock Engineering, v. 21, p. 21-51.

DERSHOWITZ, W., LEE, G., GEIER, J., FOXFORD, T., LAPOINTE, P., AND THOMAS, A., 1995, FRACMAN, Interactive Discrete Feature Data Analysis, Geometric Modeling, and Exploration Simulation—User Documentation Version 2.5: Seattle, Wash., Golder and Associates.

DERSHOWITZ, W.S., WALLMANN, P., GEIER, J.E., AND LEE, G., 1991, Preliminary discrete fracture network modeling of tracer migration experiments at the SCV site: Stripa Project Technical Report 91-23, SKB, Stockholm.

DESBARATS, A.J., 1987, Numerical estimation of effective permeability in sand-shale formations: Water Resources Research, v. 23, p. 273-286.

DESBARATS, A.J., 1990, Macrodispersion in sand-shale sequences: Water Resources Research, v. 26, p. 153-164.

DEUTSCH, C.V., AND JOURNEL, A.G., 1992, GSLIB Geostatistical Software Library and User's Guide: New York, Oxford University Press, 340 p.

DEUTSCH, C.V., AND JOURNEL, A.G., 1997, GSLIB Geostatistical Software Library and User's Guide (2d ed.): New York, Oxford University Press, 360 p.

DIMITRAKOPOULOS, R., 1990a, Permeability characterization at the Crystal Viking field, Alberta (abs.): Eos, Transactions of the American Geophysical Union, v. 71, p. 511.

DIMITRAKOPOULOS, R., 1990b, Conditional simulation of intrinsic random functions of order k1: Mathematical Geology, v. 22, p. 361-380.

DIMITRAKOPOULOS, R. AND DESBARATS, A.J., 1992, Geostatistical modeling of gridblock permeabilities for 3D reservoir simulations: Society of Petroleum Engineers Reservoir Engineering, v. 8, p. 13-18.

DOMENICO, P.A., AND SCHWARTZ, F.W., 1990, Physical and Chemical Hydrogeology: New York, John Wiley and Sons, 824 p.

DONLEY, E.M., AND DESCHAMBAULT, V.M., 1993, Environmental Software Directory: Garrison, Va., Donley Technology, 446 p.

DUBRULE, O., 1989, A review of stochastic models for petroleum reservoirs, in Armstrong, M., ed., Geostatistics: New York, Kluwer, Academic, v.2, p. 223-248.

DREYER, T., SCHEIBE, A., AND WALDERHAUG, O., 1990, Minipermeater-based study of permeability trends in channel sand bodies: American Association of Petroleum Geologists Bulletin, v. 74, p. 359-374.

DYKAAR, B.B. AND KITANIDIS, P.K., 1992a, Determination of the effective hydraulic conductivity for heterogeneous porous media using a numerical spectral approach, part 1—Method: Water Resources Research, v. 28, p. 1155-1166.

DYKAAR, B.B. AND KITANIDIS, P.K., 1992b, Determination of the effective hydraulic conductivity for heterogeneous porous media using a numerical spectral approach, part 2—Results: Water Resources Research, v. 28, p. 1167-1178.

EBRAHEEM, A.M., HAMBERGER, M.W., BAYLESS, E.R., AND KROTHE, N.C., 1990, A study of acid mine drainage using earth resistivity measurements: Ground Water, v. 28, p. 361-368.

EFRON, B., 1982, The Jackknife, the Bootstrap and Other Resampling Plans: Philadelphia, Society for Industrial and Applied Mathematics, Monograph 38.

EFRON, B., AND TIBSHIRANI, R., 1986, Bootstrap methods for standard errors, confidence intervals and other measures of statistical accuracy: Statistical Science, v. 1, p. 54-57.

ENGLUND, E.J. AND HERAVI, N., 1993, Conditional simulation—Practical application of sampling design optimization, in Soares, A., ed., Geostatistics Troia 92: Dordrecht, Kluwer, Academic Press, p. 613-624.

ENVIRONMENTAL PROTECTION AGENCY, 1993a, Environmental pathways models—Ground-water modeling in support of remedial decision-making at sites contaminated with radioactive material: U.S. Environmental Protection Agency, EPA/402/R-93/009.

ENVIRONMENTAL PROTECTION AGENCY, 1993b, Computer models used to support cleanup decision-making at hazardous and radioactive waste sites: U.S. Environmental Protection Agency, EPA/402/R-93/005.

ENVIRONMENTAL PROTECTION AGENCY, 1994, A technical guide to ground-water model selection at sites contaminated with radioactive substances, U.S. Environmental Protection Agency, EPA/402/R-94/012.

ETHRIDGE, R. G., FLORES, R.M., AND HARVEY, M.D., 1987, Recent developments in fluvial sedimentology: Society of Economic Paleontologists and Mineralogists, Special Publication 39, 389 p.

FETTER, C.W., 1994, Applied Hydrogeology (3d ed.): Englewood Cliffs, New Jersey, Prentice-Hall, 691 p.

FOGG, G.E., 1986, Groundwater flow and sand body interconnectedness in a thick multiple-aquifer system: Water Resources Research, v. 22, p. 679-694.

FOGG, G.E., 1989, Stochastic analysis of aquifer interconnectedness, Wilcox Group, Trawick area, East Texas: Texas Bureau of Economic Geology Report of Investigations 189, 68 p.

FOGG, G.E., SENGER, R.S., LUCIA, F.J., AND KERANS, C., 1993, Hierarchial heterogeneity in a grainstone outcrop—Implications for aquifer characterization and scaling: Eos, Transactions of the American Geophysical Union, v. 74, p. 134.

FORSTER, A., AND MERRIAM. D.F., 1996, Geologic Modeling and Mapping: New York, Plenum Publishing, 348 p.

FREEZE, R.A., AND CHERRY, J.A., 1979, Groundwater: Englewood Cliffs, New Jersey, Prentice-Hall, 604 p.

FREEZE, R.A., JAMES, B., MASSMANN, J., SPERLING, T., AND SMITH, L., 1992, Hydrogeological decision analysis, part 4—The concept of data worth and its use in the development of site investigation strategies: Ground Water, v. 30, p. 574-588.

FREEZE, R.A., MASSMANN, J., SMITH, L., SPERLING, T., AND JAMES, B., 1990, Hydrogeological decision analysis, part 1—A framework: Ground Water, v. 28, p. 738-766.

FRIND, E.O., SUDICKY, E.A., AND SCHELLENBERG, S.L., 1987, Micro-scale modelling in the study of plume evolution in heterogeneous media: Stochastic Hydrology and Hydraulics, v. 1, p. 241-262.

FROHLICH, R.K. AND KELLY, W.E., 1988, Estimates of specific yield with the geoelectric resistivity method in glacial aquifers: Journal of Hydrology, v. 97, p. 33-44.

GELHAR, L.W., 1986, Stochastic subsurface hydrology from theory to applications: Water Resources Research, v. 22, p. 135S-145S.

GELHAR, L.W. AND AXNESS, C. L., 1983, Three-dimensional stochastic analysis of macrodispersion in aquifers: Water Resources Research, v. 19, p. 161-180.

GELHAR, L.W., WELTY, C., AND REHFELDT, K.R., 1992, A critical review of data on field-scale dispersion in aquifers: Water Resources Research, v. 28, p. 1955-1974.

GHILARDI, P., KAI KAI, A., AND MENDUNI, G., 1993, Self-similar heterogeneity in granular porous media at the representative elementary volume scale: Water Resources Research, v. 29, p. 1205-1214.

GOGGIN, D. J., CHANDLER, M.A., KOCUREK, G., AND LAKE, L.W., 1988, Patterns of permeability in eolian deposits, Page Sandstone (Jurassic), northern Arizona: Society of Petroleum Engineers Formation Evaluation, v. 3, 297-306.

GOGGIN, D.J., CHANDLER, M.A., KOCUREK, G., AND LAKE, L.W., 1992, Permeability transects of eolian sands and their use in generating random permeability fields: Society of Petroleum Engineers Formation Evaluation, v. 7, p. 7-16.

GRAHAM, W., AND MCLAUGHLIN, D., 1989a, Stochastic analysis of nonstationary subsurface solute transport, part 1—Unconditional moments: Water Resources Research, v. 5, p. 215-232.

GRAHAM, W., AND MCLAUGHLIN, D., 1989b, Stochastic analysis of nonstationary subsurface solute transport, part 2—Conditional moments: Water Resources Research, v. 26, p. 2331-2355.

HALDORSEN, H.H., BRAND, P., AND MACDONALD, C., 1988, Review of the stochastic nature of reservoirs, in Edwards, S., and King, P., eds., Mathematics in Oil Production: Oxford, Clarendon Press, p. 109-209.

HALDORSEN, H.H., AND CHANG, D.M., 1986, Notes on stochastic shales—From outcrop to simulation model, in Lake, L.W., and Carroll, H.B., Jr. eds., Reservoir Characterization: San Diego, Academic Press, p. 445-485.

HALDORSEN, H.H., AND LAKE, L., 1984, A new approach to shale management in field-scale models, Society of Petroleum Engineers Journal, v. 24, p. 447-457.

HALDORSEN, H.H., AND MACDONALD, C.J., 1987, Stochastic modeling of underground reservoir facies (SMURF): Society of Petroleum Engineers Paper SPE 16751.

HALL, P., 1985, Resampling a coverage pattern: Stochastic Processes and Their Applications, v. 20, p. 231-246.

HARDIN, E.L., CHENG, C.H., PAILLET, F.L., AND MENDELSON, J.D., 1987, Fracture characterization by means of attenuation and generation of tube waves in fractured crystalline rock at Mirror Lake, New Hampshire: Journal of Geophysical Research, v. 92, p. 7989-8006.

HARFF, J. AND DAVIS, J.C., 1990, Regionalization in geology by multivariate classification: Mathematical Geology, v. 22, p. 573-588.

HARFF, J., EISERBECK, W., HOTH, K., AND SPRINGER, J., 1990, Computer assisted basin analysis and regionalization aid the search for oil and gas: Geobyte, v. 5, p. 11-15.

HAZEN, A., 1893, Some physical properties of sands and gravels: Massachusetts State Board of Health, 24th Annual Report.

HEGSTAD, B.K., ORME, H., TJELMELAND, H., AND TYLER, K., 1993, Stochastic simulation and conditioning by annealing in reservoir description, in Armstrong, M., and Dowd, P.A., eds., Geostatistical Simulation: Dordrecht, Kluwer, Academic Press, p. 43-55.

HEWETT, T.A., 1994, Fractal methods for fracture characterization, in Yarus, J.M., and Chambers, R.L., eds., Stochastic Modeling and Geostatistics—Principles, Methods, and Case Studies: American Association of Petroleum Geologists Computer Applications in Geology No. 3, p. 249-260.

HEWETT, T.A. AND BEHRENS, R.A., 1990, Conditional simulation of reservoir heterogeneity with fractals: Society of Petroleum Engineers Formation Evaluation, v. 5, p. 217-225.

HURST, A., AND ROSVOLL, K.J., 1991, Permeability variations in sandstones and their relationship to sedimentary structures, in Lake, L.W., Jr., Carroll, H.B., and Wesson, T.C., eds., Reservoir Characterization: San Diego, Calif., Academic Press, v. 2, p. 166-196.

HYNDMAN, D.W., AND GORELICK, S.M., 1996, Estimating lithologic and transport properties in three dimensions using seismic and tracer data—The Kesterson aquifer: Water Resources Research, v. 32, p. 2659-2670.

IMPEY, M.D., WILLIAMS, M.J., HUMM, J.P., CLARK, K.J., AND EINCHCOMB, S.J.B., 1996, The MACRO-AFFINITY code—MACRO Version 3.1, Theoretical Background: Henley-on-Thames, Oxfordshire, England, QuantiSci,Technical Report, ID4074-14 Version 1, 39 p.

INTERNATIONAL GROUND WATER MODELING CENTER, 1994, IGWMC Software Catalog: International Ground Water Modeling Center, Colorado School of Mines, 96 p.

ISAAKS, E.H., AND SRIVASTAVA, R.M., 1989, An Introduction to Applied Geostatistics: New York, Oxford University Press, 561 p.

JACOBSEN, T., AND RENDALL, H., 1991, Permeability patterns in some fluvial sandstones—An outcrop study from Yorkshire, north east England, in Lake, L.W., Jr., Carroll, H.B., and Wesson, T.C., eds., Reservoir Characterization: San Diego, Calif., Academic Press, v. 2, p. 315-338.

JAMES, B.R., AND FREEZE, R.A., 1993, The worth of data in predicting aquitard continuity in hydrogeological design: Water Resources Research, v. 29, p. 2049-2065.

JOHNSON, N.M., 1995, Characterization of alluvial hydrostratigraphy with indicator semivariograms: Water Resources Research, v. 3, no. 12, p. 3217-3228.

JOHNSON, N.M., AND DREISS, S.J., 1989, Hydrostratigraphic interpretation using indicator geostatistics: Water Resources Research, v. 25, p. 2501-2510.

JORDAN, D.W., AND PRYOR, W.A., 1992, Hierarchical levels of heterogeneity in a Mississippi River meander belt and application to reservoir systems: American Association of Petroleum Geologists Bulletin, v. 76, p. 1601-1624.

JOURNEL, A., 1983, Non-parametric estimation of spatial distributions: Mathematical Geology, v. 15, p. 445-468.

JOURNEL, A., 1988, Non-parametric geostatistics for risk and additional sampling assessment, in Keith, L., ed., Principles of Environmental Sampling: Washington, D.C., American Chemical Society, p. 45-72.

JOURNEL, A.G., AND HUIJBREGTS, CH. J., 1978, Mining Geostatistics: New York, Academic Press, 600 p.

JUERGENS, L.J., 1994, Geologic process modeling and spatial analysis of fluvial/floodplain facies for subsurface characterization: Unpub. Ph.D. Dissertation, Carnegie-Mellon University, Pittsburgh, 190 p.

KESSLER, L. G., II, AND COOPER, F.G., 1970, Channel sequences and braided stream development in the South Canadian River, Hutchinson, Roberts, and Hemphill Counties, Texas: Gulf Coast Association of Geological Societies Transactions, v. 20, p. 263-273.

KOCUREK, G., KNIGHT, J., AND HAVHOLM, K., 1991, Outcrop and semiregional three-dimensional architecture and reconstruction of a portion of the eolian Page Sandstone (Jurassic), in Miall, A.D., and Tyler, N., eds., The Three-Dimensional Facies Architecture of Terrigenous Clastic Sediments and Its Implications for Hydrocarbon Discovery and Recovery: Society of Economic Paleontologists and Mineralogists Concepts in Sedimentology and Paleontology, v. 3, p. 25-43.

KOLTERMANN, C.E., AND GORELICK, S.M., 1992, Paleoclimatic signatures in terrestrial flood deposits: Science, v. 256, 1775-1782.

KOLTERMANN, C.E., AND GORELICK, S.M., 1995, Fractional packing model for hydraulic conductivity derived from sediment mixtures: Water Resources Research, v. 31, p. 3283-3297.

KOLTERMANN, C.E., AND GORELICK, S.M., 1996, Heterogeneity in sedimentary deposits—A review of structure-imitating, process-imitating, and descriptive approaches: Water Resources Research, v. 32, p. 2617-2658.

KUEPER, B.H., 1989, The behavior of dense, non-aqueous phase liquid contaminants in heterogeneous porous media: Unpub. Ph.D. Dissertation, University of Waterloo, Ontario, Canada, 172 p.

LA POINTE, P.R., 1995, Estimation of undiscovered hydrocarbon potential through fractal geometry, in Barton, C.C., and La Pointe, P.R., eds., Fractals in Petroleum Geology and Earth Processes: New York, Plenum Press, p. 35-58.

LASSETER, T.J., WAGGONER, J.R., AND LAKE, L.W., 1986, Reservoir heterogeneities and their influence on ultimate recovery, in Lake,

L.W., and Carroll, H.B., Jr., eds., Reservoir Characterization: New York, Academic Press, p. 545-559.

LEOPOLD L.B., AND WOLMAN, M.G., 1957, River channel patterns braided, meandering and straight: U.S. Geological Survey Professional Paper 282-B, 47 p.

LOWRY, P. AND RAHEIM, A., 1991, Characterization of delta front sandstones from a fluvial-dominated delta system, *in* Lake, L.W., Carroll, H.G., Jr., and Wesson, T.C., eds., Reservoir Characterization II: New York, Academic Press, p. 665-676.

MALINVERNO, A. AND ROSSI, D.J., 1993, The method of projection onto convex sets (POCS) for the reconstruction of subsurface property maps (abs.): Eos, Transactions of the American Geophysical Union, v. 74, p. 136.

MALLET, J.L., 1997, Discrete modeling for natural objects: Mathematical Geology, v. 29, p. 199-219.

MANDELBROT, B.B., 1982, The Fractal Geometry of Nature: San Francisco, W.H. Freeman, 460 p.

MANTOGLOU, A., AND WILSON, J.L., 1981, Simulation of random fields with the turning bands method: Ralph M. Parsons Lab Report 264, Massachusetts Institute of Technology, Cambridge.

MASCH, F.D., AND DENNY, K.J., 1966, Grain-size distribution and its effect on the permeability of unconsolidated sands: Water Resources Research, v. 2, p. 665-677.

MASSMANN, J., FREEZE, R.A., SMITH, L., SPERLING, T., AND JAMES, B., 1991, Hydrogeologic decision analysis, part 2—Applications to ground-water contamination: Ground Water, v. 29, p. 536-548.

MCDONALD, B.C., AND BANERJEE, I., 1971, Sediments and bed forms on a braided outwash plain: Canadian Journal of Earth Sciences, v. 8, p. 1282-1301.

MCDONALD, M.G. AND HARBAUGH, A.W., 1988, A modular three-dimensional finite difference ground water flow model: U.S. Geological Survey Techniques in Water Resources Investigations, Book 6, Chapter A1.

MCKENNA, S.A., 1996, Geostatistical Analysis of PU-238 Contamination in Release Block D, Mound Plant, Miamisburg, Ohio: Sandia National Laboratory, SAND 97-0270, 23 p.

MIALL, A.D., 1977, A review of the braided-river depositional environment: Earth Science Reviews, v. 13, p. 1-62.

MIALL, A.D., 1985, Architectural-element analysis—A new method of facies analysis applied to fluvial deposits: Earth Science Reviews, v. 22, p. 261-308.

MICHALSKI, A., 1989, Application of temperature and electrical conductivity logging in ground water monitoring: Ground Water Monitoring Review, v. 9, p. 112-118.

MOLZ, F.J., AND BOMAN, G.K., 1995, Further evidence of fractal structure in hydraulic conductivity distributions: Geophysical Research Letters, v. 22, p. 2545-2548.

MOSKOWITZ, P.D., PARDI, R., DEPHILIPS, M.P., AND MEINHOLD, A.F., 1992, Computer models used to support cleanup decision-making and radioactive waste sites: Risk Analysis, v. 12, p. 591-621.

MOSLEY, M. P., 1982, Analysis of effect of changing discharge on channel morphology and instream uses in a braided river, Ohau River, New Zealand: Water Resources Research, v. 18, p. 800-812.

NEUMAN, S.P., 1990, Universal scaling of hydraulic conductivities and dispersivities in geologic media: Water Resources Research, v. 26, p. 1749-1758.

OLEA, R.A., 1984, Sampling design optimization for spatial functions: Mathematical Geology, v. 16, p. 369-392.

OLIVER, D.S., CUNHA, L.B., AND REYNOLDS, A.C., 1997, Markov chain Monte Carlo methods for conditioning a permeability field to pressure data: Mathematical Geology, v. 29, p. 61-90.

PANDA, M.N., AND LAKE, L.W., 1994, Estimation of single-phase permeability from parameters of particle-size distribution: American Association of Petroleum Geologists Bulletin, v. 78, p. 1028-1039.

PANDA, M.N., AND LAKE, L.W., 1995, A physical model of cementation and its effects on single-phase permeability: American Association of Petroleum Geologists Bulletin, v. 79, p. 431-443.

PAOLA, C., 1989, A simple basin-filling model for coarse-grained alluvial systems, *in* Cross, T.A., ed., Quantitative Dynamic Stratigraphy: Englewood Cliffs, New Jersey, Prentice Hall, p. 363-374.

PARK, S.K., LAMBERT, D.W., AND LEE, T-C., 1990, Investigation by DC resistivity methods of a ground water barrier beneath the San Bernardino Valley, Southern California: Ground Water, v. 28, p. 344-349.

PETERSON, C.W., 1994, Sandia National Laboratories Analysis Code Data Base: Sandia National Laboratories Report, SAND94-2177, 203 p.

PHILLIPS, F.M., AND WILSON, J.L., 1989, An approach to estimating hydraulic conductivity spatial correlation scales using geological characteristics: Water Resources Research, v. 25, p. 141-143.

PHILLIPS, F.M., WILSON, J.L., AND DAVIS, J.M., 1989, Statistical analysis of hydraulic conductivity distributions—A quantitative geological approach, *in* Molz, F., ed., Proceedings of the Conference on New Field Techniques for Quantifying the Physical and Chemical Properties of Heterogeneous Aquifers: Dublin, Ohio, National Water Well Association, p. 19-31.

PLOTNICK, R.E., AND PRESTEGAARD, K.L., 1995, Fractal and multifractal models and methods in stratigraphy, *in* Barton, C.C., and La Pointe, P.R., eds., Fractals in Petroleum Geology and Earth Processes: New York, Plenum Press, p. 73-96.

POETER, E.P., 1988, Perched water identification with nuclear logs: Ground Water, v. 26, p. 15-31.

PRICE, W. E., JR., 1974, Simulation of alluvial fan deposition by a random walk model: Water Resources Research, v. 10, p. 263-274.

RAJARAM, H., AND GELHAR, L.W., 1995, Plume-scale dependent dispersion in aquifers with a wide range of scales of heterogeneity: Water Resources Research, v. 31, p. 2469-2482.

RIPLEY, B.D., 1987, Statistical Inference for Spatial Processes: New York, Cambridge University Press, 148 p.

RITZI, R.W., DOMINIC, D.F., BROWN, N.R., KAUSCH, K.W., MCALENNEY, P.J., AND BASIAL, M.J., 1995, Hydrofacies distribution and correlation in the Miami Valley aquifer system: Water Resources Research, v. 31, p. 3271-3281.

ROSEN, L., AND GUSTAFSON, G., 1996, A Bayesian Markov geostatistical model for estimation of hydrogeologic properties: Ground Water, v. 34, p. 865-875.

ROSS, W.C., 1989, Modeling base-level dynamics as a control on basin-fill geometries and facies distribution—A conceptual framework, *in* Cross, T.A., ed., Quantitative Dynamic Stratigraphy: Englewood Cliffs, New Jersey, Prentice Hall, p. 387-397.

RUBIN, D.M., 1987, Cross-bedding, bedforms and paleocurrents: Society of Economic Paleontologists and Mineralogists Concepts in Sedimentology and Paleontology, v. 1, 187 p.

SCHEIBE, T.D., 1993, Characterization of the spatial structuring of natural porous media and its impacts on subsurface flow and transport: Unpub. Ph.D. thesis, Stanford University.

SCHEIBE, T.D., AND FREYBERG, D.L., 1990, Understanding the impacts of spatial structuring of natural porous media on the modelling of solute transport in ground water, *in* Bachu, S., ed., Parameter Identification and Estimation for Aquifer and Reservoir Characterization: Proceedings of the 5th Annual Canadian/American Conference on Hydrogeology, Banff, Alberta, Canada, National Water Well Association, p. 50.

SCHEIBE, T.D., AND FREYBERG, D.L., 1992, Characterization of the spatial structure of a geologically realistic model aquifer (abs.): Eos, Transactions of the American Geophysical Union, v. 72, p. 212.

SCHEIBE, T.D., AND FREYBERG, D.L., 1995, The use of sedimentological information for geometric simulation of natural porous media structure: Water Resources Research, v. 31, p. 3259-3270.

SCHUMM, S.A., 1968, Speculations concerning paleohydrologic controls of terrestrial sedimentation: Geological Society of America Bulletin, v. 79, p. 1573-1588.

SILLIMAN, S.E., AND ROBINSON, R., 1989, Identifying fracture interconnections between bore-holes using natural temperature profiling: Ground Water, v. 27, p. 393-402.

SMITH, N.D., 1970, The braided stream depositional environment—Comparison of the Platte River with some Silurian clastic rocks, north-central Appalachians: Geological Society of America Bulletin, v. 81, p. 2993-3014.

SPERLING, T., FREEZE, R.A., MASSMANN, J., SMITH, L., AND JAMES, B., 1992, Hydrogeologic decision analysis, part 3—Application to design of a ground-water control system at an open pit mine: Ground Water, v. 30, p. 376-389.

SRIVASTAVA, R.M., 1994, An overview of stochastic methods for reservoir characterization, *in* Yarus, J.M., and Chambers, R.L., eds., Stochastic Modeling and Geostatistics—Principles, Methods, and Case Studies, American Association of Petroleum Geologists Computer Applications in Geology No. 3, p. 3-16.

STOYAN, D., KENDALL, W., AND MECKE, J., 1987, Stochastic geometry and its applications: New York, John Wiley & Sons.

SUDICKY, E.A., AND HUYAKORN, P.S., 1991, Contaminant migration in imperfectly known heterogeneous groundwater systems: Reviews of Geophysics, v. 29, p. 240-253.

SUN, Y.F., KUO, J.T., AND TENG, Y.C., 1993, High resolution of subsurface structures and estimation of porosity and fluid content by inversion of seismic data (abs.): Eos, Transactions of the American Geophysical Union, v. 74, p. 142.

SYVITSKI, J.P.M., 1989, The process-response model in quantitative dynamic stratigraphy, *in* Cross, T.A., ed., Quantitative Dynamic Stratigraphy: Englewood Cliffs, New Jersey, Prentice Hall, p. 309-334.

TAHERI, M., 1992, Stochastic modeling of reservoir architecture for field management: Society of Petroleum Engineers Reservoir Engineering, v. 7, p. 433-438.

TAYLOR, R.W., AND FLEMING, A.H., 1988, Characterizing jointed systems by azimuthal resistivity surveys: Ground Water, v. 26, p. 464-474.

TETZLAFF, D.M., 1989a, Limits to the predictive ability of dynamic models that simulate clastic sedimentation, *in* Cross, T.A., ed., Quantitative Dynamic Stratigraphy: Englewood Cliffs, New Jersey, Prentice Hall, p. 55-65.

TETZLAFF, D.M., 1989b, SEDO, a simple clastic sedimentation program for use in training and education, *in* Cross, T.A., ed., Quantitative Dynamic Stratigraphy: Englewood Cliffs, New Jersey, Prentice Hall, p. 401-415.

TETZLAFF, D.M., AND HARBAUGH, J.W., 1989, Simulating clastic sedimentation: New York, Van Nostrand Reinhold, 202 p.

TIDWELL, V.C., AND WILSON, J.L., 1997, Laboratory method for investigating permeability up-scaling: Water Resources Research, v. 33, p. 1607-1616.

TJELMELAND, H., ORME, H., AND HEGSTAD, B.K., 1994, Sampling from Bayesian models in reservoir characterization: Technical Report in Statistics No. 2, University of Trondheim, 15 p.

TODD, D.K., 1980, Ground Water Hydrology: New York, Wiley, 535 p.

TOMPSON, A.F.B., ABABOU, R., AND GELHAR, L.W., 1989, Implementation of the three-dimensional turning bands random field generator: Water Resources Research, v. 25, p. 2227-2243.

TOMPSON, A.F.B., AND GELHAR, L.W., 1990, Numerical simulation of solute transport in three-dimensional, randomly heterogeneous porous media: Water Resources Research, v. 26, p. 2541-2562.

TSANG, C.F., HUFSCHMIED, P., AND HALE, F.V., 1990, Determination of fracture inflow parameters with a borehole fluid conductivity logging method: Water Resources Research, v. 26, p. 561-578.

TUKEY, J.W., 1958, Bias and confidence in not-quite large samples (abs.): Annals of Mathematical Statistics, v. 29, p. 614.

TURCOTT, D.L., 1986, Fractals in fluid mechanics: Annual Reviews of Fluid Mechanics, v. 20, p. 5-17.

TURCOTTE, D.L., AND HUANG, J., 1995, Fractal distributions in geology, scale invariance, and deterministic chaos, *in* Barton, C.C., and La Pointe, P.R., eds., Fractals in Petroleum Geology and Earth Processes: New York, Plenum Press, p. 1-40.

TYLER, K., HENRIQUEZ, A., AND SVANES, T., 1994, Modeling heterogeneities in fluvial domains—A review of the influence on production profiles, *in* Yarus, J.M. and Chambers, R.L., eds., Stochastic Modeling and geostatistics—Principles, Methods, and Case Studies: American Association of Petroleum Geologists Computer Applications in Geology No. 3, p. 77-89.

VAN DER HEIJDE, P.K.M., AND KANZER, D.A., 1997, Groundwater model testing—Systematic evaluation and testing of code functionality and performance: U.S. Environmental Protection Agency, National Risk Management Research Laboratory, Cincinnati, Ohio, EPA/600/R-97/007, 173 p.

VOUDOURIS, K.S., LAMBRAKIS, H.J., PAPATHEOTHOROU, G., AND DASKALAKI, P., 1997, An application of factor analysis for the study of the hydrogeological conditions in Plio-Pleistocene aquifers of NW Achaia (NW Peloponnesus, Greece): Mathematical Geology, v. 29, p. 43-59.

WEBB, E.K., 1995, Simulation of braided channel topography and topology: Water Resources Research, v. 31, p. 2601-2611.

WEBB, E.K., AND ANDERSON, M.P., 1990a, Using a sedimentary depositional model to simulate heterogeneity in glaciofluvial sediments, *in* Bachu, S., ed., Parameter Identification and Estimation for Aquifer and Reservoir Characterization, Proceedings of the 5th Annual Canadian/American Conference on Hydrogeology, Banff, Alberta, Canada, National Water Well Association, p. 23-41.

WEBB, E.K., AND ANDERSON, M.P., 1990b, Tracing contaminant pathways in sandy heterogeneous glaciofluvial sediments using a sedimentary depositional model, *in* Moltyaner, G., ed., Transport and Mass Exchange Processes in Sand and Gravel Aquifers: Field and Modelling Studies, p. 342-354.

WEBB, E.K., AND ANDERSON, M.P., 1996, Simulation of preferential flow in three-dimensional heterogeneous conductivity fields with realistic internal architecture: Water Resources Research, v. 32, p. 533-545.

WEBB, E.K., CONRAD, S.H., AND BREEDEN, R., 1993, Superfund site characterization—A probabilistic approach (abs.): Eos, Transactions of the American Geophysical Union, v. 74, p. 138.

WHEATCRAFT, S.W., AND TYLER, S.W., 1988, An explanation of scale-dependent dispersivity in heterogeneous aquifers using concepts of fractal geometry: Water Resources Research, v. 24, p. 566-578.

WILLIAMS, P.F., AND RUST, B.R., 1969, The sedimentology of a braided river: Journal of Sedimentary Petrology, v. 39, p. 649-679.

WILLIAMS, R.E., 1988, Comment on "Statistical theory of groundwater flow and transport—Pore to laboratory, laboratory to formation, and formation to regional scale," by G. Dagan: Water Resources Research, v. 24, p. 1197-1200.

WINOGRAD, I.J., AND PEARSON, F.J., 1976, Major carbon 14 anomaly in a regional carbonate aquifer—Possible evidence for mega-scale channeling, south central Great Basin: Water Resources Research, v. 12, p. 1125-1143.

WRIGHT, L.D., COLEMAN, J.M., AND ERICKSON, M.J., 1974, Analysis of major river systems and their deltas—Morphological and process comparisons: Coastal Studies Institute Technical Report No. 156, Louisiana State University.

YARUS, J.M., AND CHAMBERS, R.L., (eds.), 1994, Stochastic modeling and geostatistics—Principles, Methods and Case Studies: American Association of Petroleum Geologists Computer Applications in Geology No. 3.

ZHENG, C., 1990, A modular three-dimensional transport model for simulation of advection, dispersion and chemical reactions of contaminants in groundwater systems: Rockville, Md., S.S. Papadopulos & Associates.

USING GLACIAL TERRAIN MODELS TO DEFINE HYDROGEOLOGIC SETTINGS IN HETEROGENEOUS DEPOSITIONAL SYSTEMS

ANTHONY H. FLEMING

Indiana Geological Survey, Indiana University/Purdue University, Fort Wayne, Indiana 46805

ABSTRACT: In most glaciated basins, the arrangement of aquifers and confining units is defined by facies distributions that result directly from a variety of regional and local-scale controls on sedimentation patterns. Glacial terrain models utilize abundant sub-surface data, particularly downhole samples and geophysical logs, to document the composition and spatial variability of glacigenic sequences and attendant landscapes. When combined with water-level data and other hydraulic information, they provide a useful sedimentologic framework for interpreting hydrogeologic settings at both local and regional scales.

INTRODUCTION

Much of the glaciated region in the northern United States and southern Canada coincides with major centers of population, industry, commerce and recreation, and hosts widespread and intensive agricultural activity as well. Throughout a large part of their distribution, groundwater within these deposits constitutes *the* major water source for a variety of domestic, agricultural, commercial and in-dustrial purposes. Moreover, because these deposits are at or near the ground surface, they are the principal reposi-tory for a variety of waste disposal sites, and they are sus-ceptible to spills and a wide range of agricultural and other nonpoint pollutants. Glacial deposits also are of central importance in many other aspects of human affairs. For example, they provide inexpensive and readily available sources of mineral aggregates; act as load-bearing substrate for innumerable large structures; constitute potential sources of radon or other naturally occurring constituents that can adversely impact health and water quality; and act to either amplify or diminish the destructiveness of shear waves in regions of seismic activity, such as parts of the Midwest adjacent to the New Madrid and Wabash Valley fault systems. Many of these environmental issues are in-creasingly being addressed by diversified agencies that commonly have limited in-house expertise in hydrogeology and other earth-science topics. Consequently, the concept of hydrogeologic settings is becoming increasingly useful as a readily understandable depiction of the function of groundwater systems, particularly at the local level.

Glaciated terrains consist of characteristic assemblages of local depositional sequences and related landforms. These assemblages are elements of broader, three-dimen-sional depositional systems that constitute an episodic record of continental glaciation within a particular deposi-tional basin. The diversity of sedimentary processes typi-cal of continental glacial environments commonly results in a close commingling at both large and small scales of sediment bodies with strongly differing character that may be separated by abrupt lithologic transitions and disconti-nuities or by subtle facies changes. The potential complex-ity is made even greater by the fact that areas of the upper Midwest experienced multiple glacial events from differ-ent ice lobes, resulting in superposition or lateral juxtapo-sition of sequences having vastly different characteristics. Consequently, it is not surprising that aquifers and confin-ing units in glaciated regions exhibit a remarkably com-plex array of geometries, discontinuities and changes in physical and chemical attributes over relatively localized areas. A depositional systems-based approach, built on a foundation of modern basin analysis techniques, is thus of great utility for understanding and illustrating the distribu-tion of hydrogeologic properties in glaciated terrain.

This paper illustrates the use of a depositional systems-based approach for defining and mapping hydrogeologic settings, using the southern margin of the complex inter-lobate region of northeastern Indiana (Fig. 1) as an example. This approach, termed *glacial terrain analysis*, has evolved over the past fifteen years at the Indiana Geological Survey. It has found increasing practical application to a variety of hydrogeologic problems (Bleuer and Woodfield, 1993; Fleming et al., 1993) as well as to other relevant societal issues (Fleming, 1994b; Eggert et al., 1995). The focus of this paper is to demonstrate the conceptual benefits of a broad, depositional systems-based approach for charac-terizing glaciated terrains and to show how this approach can be used to portray the resulting three-dimensional hy-drostratigraphy in map view. A companion paper (Fleming, this volume) describes the use of the same approach to de-fine in more detail the effect of depositional environment

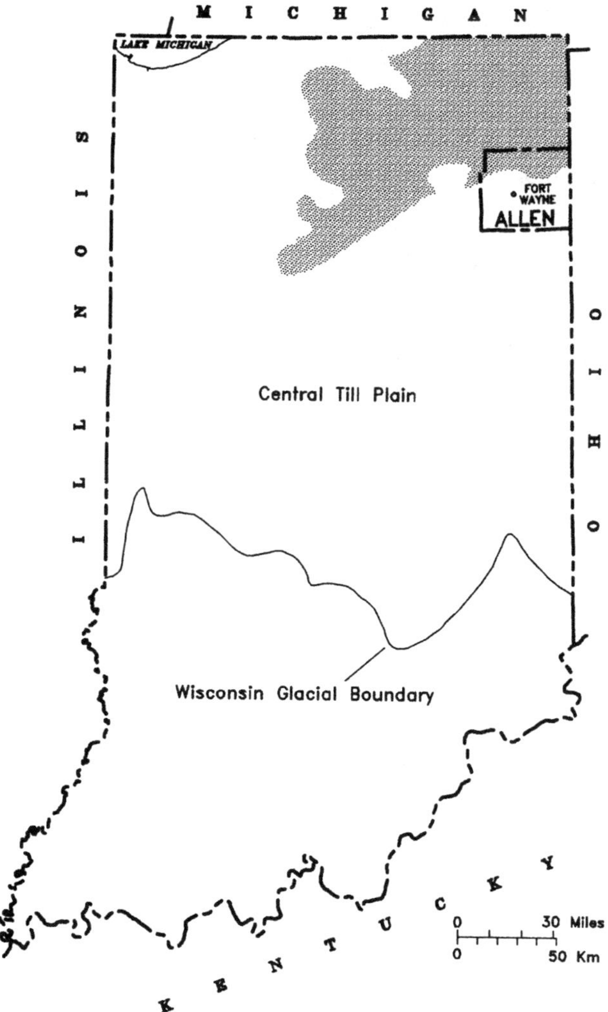

FIG. 1.—Map of Indiana showing location of hummocky supraglacial topography in northeastern Indiana (shaded), central till plain, Allen County and Fort Wayne.

on the origin and distribution of hydraulic properties and heterogeneity within a regional aquifer system and its confining units. The study of these terrains was not a theoretical exercise; rather, it was conducted on behalf of and funded by the citizens and local governmental agencies in Allen County in order to provide a holistic, practical and well-substantiated set of products to guide the development, protection and remediation of groundwater resources.

GLACIAL TERRAIN: CONCEPT AND ANALYSIS

Definition

Glacial terrains include two main elements: (1) a sequence or group of genetically related sequences that reflect deposition in a particular suite of sedimentary en-

vironments and are thus expected to possess certain ranges of physical properties and facies relations among internal sediment bodies; and (2) the overlying landscape, whose configuration typically is indicative of the nature of the underlying sequence(s). Although the practice of terrain mapping includes the varied assemblage of surface topographic forms and associated surficial materials familiar to traditional Quaternary geologic maps, "glacial terrain" map units are defined by *entire sequences* of sediments and by regionally extensive groups of related landforms. So defined, "terrain" is essentially analogous to the glacial *landsystems* model (e.g., Eyles, 1983), wherein sediment bodies are not necessarily thought of as individual, correlatable lithostratigraphic units but as components of whole depositional systems. Analysis of such systems requires the integration of a variety of subsurface data and investigative techniques to document lateral and vertical facies relationships within depositional systems as well as relations between entire systems.

The concepts of glacial terrain and hydrogeologic setting are interrelated. The particular relationship between the physical (and commonly the chemical) behavior of groundwater and the geologic sequence and landscape within which it occurs is known as a hydrogeologic setting. Sequence characteristics determine the geometry and internal properties of aquifers and confining units (hydrostratigraphy), whereas differences in surface terrain generally provide the relative elevation needed to drive groundwater flow and thus determine the locations of recharge and discharge areas. Variations in the spatial distribution of these characteristics, also known as *heterogeneity*, results in many possible configurations of hydraulic potential in near-surface groundwater systems. The saturated, coarse-grained elements of depositional sequences are the aquifers whose characteristics are of interest, whereas surface terrain affects the relative rates and amount of recharge (versus surface runoff) into the sequences. Similarly, the nature of facies relations within sequences affects the amount of hydraulic interconnection between different aquifers at various positions, and thus determines the nature of recharge to deeper aquifers. Fine-grained elements of the sequences affect the directions of movement and travel times of water and contaminants through the sequences and into or around specific aquifers. At many places, moreover, fine-grained sediment bodies at or near the surface are themselves saturated and interact with various surface waters, such as ecologically sensitive wetlands or riparian habitats. The relative importance and interaction of these hydrogeologic elements are determined by the character of the whole sequences and landscapes in which they occur; for all these reasons, glacial terrains are, in fact, representations of hydrogeologic settings.

Process

The cornerstone of glacial terrain analysis is the extensive use of downhole geophysical logs to establish sequences over broad depositional basins, from which many log-defined facies and examples of local vertical sequences are developed (Fig. 2). The bulk of such data is acquired by collecting mud-rotary cuttings during the construction of domestic water wells, and logging the completed wells by natural gamma ray. These data are locally supplemented with downhole logs from various environmental sites and public or private well fields, as well as detailed descriptions of large exposures. Downhole data provide an unbiased picture of glacial sequences and hydrostratigraphic relationships and afford numerous local examples of the nature and variation of the larger terrains. Interpretation of glacial sequences by logging is covered only superficially here; a more comprehensive treatment of this subject is given by Bleuer (1985). Other types of subsurface information also are employed, such as water well records, geotechnical borings and oil and gas wells, but their often crude, restrictive, inconsistent or uncertain descriptive attributes limit their usefulness in holistic interpretations of depositional systems. However, these other data do provide some indication of gross lithologic properties (e.g., "sand," "clay" and so on) and can be used to infer the presence, continuity and thickness of sequences or lithofacies between locations of more reliable downhole data.

In analyzing glacial terrains, conventional lithostratigraphic units, such as individual units of "clay," "sand" and the like, assume less significance than the overall sequence, or depositional setting, in which they occur. Instead, the nature of facies variation is of paramount importance because it determines the relationships among bodies of contrasting hydraulic properties, and the nature of these relationships ultimately controls the general character of groundwater movement that defines a particular hydrogeologic setting. Traditional lithostratigraphic elements (e.g., extensive till sheets or sand and gravel bodies) do exist in some, if not most, of these settings; such units commonly form distinct aquifers or confining units and locally constitute major parts of sequences that can be identified over broad areas. Such units, however, are only one element of a local sequence, and knowledge about the larger relations within the depositional system often allows more

definitive interpretations to be made about the geometry, range of physical properties and other essential features of individual aquifers and confining units. Moreover, some vertical sequences themselves (or characteristic facies thereof) form regionally extensive and identifiable stratigraphic entities, often expressed as consistent "motifs" on geophysical logs (e.g., Figs. 2, 6 and 7), that can provide the critical linkage between disparate parts of large depositional systems.

The glacial terrain models and map units that result from this process are characterized by distinct and documented ranges of surface and subsurface conditions. Analysis of these terrains, and the larger depositional systems of which they are part, is the foundation for various derivative products, such as maps showing geometry and physical properties of specific aquifer systems and confining units, corresponding water levels and recharge-discharge relations, and configurations of geotechnically significant buried surfaces. As noted above, these map units also constitute discrete hydrogeologic settings. Although somewhat generalized, hydrogeologic settings are nonetheless useful representations of the basic groundwater regime in a particular geologic terrain. They are especially valuable for illustrating the concept of natural geologic variability, especially to the many non-earth scientists who directly or indirectly deal with hydrogeologic issues (e.g., health officials, planning commissions, water departments and the like). Hydrogeologic settings also have proven to be an important tool for hydrogeologists in developing conceptual models of an area or site and in planning regional groundwater flow or geochemical studies.

DEPOSITIONAL SYSTEMS

Regional Setting

In northeastern Indiana much of the area immediately north of Fort Wayne was alternately invaded by ice from the east (Huron-Erie Lobe), north (Saginaw Lobe) and southeast (Erie Lobe) during the late Wisconsin Age and probably during the pre-Wisconsin as well. At many places a predominant process appears to have been the overriding and partial burial of stagnant ice and moderate- to high-relief constructional surfaces of earlier sequences by subsequent ice advances. In essence, each successive ice sheet encountered a landscape actively undergoing deglaciation from the preceding advance(s). Consequently, a variety of landscape elements formed whose particular expressions or locations are not directly related to the most

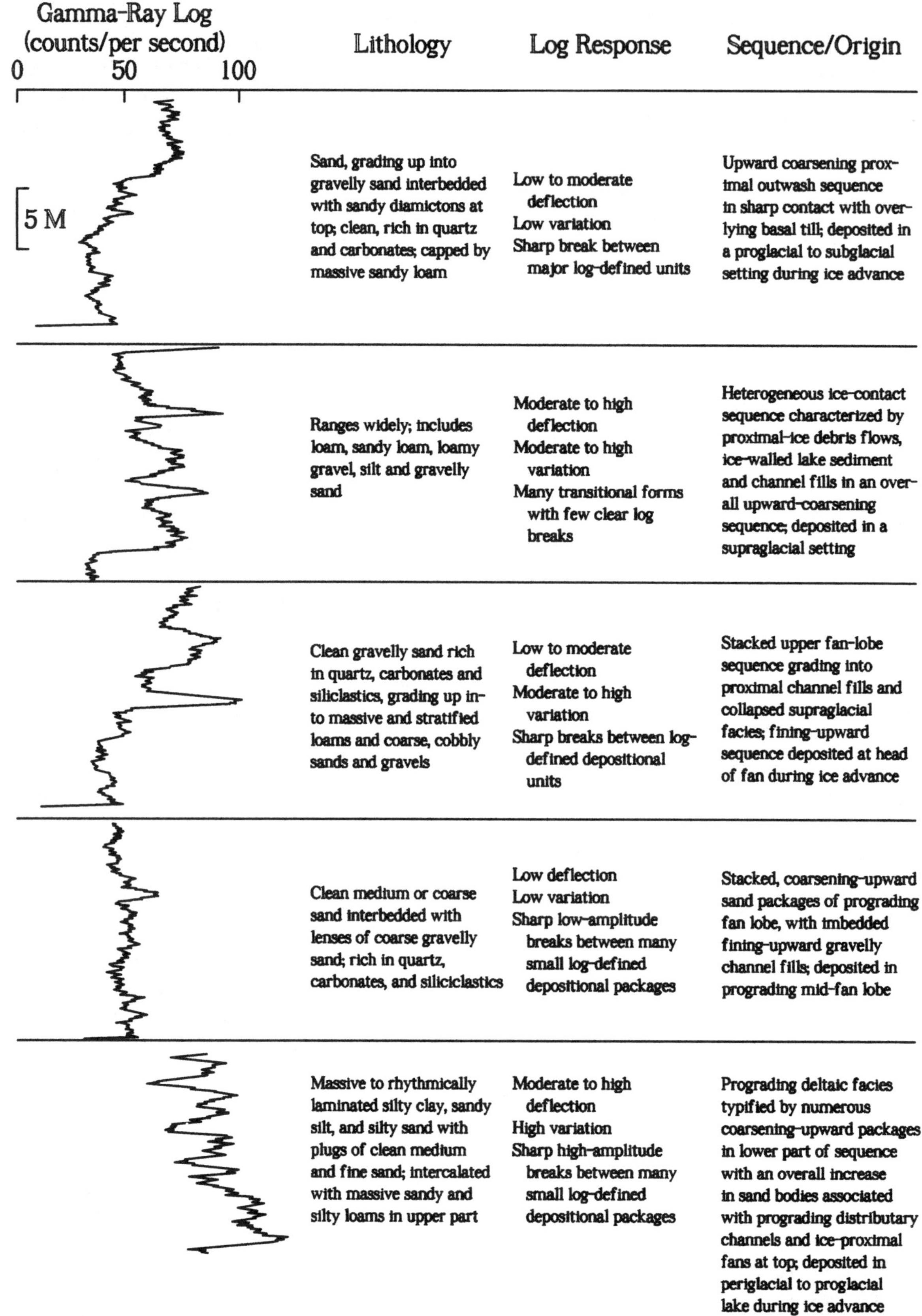
Gamma-Ray Log
(counts/per second)
0   50   100

Lithology

Log Response

Sequence/Origin

5 M

Sand, grading up into
gravelly sand interbedded
with sandy diamictons at
top; clean, rich in quartz
and carbonates; capped by
massive sandy loam

Low to moderate
deflection
Low variation
Sharp break between
major log-defined units

Upward coarsening prox-
imal outwash sequence
in sharp contact with overlying basal till; deposited in
a proglacial to subglacial
setting during ice advance

Ranges widely; includes
loam, sandy loam, loamy
gravel, silt and gravelly
sand

Moderate to high
deflection
Moderate to high
variation
Many transitional forms
with few clear log
breaks

Heterogeneous ice-contact
sequence characterized by
proximal-ice debris flows,
ice-walled lake sediment
and channel fills in an overall upward-coarsening
sequence; deposited in a
supraglacial setting

Clean gravelly sand rich
in quartz, carbonates and
siliclastics, grading up into massive and stratified
loams and coarse, cobbly
sands and gravels

Low to moderate
deflection
Moderate to high
variation
Sharp breaks between log-
defined depositional
units

Stacked upper fan-lobe
sequence grading into
proximal channel fills and
collapsed supraglacial
facies; fining-upward
sequence deposited at head
of fan during ice advance

Clean medium or coarse
sand interbedded with
lenses of coarse gravelly
sand; rich in quartz,
carbonates, and siliciclastics

Low deflection
Low variation
Sharp low-amplitude
breaks between many
small log-defined
depositional packages

Stacked, coarsening-upward
sand packages of prograding
fan lobe, with imbedded
fining-upward gravelly
channel fills; deposited in
prograding mid-fan lobe

Massive to rhythmically
laminated silty clay, sandy
silt, and silty sand with
plugs of clean medium
and fine sand; intercalated
with massive sandy and
silty loams in upper part

Moderate to high
deflection
High variation
Sharp high-amplitude
breaks between many
small log-defined
depositional packages

Prograding deltaic facies
typified by numerous
coarsening-upward packages
in lower part of sequence
with an overall increase
in sand bodies associated
with prograding distributary
channels and ice-proximal
fans at top; deposited in
periglacial to proglacial
lake during ice advance

recent glacial event (i.e., that represented by the sequence immediately underlying the modern landscape) but to characteristics associated with older buried sequences. Parts of the region thus constitute large-scale *palimpsest* landscapes, with different terrains and terrain elements having been cored by various combinations of sequences from different sources. These composite terrains are fundamentally unlike those in most other parts of the state because their architecture reflects the superposition and interaction of two or more large depositional systems.

Allen County is situated at the western apex of the Erie Lowland, or Maumee Lacustrine Plain (Malott, 1922), a regionally extensive basin of minimal internal relief whose extent is limited by bedrock uplands to the south and a massive pre-Wisconsin highland to the north (Fleming, 1995). The broadly lobate shape of the basin generally reflects its position along the axis of ice flow out of ancestral Lake Erie. The depositional surface of the basin appears to have been as much as 50 to 100 m below the surrounding uplands at various times during the late Wisconsin. Consequently, the basin frequently was the site of large proglacial lakes that formed when eastern-source ice blocked outlets to the east. Furthermore, it generally acted as the central depositional basin throughout the late Wisconsin not only for eastern-source ice but also for northern-source ice and meltwater that were directed down its northern slope.

Stratigraphic Background

The series of late Wisconsin glacial events that affected Allen County can be distilled into three major episodes (Fleming, 1994a). From oldest to youngest, these episodes consist of:
1. multiple advances of the Huron-Erie Lobe, beginning about 21.5 ka, that deposited loam-textured diamicts and associated glaciofluvial units of the Trafalgar Formation (Wayne, 1963; Bleuer and Moore, 1972; Bleuer, 1974; Gray, 1989; Fleming, 1994a) across most of the northern two-thirds of Indiana and into eastern Illinois;
2. a major period of Saginaw Lobe activity that produced sequences of generally sandy aspect. Initially defined

as an "unnamed member" of the Lagro Formation (Wayne, 1963; Bleuer and Moore, 1974), the Saginaw Lobe deposits in Allen County have more recently been referred to as the "Huntertown megasequence" (Fleming, 1992) and Huntertown Formation (Fleming, 1994a); and
3. two or more advances of the Erie Lobe that took place from approximately 15.3 to 14 ka and produced the classical series of looping moraines broadly concentric about the Maumee Lacustrine Plain in northeastern Indiana and adjacent states (e.g., Leverett and Taylor, 1915; Gooding, 1973; Lineback et al., 1983; Gray, 1989). Associated sequences consist chiefly of a variety of clay-rich basal tills[1], waterlain diamicts and supraglacial sediments known as the Lagro Formation in Indiana (Wayne, 1963; Gray, 1989; Fleming, 1994a).

The Trafalgar and Lagro Formations are generally described as being of "eastern source," whereas the Saginaw Lobe deposits are regarded as "northern source." At most places in northeastern Indiana, deposits from these two source areas are discriminated by differences in clast lithology, heavy mineral composition, carbonate content and other features.

The terrains of northern Allen County are largely the product of the second and third episodes above, and their structure reflects various relationships between Saginaw and Erie Lobe sequences. The older Huron-Erie Lobe sequences are an increasingly significant element within terrains in central Allen County southward into the central till plain; in northern Allen County, however, they generally constitute a thick stack of severely overconsolidated diamictons that lies at considerable depth below the modern land surface. Their significance in that area derives from the fact that they comprise "basement" for the depositional systems represented by the overlying Saginaw and Erie Lobe sequences. Overall basin shape and internal morphology of the depositional surface are closely approximated by the surface of the Trafalgar Formation, which, for simplicity, is referred to herein as the "Trafalgar surface."

FIG. 2 (left).—Downhole gamma ray log responses of common glacigenic sequences in northern Allen County. Background gamma radiation emitted from glacial sediments is determined chiefly by the presence of potassium-40, thorium and uranium; hence, sediments rich in clay or black shale fragments tend to produce the highest counts in a log. Different types of glacial sequences also tend to exhibit a typical pattern, or motif, that is indicative of the environment of deposition.

1. In this paper, the term "till" is used in a highly restrictive sense to include only those unsorted, generally massive diamictons that have undergone essentially no mass movement or washing by meltwater since their deposition directly from ice. Such conditions are generally found only in the subglacial environment of terrestrial glaciers; that is, the definition employed here is genetic and refers specifically to basal till of lodgment or meltout origin. This usage tends to exclude a great variety of otherwise "till-like" sediments, such as those deposited subaqueously or supraglacially. To avoid confusion and provide a clearer understanding of sedimentary sequences and their environmental interpretations, the more general terms "diamict" and "diamicton" are used for these other deposits and for the many transitional forms in which a clear genetic interpretation is problematic.

Saginaw Lobe Sequences

The Saginaw Lobe deposits in northern Allen County comprise three regional facies associations (Fig. 3): (1) a lower, commonly till-capped basal outwash association generally deposited in a proglacial to subglacial setting during ice advance; (2) a distal assemblage of predominantly glaciolacustrine aspect that exhibits a lateral facies relationship to the basal outwash association and that contains both periglacial and proglacial elements. This assemblage is especially well-developed over a broad area in central Allen County along and outboard of the inferred Saginaw Lobe terminus; and (3) an upper, northward-thickening complex of supraglacial aspect composed of a variety of intimately commingled sandy and silty diamicts and small to very large granular units, all with a hummocky upper surface.

Most of the sand and gravel units within this depositional system are hydraulically interconnected, and they exhibit a well-defined groundwater flow system. They constitute the predominant source of groundwater in northern Allen County and are collectively referred to as the Huntertown aquifer system (Fleming, 1994a).

Erie Lobe Sequences

Erie Lobe deposits in Allen County are decidedly more fine-grained, and the depositional system generally is char-

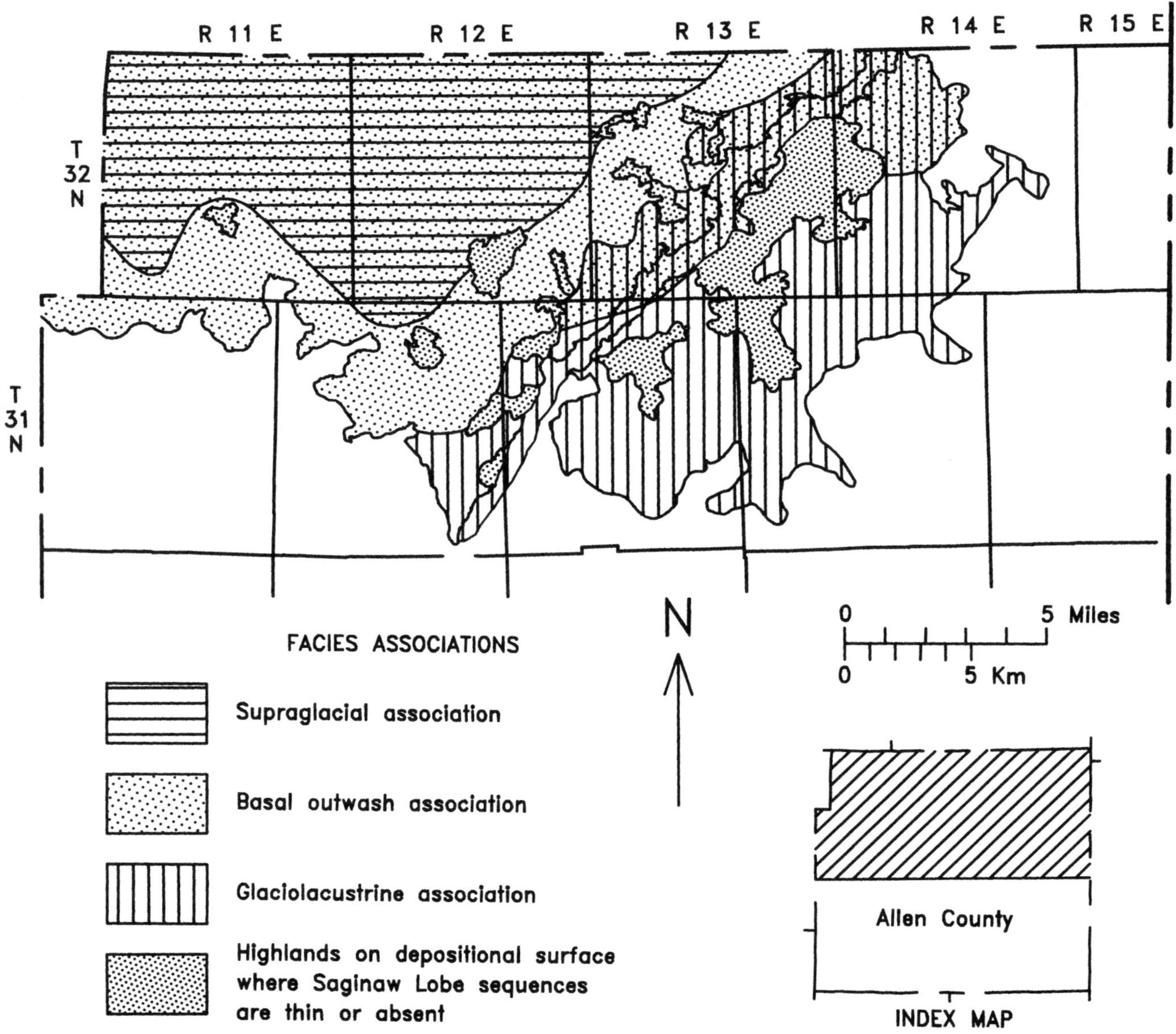

Fig. 3.—Map of northern Allen County showing distributions of three regional facies associations in the Saginaw Lobe depositional system (Huntertown megasequence).

acterized by less conspicuous differences in regional facies associations than Saginaw Lobe deposits. In most places, Erie Lobe sequences contain much more clay than underlying Saginaw Lobe sequences and consist dominantly of massive or poorly stratified diamictons. The distinction is readily apparent on logs (e.g., Figs. 6 and 7) and in most water well records, wherein drillers commonly refer to the Erie Lobe diamictons as "waxy clay" or "soft clay." Underlying diamictons are generally called "sandy clay" and "mixed sand and clay" (Huntertown megasequence) and "hardpan" or "hard clay" (Trafalgar Formation).

Three distinct depositional assemblages are recognized within the fine-grained elements of the Lagro megasequence in northern Allen County (Fig. 4): (1) an outer, partly palimpsest sequence of thin, mostly resedimented clay-loam diamicts intercalated with a heterogeneous group of sandy, silty and clayey kettle fills, and small glaciofluvial units; (2) an inner, constructional sequence of thick, repetitively sheared and stacked, massive diamicts of predominantly silty-clay to silty-clay-loam texture; and (3) a glaciolacustrine assemblage of thin to thick waterlain diamicts, laminated lake muds and subaqueous meltwater deposits underlying the Maumee Lacustrine Plain and immediately adjoining parts of neighboring terrains.

The Erie Lobe depositional system also includes three significant glaciofluvial sequences in northern Allen County (Fig. 4): (1) a sandy, fining-upward valley train sequence that occupies the St. Joseph River Valley, the narrow, entrenched marginal channel of the Fort Wayne Moraine; (2) a fining-upward sequence of medium to coarse sand and gravel found within the confines of Cedar Creek Canyon, a deeply entrenched tunnel valley across the northern limb of the Wabash Moraine; and (3) a combination of sandy to gravelly valley-train sequences associated with the Eel River Valley, a broad, arcuate sluiceway partly ice-marginal to the Wabash Moraine. The Eel also contains proximal and distal fan elements associated with the mouth of Cedar Creek Canyon. At places, all three of these glaciofluvial sequences are entrenched into older glaciofluvial and glaciolacustrine sequences of the Saginaw Lobe, resulting in locally thick, composite complexes of sand and gravel.

GLACIAL TERRAINS

Distribution and General Characteristics

In general terms, glacial terrains in northern Allen County are defined by the relationships between specific kinds of vertical sequences and their associated landscape(s). The fine-grained Erie Lobe sequences and their glaciofluvial counterparts represent the visible surface expressions of these terrains; i.e., they compose most of the individual landscape elements and thus form the immediate substrate for human endeavors. The geographic distribution of these sequences within the overall Erie Lobe depositional system thus constitutes a natural basis for defining major terrain boundaries at the land surface. So defined, the outlines of a glacial terrain map (Fig. 4), in many respects, resemble earlier maps of surficial geology and major geomorphic features for this area (e.g., Malott, 1922; Wayne, 1958; Johnson and Keller, 1972).

At many places in Allen County and elsewhere, Erie Lobe sequences are sufficiently thick (10 to 25 m or more) that the information provided by terrain units based purely on the Erie Lobe sequences likely would satisfy a variety of typical map applications (e.g., soil characteristics and parent materials, geotechnical conditions, shallow groundwater investigations, wetlands). However, the nature and distribution of subsurface heterogeneities and discontinuities in northern Allen County in large part reflect the characteristics and interaction of more than one depositional system, and the glacial terrains in this area ultimately are manifestations of these superposed systems. Northwestward across the county, for example, the surface configurations and internal structure of terrains increasingly reflect the effect of Saginaw Lobe deglaciation on Erie Lobe ice flow and sedimentation. In contrast, this particular dynamic is less pronounced to the southeast, where at places a strong relation is apparent between certain parts of Erie Lobe moraines and buried uplands on the Trafalgar surface (Bleuer, 1974; Fleming, 1994a). In hydrogeologic terms, the major groundwater resources are associated with Saginaw Lobe deposits (Huntertown aquifer system), whereas the eastern-source sequences above and below constitute regional confining units. In order to maximize their interpretive value, therefore, the glacial terrains of Allen County are, by necessity, defined by the nature of relationships between *entire depositional systems*, while the descriptions and explanations of glacial terrain map units provide information directly concerning the entire vertical sequences that characterize these terrains.

Because they are based on a combination of sedimentary characteristics (hydrostratigraphy) and surface morphology (hydrology), each glacial terrain is generally representative of a certain range of hydrogeologic conditions. By extension, the boundaries between terrains are commonly indicative of hydrogeologic discontinuities or

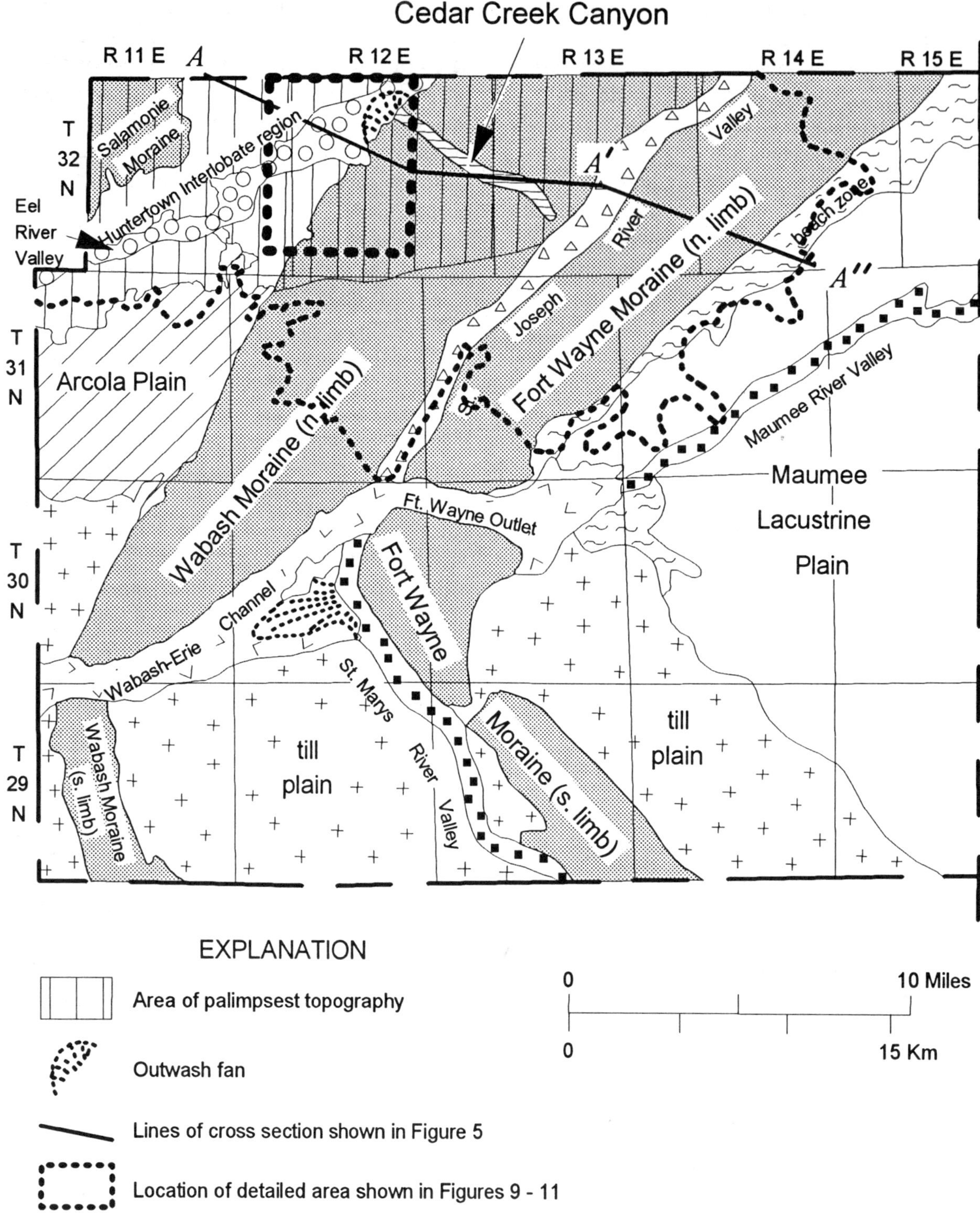

FIG. 4.—Map showing glacial terrain regions of Allen County. With the exception of river valleys and meltwater channels, the vertical sequences beneath these terrains are capped by one of three types of fine-grained Erie Lobe assemblages: (1) an inner constructional sequence forms the Fort Wayne and Wabash Moraines, as well as the till plains inboard of their southern limbs; (2) an outer sequence lies outboard of the Wabash Moraine and is partly palimpsest over hummocky Saginaw Lobe deposits in the Huntertown interlobate region; or (3) a dominantly glaciolacustrine sequence underlies the Maumee Lacustrine Plain and adjoining parts of the Fort Wayne Moraine and till plain.

transitions, such as a change in the character of a confining unit or a facies change in the aquifer system. In Allen County the distributions of several key hydrogeologic features and conditions closely match certain glacial terrains. In particular, the distribution of hydraulic potential in the Huntertown aquifer system is strongly affected by relationships between the top of the Saginaw Lobe depositional system and its overlying Erie Lobe confining units across different terrains. Consequently, the terrains also represent distinct hydrogeologic settings at different points within the regional hydrogeologic system, each typified by a characteristic range of subsurface conditions, surface terrain and resulting hydraulic head configuration.

Figure 5 conceptually illustrates several of the major glacial terrains in terms of (1) structural relationships between the Saginaw Lobe and Erie Lobe depositional systems; (2) distributions of major facies associations and internal sequences within these systems, and their relationship to the surface of the depositional basin; and (3) general profile of the potentiometric surface of the Huntertown aquifer system. Progressing from southeast to northwest across northern Allen County, the following sections summarize and contrast the central features of each of these terrains, including their identities as hydrogeologic settings. It is important to note that all of these regional-scale terrains can be further subdivided into distinct local subregions, or *terrain elements*, based on specific, mappable characteristics that differ somewhat from the mode. Space limitations preclude a detailed discussion of all possible variations, but some examples are given in the discussion of terrain maps and map units.

Northern Maumee Lacustrine Plain

The northern Maumee Lacustrine Plain (Fig. 4) is the core of the depositional basin. This landscape represents a former lake bottom and is thus virtually flat and very poorly drained. The surface of this terrain is blanketed by postglacial sediments deposited in Glacial Lake Maumee. A veneer of laminated lake mud is widespread over the southern two-thirds of the terrain, whereas a variety of sandy beach and nearshore deposits occur in the higher energy shoreline area to the north. The northern edge of the lake plain is prominently defined by a set of beach ridges composed of as much as 10 m of sand and gravel deposited atop wave-scoured benches incised into the lakeward edge of the Fort Wayne Moraine.

The postglacial sediments overlie older, superposed glaciolacustrine sequences that include (1) a variety of proglacial lacustrine facies of the Lagro Formation, which are generally thickest toward the core of the lake basin and over low-lying areas of the Trafalgar surface; and (2) a northward-thickening wedge of distal, predominantly northern-source sediments, including several sizable sand bodies of deltaic aspect that were deposited in an earlier interlobate phase of Lake Erie. All these sediments overlie a subdued, low-lying surface on loamy to silty diamicts of the Huron-Erie Lobe, some of which also appear to be of glaciolacustrine aspect. The depositional surfaces that separate these sediment packages represent the bottoms of various lake phases and are mostly of very low relief, similar to the modern landscape.

The glaciolacustrine sequences beneath the northern lake plain exhibit considerable variety from place to place (Fig. 6), which is probably related to variable water depths and distance from ice margins during deposition. Some are typified by numerous facies changes and discontinuities between medium- and fine-grained units, which may reflect sudden changes in sediment input to certain parts of the basin. Isolated, pod- or ribbon-like bodies of sand are embedded within finer grained units at a variety of positions within the glaciolacustrine sequences and probably mark the locations of distributary channels within a dominantly fine-grained, distal delta system.

Coalesced delta lobes consisting of fine to medium sand in sequences as thick as 20 m are present at depth beneath as much as 40 km^2 along the northern edge of the lake plain. These sands appear to grade northward into thinner, coarser and locally debris-flow-capped bodies of gravelly sand at depth beneath parts of the Fort Wayne Moraine (Fig. 6). Collectively, all these units appear to constitute a fan-delta system associated with the approach of the Saginaw Lobe to its terminal position.

The surface of the Maumee Lacustrine Plain lies as much as 30 m or more below the crest of the Fort Wayne Moraine and is generally the lowest lying area throughout the entire region. Consequently, the terrain constitutes a major regional groundwater discharge area, and water-level elevations in most aquifers are commonly near the land surface. Upward seepage of groundwater maintains numerous small creeks and contributes to the perennially wet surface conditions. However, the beach ridges along the northern margin are significantly elevated above the general surface of the plain and act as local recharge areas, especially where they are incised into the deltaic sands, which are the principal aquifers in this terrain.

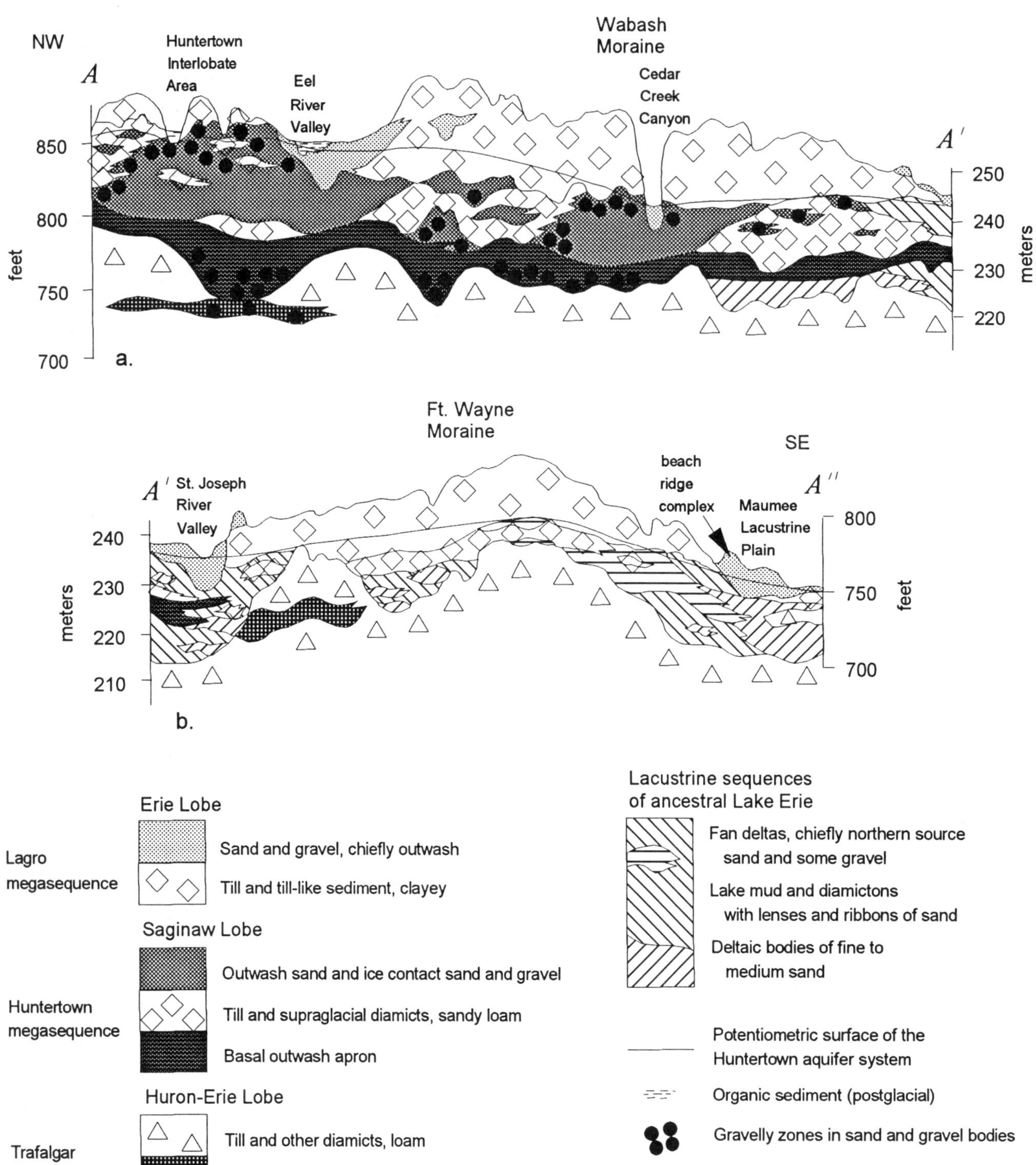

Fɪɢ. 5.—Schematic sections showing structure of the glacial terrains in northern Allen County and generalized potentiometric profile of the Huntertown aquifer system. A–A′: terrains northwest of the St. Joseph River Valley. A′-A″: terrains to the southeast. General location of section line is shown in Figure 4.

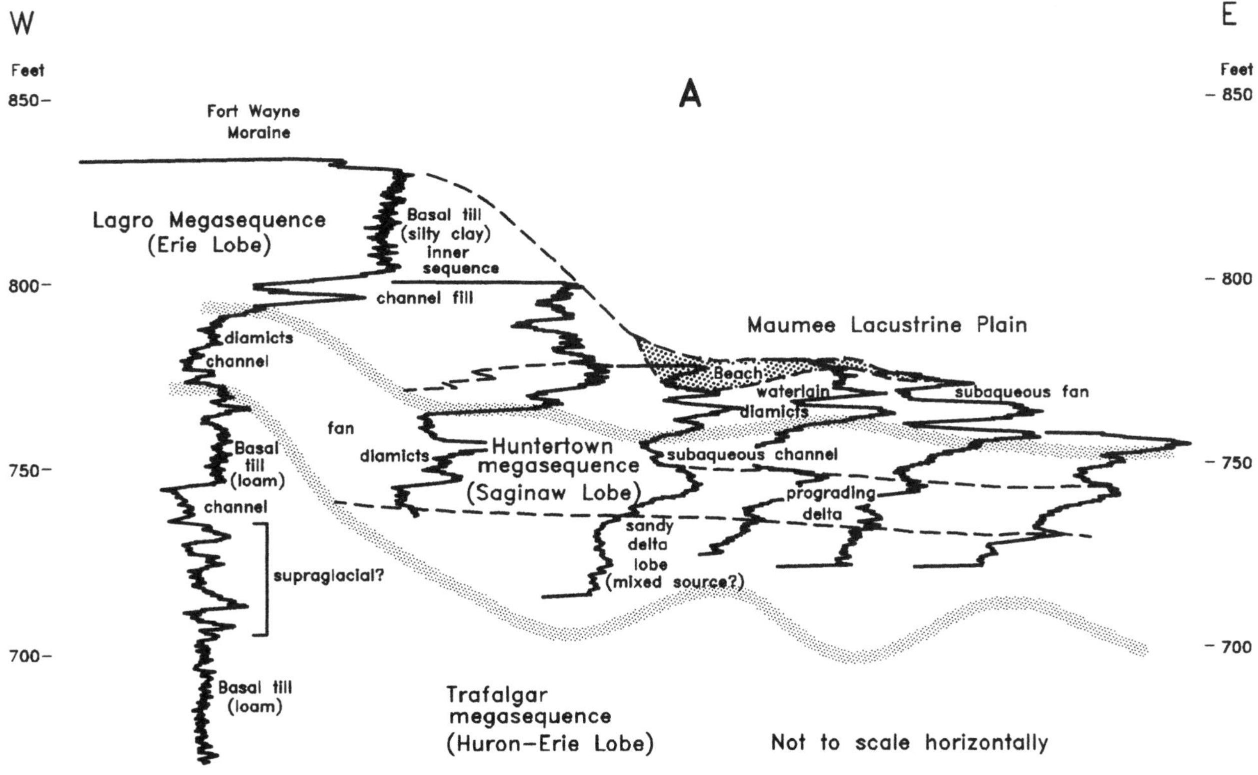

W
E
Feet
850—
Feet
— 850
A
Fort Wayne Moraine
Lagro Megasequence (Erie Lobe)
Basal till (silty clay) inner sequence
800—
— 800
channel fill
Maumee Lacustrine Plain
diamicts
channel
Beach
waterlain diamicts
subaqueous fan
fan
diamicts
Huntertown megasequence (Saginaw Lobe)
subaqueous channel
Basal till (loam)
750—
— 750
channel
prograding delta
sandy delta lobe (mixed source?)
supraglacial?
700—
— 700
Basal till (loam)
Trafalgar megasequence (Huron–Erie Lobe)
Not to scale horizontally

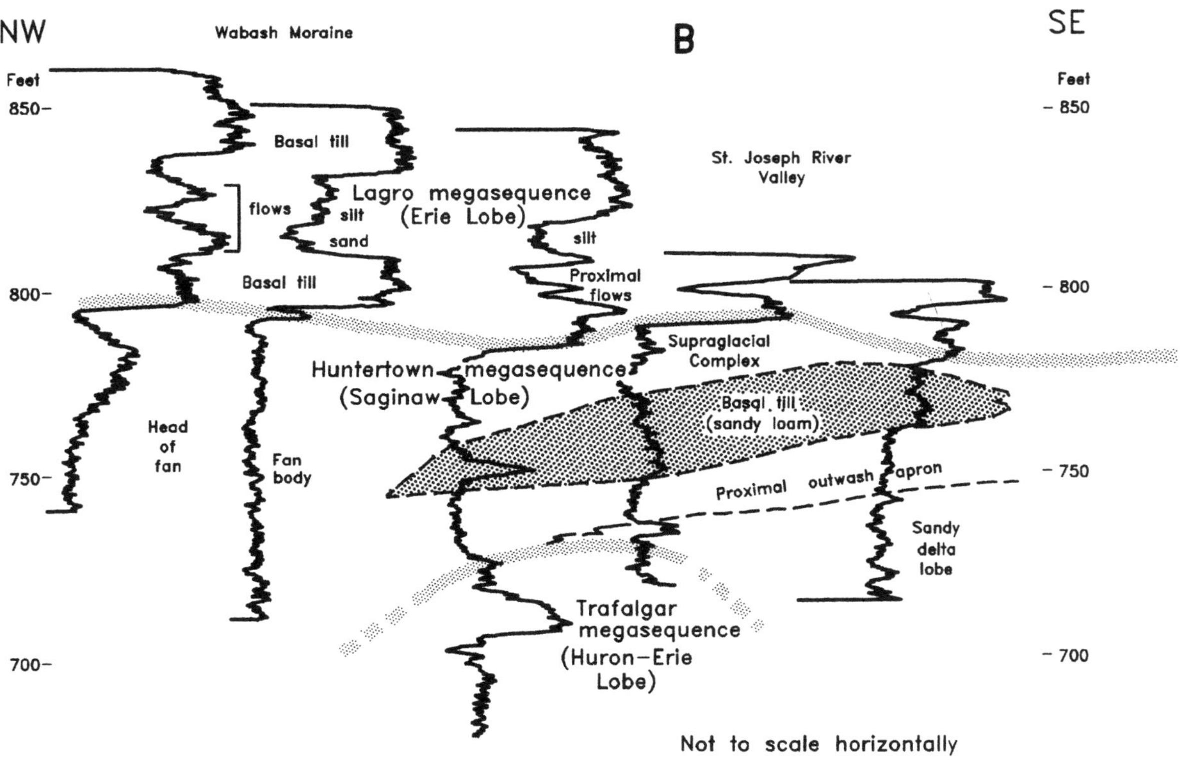

NW
SE
Wabash Moraine
B
Feet
850—
Feet
— 850
Basal till
St. Joseph River Valley
flows
silt
Lagro megasequence (Erie Lobe)
sand
silt
800—
Basal till
Proximal flows
— 800
Supraglacial Complex
Huntertown megasequence (Saginaw Lobe)
Basal till (sandy loam)
Head of fan
Fan body
750—
Proximal outwash apron
— 750
Sandy delta lobe
Trafalgar megasequence (Huron–Erie Lobe)
700—
— 700
Not to scale horizontally

Fort Wayne Moraine (North Limb)

The north limb of the Fort Wayne Moraine (Fig. 4) constitutes the inner core of the wide belt of morainal topography that characterizes northern Allen County. It is composed of a stack of silty-clay to silty-clay loam diamictons that collectively range from 10 m to more than 25 m thick. It is separated from the north limb of the Wabash Moraine only by the relatively narrow valley of the St. Joseph River, which acted as the ice-marginal channel when the moraine was constructed. The moraine generally forms a broad upland as wide as 7 km with a well-defined central ridge crest; the land surface typically is strongly rolling, but becomes increasingly hummocky to the northeast. Crestal elevation rises gradually to the northeast; the greatest elevations are associated with the highest uplands on the buried Trafalgar surface. Thus, parts of the moraine appear to be palimpsest over parts of an older, Huron-Erie Lobe morainal highland, a relationship previously noted for several Erie Lobe Moraines in the Lake Erie Basin (see Bleuer, 1974, for a summary). The morainal topography of the buried Trafalgar surface in this part of the county is punctuated by scattered embayments and other local basins. The latter features comprise the irregular northern fringe of the larger basin of ancestral Lake Erie, and they almost universally contain deltaic sequences of both northern and eastern source. The glaciolacustrine

---

Fig. 6 (previous page).—Sections illustrating gamma-ray log-defined facies of the basal outwash association, supraglacial complex and glaciolacustrine sequences of the Saginaw Lobe, and fine-grained confining diamicts of the overlying Lagro megasequence (Erie Lobe). Heavy stippled lines represent regional bounding surfaces between major depositional systems. The deltaic sequence shown in (A) exhibits a persistent log signature that is readily traceable across the eastern part of the section, whereas the Lagro Formation shows a sharp facies change between the blocky basal tills that cap the moraine and the heterogeneous waterlain assemblages of the lake plain. Note also the nearly 80 ft of relief on the buried surface of the Trafalgar megasequence and the massive, diamicton-dominated vertical sequence that typically cores the highlands on this surface beneath the Fort Wayne Moraine. In (B) the contact between the basal outwash apron and the overlying till sheet in the Saginaw Lobe sequences is abrupt, suggesting it is erosional and perhaps a result of rapid ice advance. The repetitive, blocky-logging nature of the inner, constructional Erie Lobe sequence is evident in the logs from the Wabash Moraine, suggesting a stack of clayey lodgment tills. The Saginaw Lobe tills exhibit a similar blocky pattern, but their much lower response reflects their sandy texture. One of several massive outwash fans of the Saginaw Lobe that underlie the Erie Lobe diamictons in the northern "interlobate" part of the moraine is evident in the northwestern part of (B). Logs in (A) generally lie along section line A'-A'' shown in Figures 4 and 5; logs in (B) are from just west of point A'.

sequences contain both proglacial and periglacial facies and thus exhibit considerable internal variability at places. They also contain most of the aquifers that occur below this terrain.

With some exceptions, buried morainal highlands on the Trafalgar surface are generally composed of thick sequences of loam-textured diamicts that contain only a few, small, widely scattered sand intercalations (Fig. 6). Most of these paleouplands are inferred to have stood above the general level of sedimentation that took place in this part of ancestral Lake Erie, hence most northern-source deposits tend to be of limited extent. Thus, the superposition of the clayey Erie Lobe tills on these buried highlands results in very thick, relatively homogeneous, composite diamict sections that commonly extend to bedrock at depths of 50 to 75 m. At many places, few aquifers of any size occur within these sections, and the buried highlands locally are areas of very limited groundwater availability.

Parts of the Fort Wayne Moraine are internally drained, but much of this rolling landscape is moderately well-drained by surface runoff. The moraine is associated with relatively high water levels in underlying aquifers and appears to act as a groundwater flow divide for shallow aquifers. A relatively active zone of shallow groundwater probably circulates under water-table conditions in the weathered and fractured till within a few meters of the surface; however, the thick sequence of clayey till appears to greatly retard any interchange between this shallow system and subjacent aquifers. Moreover, the predominantly glaciolacustrine aquifers are relatively fine-grained and are commonly isolated from one another by muddy facies. These hydrostratigraphic characteristics suggest that groundwater beneath this terrain is likely to be characterized by long residence times; this conclusion appears to be borne out by the distinctly more mineralized groundwater in this area, relative to other terrains to the northwest (Fleming and Yarling, 1994).

St. Joseph River Valley

The St. Joseph River Valley (Fig. 4) is the narrow, ice-marginal sluiceway that parallels the distal side of the Fort Wayne Moraine and separates that moraine from the more massive Wabash Moraine to the northwest. The valley bottom is underlain by 5 to 20 m of Erie Lobe outwash and is flanked by extensive, high-level outwash terraces that stand 15 to 20 m above the modern flood plain. The outwash consists chiefly of sand within an overall, fining-upward valley-train sequence that is inset into older sequences.

Several significant features of this valley not readily apparent from its surficial expression contribute to considerable subsurface heterogeneity and numerous discontinuities among bodies of contrasting physical properties (Figs. 5 and 6; see also Fig. 6 of Fleming, this volume). For example, the sluiceway overlies a major transition in the subjacent Saginaw Lobe depositional system, from predominantly subaerial sequences (basal outwash association and supraglacial complex) northwest of the river, to subaqueous sequences (glaciolacustrine association) to the southeast. Consequently, sequences below the valley-train outwash may consist mainly of glaciolacustrine sediments, basal outwash and sandy diamictons, or contain elements of both facies associations. A prominent upward-coarsening element is evident in many of these local sequences.

These relations are further complicated by the irregular nature of the Trafalgar surface, which contributes to numerous local facies changes in the depositional system. In yet another variation, some protuberances of the buried surface are sufficiently elevated to have caused surface outwash locally to be deposited directly over thick sections of hard Trafalgar diamicts. The deceptively simple surface form of the river valley thus belies a relatively complicated terrain in which the Erie Lobe outwash is inset into a remarkable assortment of older sequences in places.

The St. Joseph River Valley is the regional discharge area of the Huntertown aquifer system, and upward gradients are present across the sequences below the surface outwash. The elevation of the potentiometric surface equals or exceeds river level at most places, indicating that the aquifer system is, in fact, discharging upward into the river or into the outwash below. An average of 14 million gallons (53 million L) per day are discharged along only one 10-mile (16 km) stretch of the river (Arvin, 1989). Much of the discharge probably is focused where Erie Lobe outwash is incised into fan-deltas or other sizable aquifers, creating a direct hydraulic connection with the river. Although primarily a discharge area, the valley is susceptible to rapid, direct recharge through the sandy outwash in the valley bottom and fringing terraces. Most of this recharge probably flows directly to the river, although the presence of tritiated water in several wells developed in aquifers below the surface outwash indicates that some recharge reaches the upper parts of the aquifer system.

Wabash Moraine

The northern limb of the Wabash Moraine (Fig. 4) constitutes the outer part of the broad belt of constructional Erie Lobe topography that dominates northern Allen County. It forms a ridged upland up to 15 km wide with a prominent crest that locally stands some 20 to 35 m above outlying terrains. It is cored by thick (up to 35 m), repetitively sheared sequences of massive, silty-clay to silty-clay-loam diamicts that are locally intercalated with a variety of silty pond sediments, small channel fills and several much larger, wedge-shaped or irregular sand bodies that pierce the diamicton sequence in places.

The character of the moraine gradually changes northward across Allen County as it becomes increasingly superposed upon hummocky supraglacial topography of the Saginaw Lobe. The crestal area loses its distinctly ridged aspect, giving way to an increasingly broad and somewhat amorphous belt of hummocky topography that appears to be partly inherited from the buried Saginaw Lobe surface. Enclosed depressions and abandoned meltwater channels also become more prevalent, resulting in a preponderance of internally drained areas and peatlands. The northward change in surface morphology coincides with somewhat greater compositional variability of the Erie Lobe sequences. In places they contain a significant proportion of intercalated sand, silt and stratified diamictons, whereas well-developed massive basal till sequences are less common and appear to occur in discrete streamlined ridge-forms. All of the Erie Lobe sequences in the northernmost segment of the moraine overlie an extensive supraglacial sequence of the Saginaw Lobe that contains several large ice-contact fans (Figs. 5, 6 and 7). Basal outwash facies and associated diamictons are widespread beneath the supraglacial complex in most of this terrain.

The Huntertown aquifer system is extremely well confined below virtually all of the moraine. A very shallow zone of saturation typical in the confining unit interacts with small streams and numerous depressional bogs in places. No physical or geochemical evidence has, however, been found to suggest that any appreciable recharge is occurring through the 20 to 35 m of clayey diamicts in the moraine. During the construction of several wells, cuttings of basal till and other diamicts taken from depths of 15 to 30 m appeared hard and unsaturated, suggesting that little vertical flow is occurring. Moreover, the considerable rise in surface elevation associated with the moraine appears to have no corresponding effect on the potentiometric surface of the aquifer system, whose southeastward gradient toward the St. Joseph River continues unaffected beneath the moraine. Some recharge may be occurring at places in the slightly more heterogeneous northernmost part of the moraine, chiefly beneath high-level meltwater channels and wedge-shaped sand bodies, where the confining unit is thinned or pierced and the surface of the aquifer system is relatively elevated.

Cedar Creek Canyon

Cedar Creek Canyon (Fig. 4) is a deeply entrenched meltwater channel that cuts straight across the north limb of the Wabash Moraine between the Eel River Valley and the St. Joseph River. It consists of two segments of differing morphology. The upper segment, closest to the Eel River Valley, forms a perfectly straight, narrow, gorge-like channel about 7 km long whose flat floor is as much as 30 m below the adjoining moraine crest. The more irregular lower segment is characterized by less imposing slopes and several sharp bends that impart a prominent rectilinear drainage pattern.

The canyon is floored by a fining-upward sequence of sand and gravel that is thickest in the upper segment nearest the outlet. In the upper canyon the Erie Lobe outwash is superposed atop several large outwash fans and other supraglacial deposits of the Saginaw Lobe (Fig. 5), giving rise to considerable variability in total sand and gravel thicknesses. By contrast, the regional southeastward facies change within the Saginaw Lobe system leads to an increasing number of fine-textured diamicts and lacustrine sediments underlying the surface outwash in the lower segment. Total sand and gravel thicknesses range from as little as 1 m in the lower canyon to more than 25 m in parts of the straight upper segment. The outlet of the canyon is at the head of a coarse outwash fan that lies within the southeast side of the Eel River Valley (Fig. 4).

Cedar Creek is primarily a local discharge area for the Huntertown aquifer system; the potentiometric surface forms a small valley around the canyon, indicating inward flow. Most of the discharge is inferred to occur in the lower segment, where the facies change between the thick, transmissive outwash aquifers upgradient to the northwest and the thinner, less continuous lacustrine and distal outwash aquifers to the southeast may create zones of upwelling.

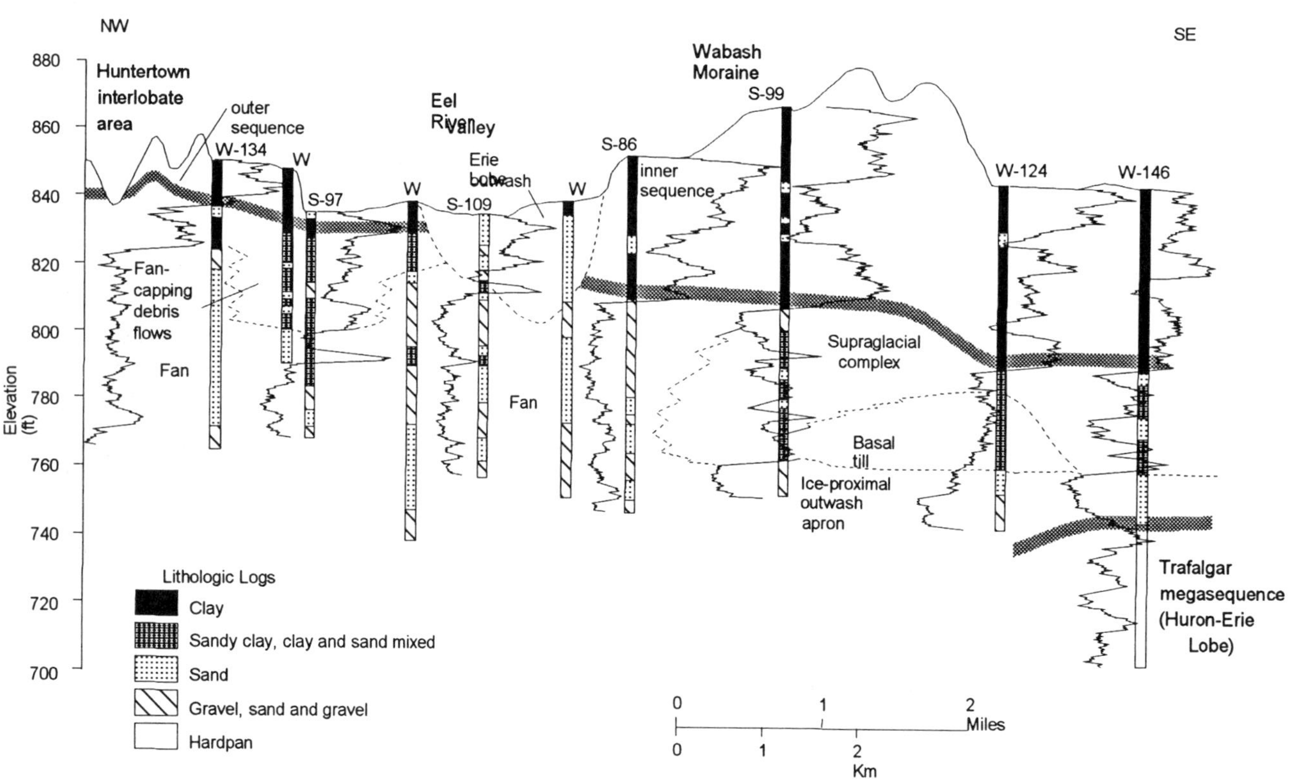

FIG. 7.—Subsurface profile from northwestern Allen County showing gamma-ray log-defined proglacial, fan and supraglacial facies of the Huntertown megasequence, and architecture of the inner Erie Lobe sequence in the Wabash Moraine. Heavy stippled lines mark regional bounding surfaces between major depositional systems. Lighter dashed lines are facies boundaries within each sequence. Lithologies reported in several well logs are also shown for comparison. The large fan in the western half of the diagram is as much as 30 m (100 ft) thick and fills a massive "hole," or local basin, in this part of the depositional system. See Figure 9 for location of section line.

Huntertown Interlobate Area

The Huntertown interlobate area (Fig. 4) is the most heterogeneous terrain in Allen County. It forms a broad, northeast-to-southwest-trending belt of rolling, hummocky topography that is situated outboard of the Wabash Moraine. It is bisected down its long axis by the closely related Eel River Valley, which is discussed in the following section. The landscape is almost entirely internally drained, being marked by numerous enclosed depressions, steep-sided irregular ridges and hummocks and abandoned meltwater channels of various sizes.

Uplands in this terrain are veneered by the outer Erie Lobe sequence, which is rarely more than several meters thick and exhibits many discontinuities (Figs. 5 and 8). Local sequences consist of a variety of clayey to loamy diamictons, as well as small intercalated silt units and sandy or gravelly channel fills. The Erie Lobe sequences are draped over a relict supraglacial landscape of the Saginaw Lobe that is generally cored by a variety of thin diamictons and intercalated ice-contact stratified units and channel deposits; several extensive ice-contact fans that outline former ice margins also are present beneath this region. The supraglacial complex is exposed in the bottoms and sides of numerous depressions and small ravines, where Erie Lobe sequences are absent. Many of the depressions extend below the water table and thus contain wetlands atop sections of peat and muck. The basal outwash association forms the base of the Saginaw Lobe system over much of the terrain, although its thickness ranges widely.

The Huntertown interlobate area is characterized by a distinctive hydrogeologic regime (Fig. 8) that reflects a close relationship among surface drainage, wetlands and groundwater recharge (Fleming, 1994b). Hummocks are typified by steep slopes and poorly drained soils that lead to considerable surface runoff into adjacent depressions. It is not uncommon to see as much as 1 to 2 m of water standing in depressional wetlands following major spring and summer storms. In contrast, hydraulic head in deeper parts of the aquifer system is as much as 2 to 5 m lower than the

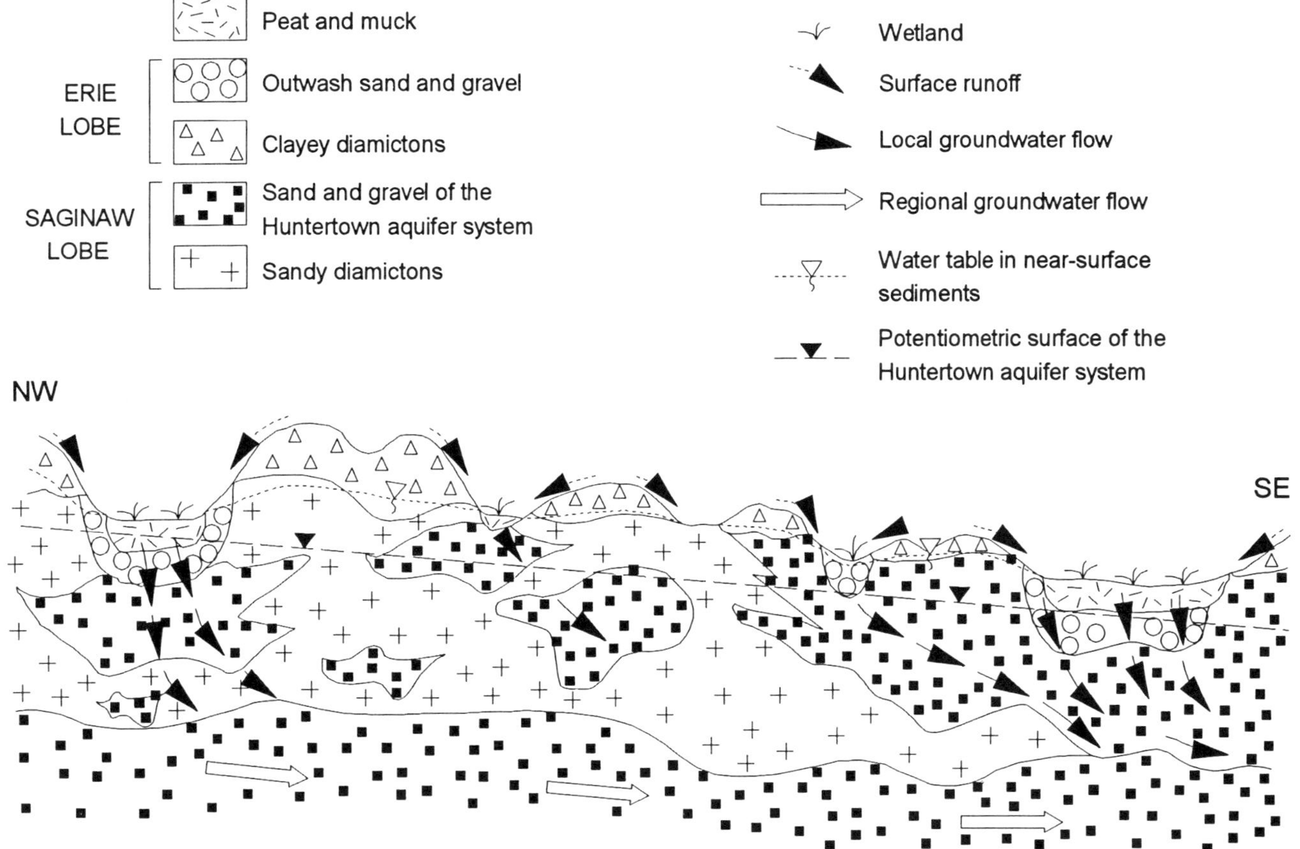

FIG. 8.—Schematic diagram illustrating hydrogeologic components of the Huntertown interlobate area.

water table represented by these wetlands, resulting in a rather pronounced downward gradient in many parts of the terrain. Ponded water seeps into the underlying aquifer system largely unimpeded; consequently, many depressions drain within several days or less. Together with the Eel River Valley, the location of this terrain coincides with a prominent flat region of the potentiometric surface, and the aquifer system contains tritiated groundwater at many places (Fleming and Yarling, 1994). These relations indicate that this terrain functions as a regional recharge area for the Huntertown aquifer system.

Eel River Valley

The Eel River Valley (Fig. 4) lies along the axis of the Huntertown interlobate region. It is a polygenetic meltwater channel that experienced a complex drainage history involving outwash alluviation from both northern and eastern sources as well as subsequent stream piracy and channel abandonment. Today the Eel River "Valley" of northwest Allen County constitutes an abandoned high-level sluiceway that is crossed by the drainage divide separating Wabash (Mississippi) River drainage from Great Lakes drainage. Most of this terrain contains no natural through-flowing streams (although extensive peatlands are abundant), and the surface elevation of its "bottom" is barely below that of surrounding areas.

Typical sequences below the valley surface are composed of 3 to 12 m of gravelly Erie Lobe sand that is commonly superposed atop Saginaw Lobe fan and supraglacial sequences (Figs. 5 and 7). The Erie Lobe outwash was derived mostly from the outlet of Cedar Creek Canyon, where it forms coarse-grained, proximal (cobbly) and medial (gravelly) fan facies. The outwash becomes somewhat finer, although still somewhat gravelly, to the southwest, where it occurs in a variety of streamlined bedforms typical of a high-energy, moderately confined sluiceway. The northeastern half of the sluiceway overlies medial and distal sequences of a large Saginaw Lobe fan. These fan sequences commonly exhibit an upward-coarsening pattern typical of prograding fan lobes, locally interrupted by short, fining-upward intervals suggestive of abandoned distributaries (Fig. 7). Unbroken sand and gravel sections in excess of 30 m are locally present below this part of the valley, and the northern-source fan sequences extend a considerable distance in either direction beneath capping sediments of the adjacent Huntertown interlobate terrain. Discontinuous lenses of sandy diamictons locally present near the top of the fan separate it from the overlying valley train

outwash; diamict bodies are larger and more common in the supraglacial sequences underlying the valley train elsewhere in the valley. The glaciofluvial sequences locally are capped by up to 10 m of peat, which filled the former sluiceway channel following its abandonment.

The Eel River Valley represents the most productive region of the Huntertown aquifer system, but it is also the most sensitive to contamination. Large, properly constructed wells in or adjacent to the valley commonly are capable of producing a million gallons per day or more. The combination of thick, unconfined sections of outwash, shallow water table and a mostly flat landscape result in high recharge potential, a condition manifested by the virtually flat potentiometric surface. A sharp downward gradient is evidenced in places by a 1- to 2 m hydraulic head difference between the lower parts of the sand and gravel and the water table. A considerable amount of recharge does, in fact, appear to reach deeper parts of the aquifer system because some wells as deep as 35 m exhibit moderately tritiated water (10 to 30 T.U.).

Several sizable peatlands in the valley receive runoff from small streams and ravines that drain adjacent hummocks within the Huntertown interlobate area. Like the depressional wetlands in the adjacent region, these peatlands also may act as areas of focused recharge. Considerable quantities of chemical fertilizers and pesticides are applied to the predominantly agricultural land in much of the Eel Valley and adjacent Huntertown interlobate area. Yet these compounds have not been detected by extensive sampling of wells in the area (Indiana Dept. Environmental Management, unpub. data), despite the high potential for contamination (e.g., Aller et al., 1987). This somewhat surprising relationship suggests that processes within the wetlands, which play such a central role in the recharge mechanism of this area, also may mitigate the effects of agricultural chemicals carried by surface runoff.

DISCUSSION

In its most elemental form, a glacial terrain map should convey the intimate relationships in the glacial and near-glacial environment among process (e.g., geologic history and modern surface processes), material (depositional sequence) and form (landscape). The area shown in Figure 9 is located in the northwestern part of the county, along the juncture of the Huntertown interlobate area, Eel River Valley and interlobate segment of the Wabash Moraine. Abrupt facies changes and diverse sequences are characteristic of these terrains (Fig. 7). Landforms and near-surface materials also are highly varied (Fig. 10), reflecting the wide range of subsurface conditions as well as a host of meltwater events during the latest glacial episodes.

Due to this great diversity, map units may be most meaningful when designed to illustrate the *range of conditions* present within their boundaries. Consequently, each terrain unit denotes a particular set of landscape elements, vertical sequences and facies associations known or inferred to occur within its mapped bounds. In general, terrain map units may be defined geographically, morphologically and (or) genetically, depending on the defining features of the terrains and on the kinds of information the map is attempting to convey. The names of most regions typically involve a geographic location combined with either a genetic or morphologic descriptor (e.g., Huntertown interlobate region, Eel River Valley), although the geographic name may be omitted (e.g., interlobate morainal regions) when the terrain occurs in two or more disjunct places. In this example, a hierarchical system of map units is used to illustrate major terrain regions as well as relatively distinctive subregions (i.e., terrain elements) within each region.

Each terrain unit mapped in Allen County not only describes a set of morphologic and sequence elements but also connotes a specific series of events that differ in one or more fundamental respects from events in neighboring terrains. The Huntertown interlobate area, for example, defines a broad region wherein the Erie Lobe initially overrode large-scale ice-contact stratified deposits and stagnant ice of the Saginaw Lobe. Subsequent collapse of the buried ice resulted in abundant coalesced depressions and irregular meltwater channels that characterize the region, as well as the discontinuous cap of Erie Lobe sediments that mantles hummocks and other uplands. In contrast, the Eel River Valley represents an abandoned high-level sluiceway that acted as a marginal channel for both the Saginaw and Erie Lobes. Torrential meltwater discharges from the Erie Lobe during its latest advance scoured the valley, leaving a thin veneer of Erie Lobe outwash over a very thick, exhumed fan complex deposited along a previous Saginaw Lobe margin.

Consistent variations on the general theme can be mapped within all the terrains (Fig. 9). These subregions have experienced the same basic history as the rest of the terrain, but they differ in one or more important respects. Most commonly, subregions are characterized by the absence (e.g., unit Ho) or a significant thickening (e.g., units Hr, MHr) of an important part of the sequence, but they may also be identified by morphologic expression (e.g., unit Mm) or by a significant but localized departure from the expected sequence. In any event, each subregion is a distinct mappable unit but is not sufficiently different in overall aspect to constitute a separate terrain.

The depositional systems-based foundation of terrain mapping provides insights into the geometry and sedimentologic characteristics of both surface and buried lithofacies that might not be evident in traditional forms of mapping. Previous surficial mapping in this area, for example, identified the Eel River Valley as a valley train (Johnson and Keller, 1972; Gray, 1989), which apparently is only the latest, uppermost part of the sequence there. In view of the history of the Eel River Valley inferred from the underlying sequences, however, the sand and gravel within the valley actually consists of two or more superposed units that can be expected to have very different geometries and internal properties. The relatively thin, upper (Erie Lobe) outwash does, in fact, form terraces and channel deposits confined to the valley. In contrast, the lower, exhumed (Saginaw Lobe) unit, which forms the great bulk of the sand and gravel, extends northward in the subsurface from the valley wall for a distance of at least 3 to 4 km, rising sharply in elevation to the buried head of the fan along a former Saginaw Lobe margin. A similar comparison arises in the Huntertown interlobate area, most of which traditionally has been shown as "Erie Lobe till" on regional geologic maps (Johnson and Keller, 1972; Gray, 1989). Over large parts of this "till" area, as much as 50 percent of the surface is underlain by various combinations of Saginaw Lobe deposits, Erie Lobe outwash and peat, albeit mostly in very small tracts exposed through discontinuities in the Erie Lobe "till" cover (Fig. 10). Likewise, much of the "till" within parts of the region actually underlain by fine-grained Erie Lobe sequences is in reality a varied assemblage of debris flows, pond sediments and small glaciofluvial units that exist in complicated facies relations with one another and with true tills according to the specific position in the landscape. Recognizing these relationships is critical to understanding the history, structure and hydrogeology of this area, and points out the complications of basing hydrogeologic interpretations on a strictly surficial materials-mapping approach. On the other hand, mapping terrains and terrain elements represents a practical alternative for showing the range of conditions present and which of those conditions likely are associated with different parts of the landscape.

Terrain maps and their derivatives have a wide range of practical applications, hydrogeologic and others. The greatest value of this approach may derive from its emphasis on defining natural variation in geologic systems and the

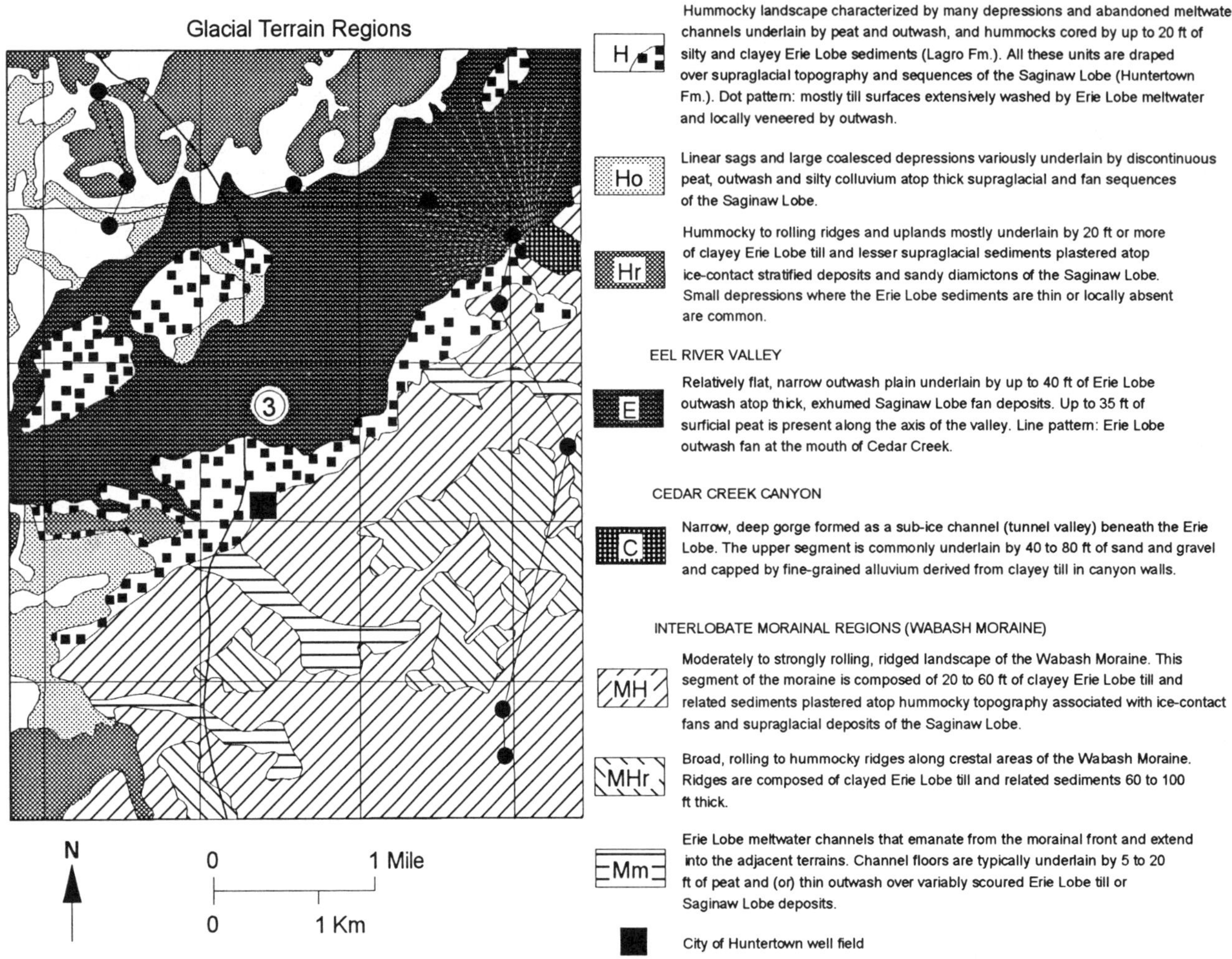

FIG. 9.—Map showing details of glacial terrains and their explanations in a small area in northwestern Allen County along the juncture of the Wabash Moraine, Eel River Valley and Huntertown interlobate area. Location of map area is shown in Figure 4.

ability to depict heterogeneities. This certainly is true for regional studies, such as those encompassing an entire depositional or hydrogeologic system, wherein large-scale trends and features are depicted in the form of maps of terrain and hydrogeologic settings (Figs. 9 and 11).

However, these benefits also commonly apply on a much more localized basis. Numerous examples of local sequences and conditions derived from analysis of the depositional system illustrate important details that may be critical to proper understanding of a specific issue or site. This can be illustrated by conditions around the Huntertown municipal well field (Fig. 12). The well field, situated at the toe of the Wabash Moraine, is developed in basal outwash of the Saginaw Lobe sequence. If a capture-zone delineation and wellhead protection strategy were to be based purely on the geological characteristics evident from boreholes at the well field (a common procedure), it is uncertain whether the outcome would reflect the full hydro-geologic setting. The well field is situated where the basal outwash aquifer is relatively well confined, as is most of the area affected by the immediate zone of influence of the well field. Analysis of depositional systems in this area indicates that the same aquifer is unconfined as little as 0.5 km to the north in the Eel River Valley, where the combination of landscape and hydraulic characteristics have created a regional recharge area. The long-term (10+ years) zone of contribution for the well field extends northward into the recharge area; it is clear then, based on the hydrogeologic settings and probable flow paths, that equal or greater effort should be directed at protecting this more

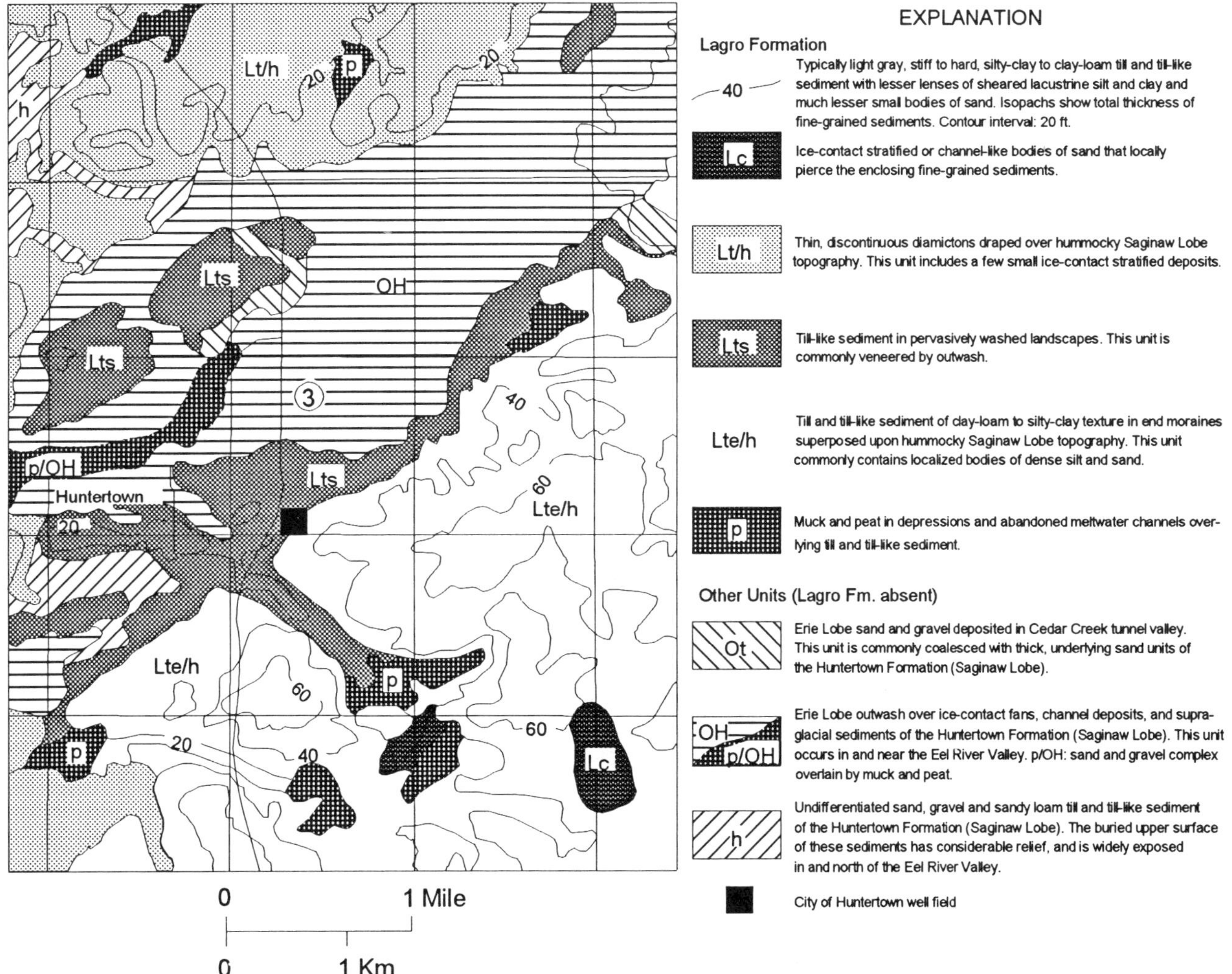

FIG. 10.—Map derived from terrain map of Figure 9, showing near-surface glacial and postglacial sequences.

distant, but much more sensitive, recharge area than the better confined parts of the aquifer close to the well field.

SUMMARY AND CONCLUSIONS

Glacial terrains are composed of distinctive groups of landforms and the underlying vertical sequences that constitute those forms. Glacial terrains include the characteristic assemblage of surface topographic forms and associated surficial materials familiar to traditional Quaternary geologic maps, but they are ultimately defined by the relationships within and between entire sedimentary sequences. As elements of larger depositional systems, terrains and express the different relations within those systems among process, composition and form. Individual lithostratigraphic units tend to be less important in the definition of terrains than the total facies associations within

which such units occur. The mapping of terrain emphasizes the nature of facies variation and its relationships to landscape elements and to the distribution of heterogeneity and discontinuities in internal properties of depositional sequences. The cornerstone of glacial terrain analysis is the extensive use of downhole geophysical logs and associated sample sets to establish sequence over broad depositional basins.

Glacial terrains in northern Allen County represent the transitional southern terminus of the interlobate region of northeastern Indiana. Individual terrains reflect the superposition and interaction of two large, latest Wisconsin depositional systems, and the relationship of those systems to the depositional basin, as represented by the surface of a third depositional system. Superposition of these systems results in a series of composite terrains whose overall characteristics depend on the distributions of facies within each

43

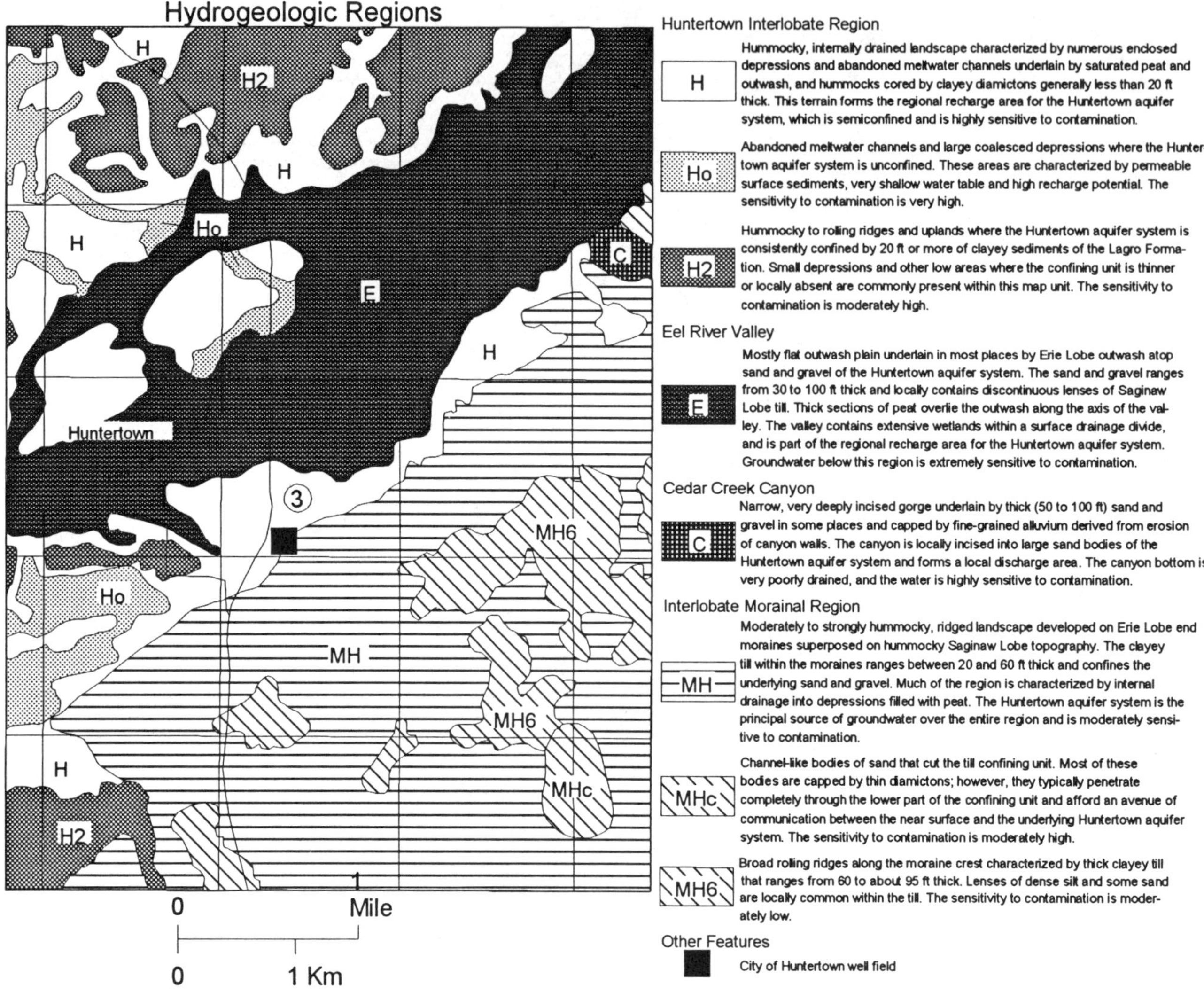

FIG. 11.—Map showing hydrogeologic settings derived from terrain units of Figure 9.

system as well as the effects of buried or recently deglaciated surfaces of one or more systems on the behavior of succeeding ice sheets. The key to understanding and depicting these terrains depends on recognition of fundamentally different types of sequences and physical properties associated with depositional systems from northern- and eastern-source areas.

The main elements that constitute a glacial terrain also are the major determinants of the local hydrogeologic regime. Sequence characteristics control the presence, hydraulic properties and configurations of particular aquifers and confining units and thus determine local hydrostratigraphy. Likewise, surface terrain, which provides the elevation differential that helps drive groundwater flow, thus affects the distribution of recharge and discharge areas as well as the specific mechanisms involved in these processes.

All these hydrogeologic elements are dependent on the character of entire sequences and landscapes; hence, glacial terrains are representations of hydrogeologic settings.

Terrain maps are designed to illustrate heterogeneity in local vertical sequences associated with specific terrains as well as across entire depositional systems. They are useful for a variety of applications at both regional and local scales, and they serve as a descriptive and practical foundation for many kinds of derivative maps. Terrain maps and their derivatives are especially useful for analyses of regional aquifer systems, such as the Huntertown aquifer system presented in the next paper (Fleming, this volume). Understanding the distributions of hydrofacies and their internal variation in such systems is essential for identifying the function of hydrogeologic settings. On the other hand, terrain maps also provide hydrogeologic insight in

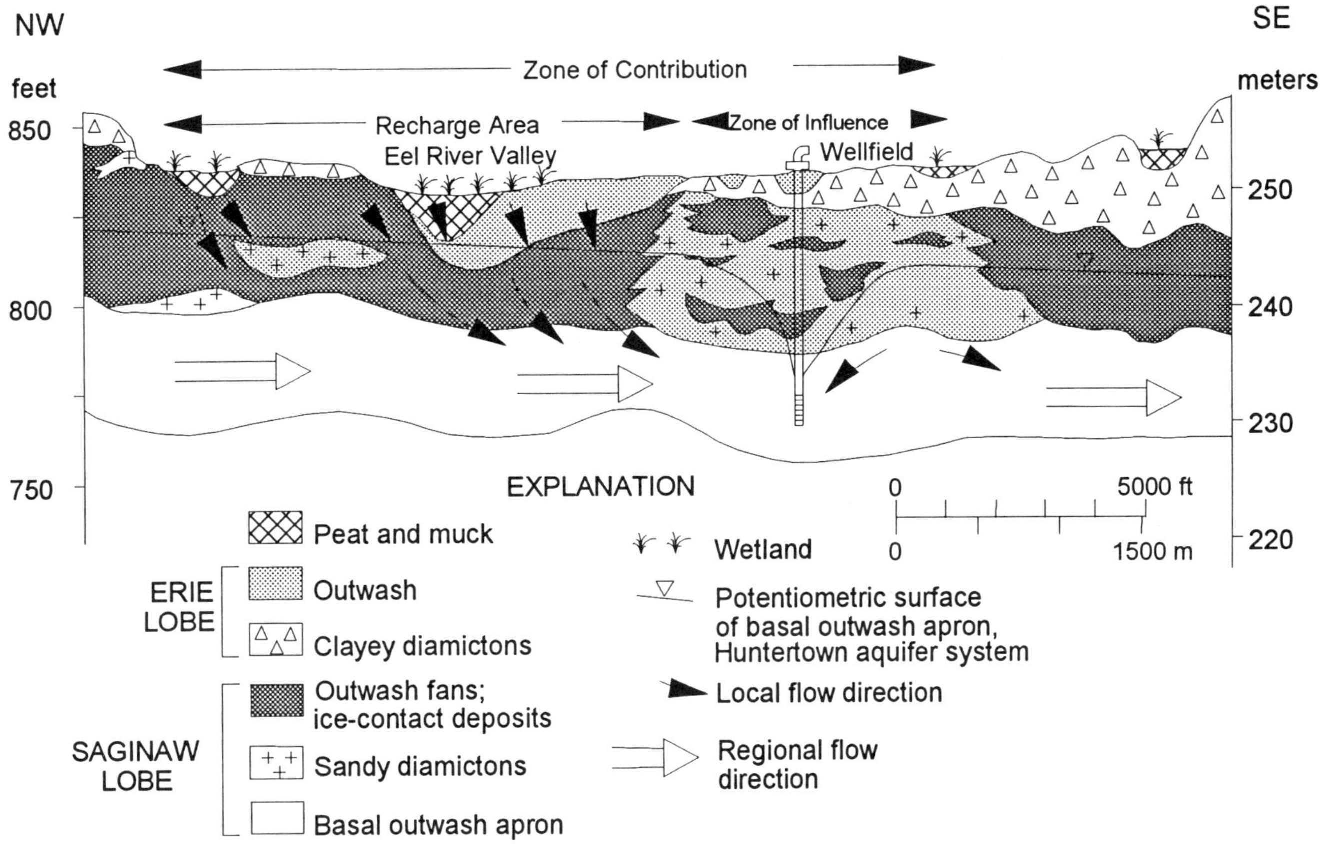

F I G. 12.—Cross section showing hydrogeologic characteristics in the vicinity of the City of Huntertown public well field.

specific areas. The range of conditions represented by terrain map units can provide a regional, three-dimensional context for site evaluations, and may help in designing drilling programs and maximizing interpretation of data from isolated boreholes.

ACKNOWLEDGMENTS

Many of the results presented herein were derived from a hydrogeologic study of Allen County funded in large part by the Allen County Board of Commissioners and coordinated by the Allen County Department of Planning Services. I particularly wish to acknowledge Kimberly Bowman and Dennis Gordon of the latter agency; their interest and initiative in understanding and protecting the county's groundwater resources directly led to this study and to the idea for developing hydrogeologic settings from glacial terrain units. Many of the ideas concerning glacial terrains of northern Indiana and the techniques for their characterization evolved through numerous discussions with N.K. Bleuer of the Indiana Geological Survey, who pioneered the application of downhole geophysical logging to glaciated terrains in Indiana. I am also grateful to Michael Yarling of the Indiana Department of Environmental Management, who shared geochemical and isotopic data from the Huntertown aquifer system. Finally, I thank the many water well contractors who so generously shared their expertise and allowed us to sample and log numerous wells.

REFERENCES

ALLER, L., BENNETT, T., LEHR, J., PETTY, R., AND HACKETT, G., 1987, DRASTIC—A standardized system for evaluating ground water pollution potential using hydrogeologic settings: U.S. Environmental Protection Agency Report EPA 600/2-87-035, 622 p.

ARVIN, D.V., 1989, Statistical summary of streamflow data for Indiana: U.S. Geological Survey Open-File Report 89-62.

BLEUER, N.K., 1974, Buried till ridges in the Fort Wayne area, Indiana, and their regional significance: Geological Society of America Bulletin, v. 85, p. 917-920.

BLEUER, N.K., 1985, The nature of some glacial and manmade sedimentary sequences and their downhole logging by natural gamma ray, *in* West, T.L., ed., Building On and With Sedimentary Bedrock: Proceedings of the 36th Annual Highway Geology Symposium, p. 204-219.

BLEUER, N.K., AND MOORE, M.C., 1972, Glacial stratigraphy of the Fort Wayne, Indiana, area and the draining of Glacial Lake Maumee: Proceedings of the Indiana Academy of Science, v. 81, p. 195-209.

BLEUER, N.K., AND MOORE, M.C., 1974, Buried pinchout of Saginaw Lobe drift in northeastern Indiana: Proceedings of the Indiana Academy of Science, v. 84, p. 362-372.

BLEUER, N.K., AND WOODFIELD, M.C., 1993, Glacial terrain, sequences, and aquifer sensitivity, Porter County, Indiana: Indiana Geological Survey Open-file Report 93-2, 90 p.

EGGERT, D.L., BLEUER, N.K., HARPER, D., AND WOODFIELD, M.C., 1995, Terrain mapping and seismic hazard analysis, Evansville, Indiana, USA, *in* Proceedings, Pacific Conference on Earthquake Engineering: Australian Earthquake Engineering Society and New Zealand National Society for Earthquake Engineering, Melbourne, Australia.

EYLES, N., 1983, Glacial geology—A landsystems approach, *in* Eyles, N., ed., Glacial Geology—An Introduction for Engineers and Earth Scientists: Oxford, Pergamon Press, p. 1-18.

FLEMING, A.H., 1992, The hydrogeologic framework of Allen County, Indiana—Hydrostratigraphic atlas emphasizing subsurface sequence stratigraphy and ground-water contamination potential: Indiana Geological Survey Open-File Report 92-14, 55 p.

FLEMING, A.H., 1994a, The hydrogeology of Allen County, Indiana— A geologic and ground-water atlas: Indiana Geological Survey Special Report 57, 111 p.

FLEMING, A.H., 1994b, Origin and hydrogeologic significance of wetlands in the interlobate region of northwest Allen County: Proceedings of the Indiana Academy of Science, v. 103, nos. 3-4, p. 147-166.

FLEMING, A.H., 1995, Glacial geology of the Maumee River Basin: Unpub. report prepared for the Indiana Department of Natural Resources, Division of Water.

FLEMING, A.H., BROWN, S.E., AND FERGUSON, V.R., 1993, The hydrogeologic framework of Marion County, Indiana—Atlas emphasizing hydrogeologic terrain and sequence: Indiana Geological Survey Open-File Report 93-5, 67 p.

FLEMING, A.H., AND YARLING, M., 1994, Facies distributions, recharge-discharge relations, and aquifer sensitivity in a glacial aquifer system, northeastern Indiana (abs.): Geological Society of America, Abstracts with Program, v. 26, no. 5, p. 15.

GOODING, A.M., 1973, Characteristics of Late Wisconsinan tills in eastern Indiana: Indiana Geological Survey Bulletin 49, 28 p.

GRAY, H.H., 1989, Quaternary geologic map of Indiana: Indiana Geological Survey Miscellaneous Map 49.

JOHNSON, G.H., AND KELLER, S.J., 1972, Geologic map of the 1° × 2° Fort Wayne Quadrangle, Indiana, Michigan, and Ohio, showing bedrock and unconsolidated deposits: Indiana Geological Survey Regional Geologic Map 8.

LEVERETT, F., AND TAYLOR, F.B., 1915, The Pleistocene of Indiana and Michigan and the history of the Great Lakes: U.S. Geological Survey Monograph 53, 529 p.

LINEBACK, J.A., BLEUER, N.K., MICKELSON, D.M., FARRAND, W.R., AND GOLDTHWAIT, R.P., 1983, Quaternary geologic map of the Chicago 4° × 6° Quadrangle: U.S. Geological Survey Miscellaneous Investigations Series Map 1–1420 (NK-16).

MALOTT, C.A., 1922, The physiography of Indiana, *in* Handbook of Indiana Geology: Indiana Department of Conservation Publication 21, part 2, p. 59-256.

WAYNE, W.J., 1958, Glacial geology of Indiana: Indiana Geological Survey Atlas of Mineral Resources of Indiana, Map 10.

WAYNE, W.J., 1963, Pleistocene formations in Indiana: Indiana Geological Survey Bulletin 25, 85 p.

USING GLACIAL TERRAIN MODELS TO CHARACTERIZE AQUIFER SYSTEM STRUCTURE, HETEROGENEITY AND BOUNDARIES IN AN INTERLOBATE BASIN, NORTHEASTERN INDIANA

ANTHONY H. FLEMING

Indiana Geological Survey, Indiana University/Purdue University, Fort Wayne, Indiana 46805

ABSTRACT: Aquifers and confining units in glaciated basins commonly exhibit a remarkable array of geometries, internal discontinuities and changes in physical and chemical properties at a variety of scales. These heterogeneities commonly control the distributions, both locally and regionally, of productive zones in an aquifer system as well as the locations of preferred pathways for contaminant migration. Glacial terrain models emphasize the nature of variation *within* individual sediment bodies as well as between entire depositional sequences; thus, they provide a useful sedimentologic framework for interpreting hydrostratigraphic complexity at both local and regional scales. The usefulness of these models is illustrated by the Huntertown aquifer system of northern Allen County, which consists of several distinct proglacial, supraglacial and glaciolacustrine hydrogeologic facies tracts whose distribution and internal variability reflect the interaction of the Saginaw Lobe and its meltwaters with contrasting morphologic regions of the depositional surface. Aquifer system properties, such as hydraulic conductivity and transmissivity, and the character of internal and regional system-bounding confining units vary systematically across the depositional basin in response to changes in sedimentary architecture. The development of a glacial terrain model for the depositional systems of northern Allen County provides the essential regional context necessary for interpreting, depicting and forecasting the nature and hydraulic significance of heterogeneities that characterize different parts of this aquifer system and its confining units.

INTRODUCTION

Depositional sequences of continental ice sheets in the upper Midwest often are typified by a close commingling at both large and small scales of sediment bodies with strongly contrasting hydraulic properties and a resulting abundance of lithologic transitions and discontinuities. Resulting aquifer systems are characterized by complex boundary conditions, internal discontinuities and other forms of physical and chemical heterogeneity over relatively localized areas. In many cases such heterogeneities commonly control the distribution of productive zones in an aquifer system as well as the locations of preferred pathways for contaminant migration. Consequently, a wide variety of issue- and site-specific tasks related to the characterization, protection and remediation of groundwater resources can benefit from the availability of models that document the nature and spatial distributions of heterogeneities in an aquifer system and that provide a basis for forecasting their impact on specific parts of the hydrogeologic system.

Until relatively recently, however, few systematic attempts had been made to characterize the distribution and origins of regional heterogeneity in glacigenic aquifer systems, and relatively little information was available concerning the architecture of entire depositional systems and attendant variability. The marked growth in efforts to identify, protect and remediate groundwater resources during the past decade has created a strong impetus for developing new approaches to better define the subsurface in glaciated basins. Definitive studies of groundwater flow and transport processes at both local and regional scales, for example, require information about the three-dimensional relationships among sediment bodies within a sequence and how the distributions of hydraulic properties are affected (Anderson, 1989; 1991; Webb and Anderson, 1991; Straw et al., 1993; and other papers in this volume). To meet this need, facies models, sequence stratigraphy, statistical analyses and other techniques of modern basin analysis long utilized by sedimentologists in other fields (e.g., the petroleum industry) are beginning to find increased application in studies of Pleistocene and older glaciated terrains (Eyles et al., 1985; Ashley et al., 1985; Eyles and Miall, 1984; Levell et al., 1988; Shaw and Ashley, 1988; Bleuer, 1985).

Glacial terrain models utilize extensive downhole geophysical logging and sampling to document relations between local sedimentary sequences and associated landscape elements, which together constitute regional-scale, genetically based glacial-depositional systems, or "megasequences." Terrain models emphasize the nature of variation *within* map units as well as between entire depositional sequences. The concept of glacial terrain analysis and its mapping are presented in a companion paper (Fleming, this volume), which focuses on the relationship between glacial terrain map units and hydrogeologic settings. This paper further illustrates the use of glacial terrain models to derive and depict large-scale facies trends and heterogeneity in a glacigenic aquifer system, and how the origin and distribution of these heterogeneities can be traced back to the depositional environments of the aquifer system and its regional confining (bounding) units. The example presented herein is from the Huntertown aquifer system of northern Allen County (Fig. 1), whose hydraulic properties, recharge-discharge relations, geochemistry and other characteristics are intimately related to the depositional systems that operated in this area and to local differences in resulting terrain (Fleming, 1994; Fleming and Yarling, 1994).

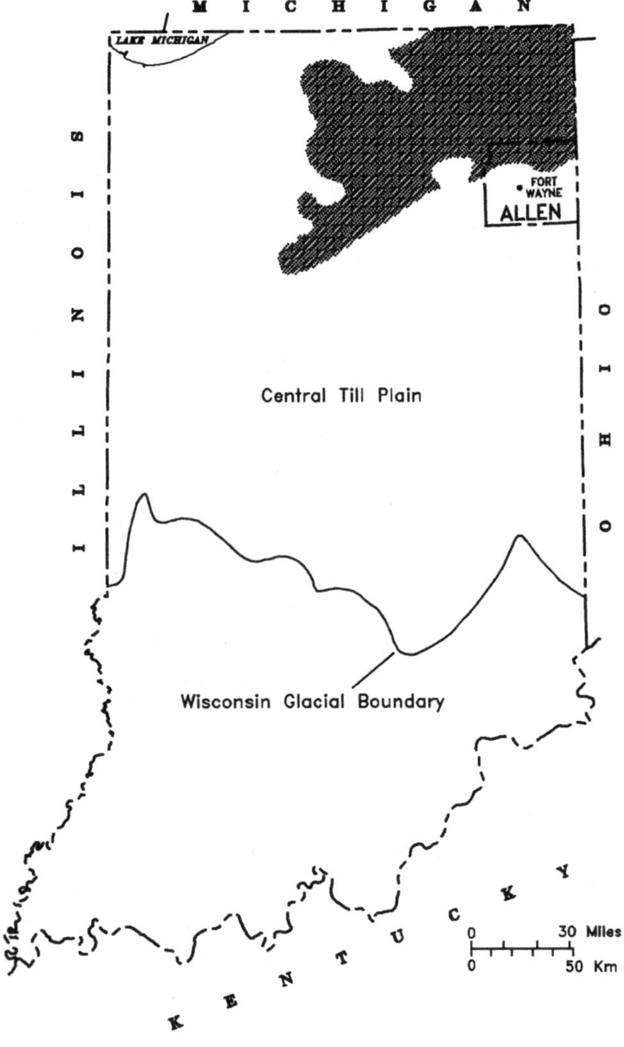

FIG. 1.— Map of Indiana showing location of hummocky supraglacial topography of the northeastern Indiana interlobate area (shaded), central till plain, Allen County and Fort Wayne.

REGIONAL HYDROGEOLOGIC SETTING

General Geologic Setting

Allen County is situated along the transition between the nearly flat central Indiana till plain and the more topographically and geologically diverse interlobate region of northeastern Indiana (Fig. 1). The latter region is characterized by as much as 50 to 70 m of local relief in some areas and is dotted with numerous lakes, wetlands and depressions. The physiographic change generally is mirrored by greater complexity of depositional sequences as well as a concommitant increase in the abundance, size and variety of sand and gravel units within the sequences. These characteristics reflect a predominance of supraglacial processes that operated in the region, which may be attributed partly to the interaction between ice advancing from the east (Erie Lobe) and from the north (Saginaw Lobe) at different times. Northward from Fort Wayne, Allen County is generally typified by this complex interlobate geology and by glacigenic aquifers and aquifer systems that interact directly with surface waters.

Late Wisconsin glacigenic sequences in northern Allen County result from three distinct but somewhat overlapping episodes, each marked by one or more advances of a particular ice lobe (Fleming, 1994, and this volume). The Huron-Erie Lobe was the earliest ice to enter the area; several advances produced a thick sequence of loamy diamictons and associated glaciofluvial units known as the Trafalgar Formation (Wayne, 1963). This was followed by a period of Saginaw Lobe activity that resulted in a markedly sandy sequence containing numerous small to large aquifers that are the focus of this paper. The final episode consisted of two or more advances of the Erie Lobe that left a distinctive suite of very clayey diamictons and stratified sediments known as the Lagro Formation (Wayne, 1963; redefined by Fleming, 1994) and associated end moraines.

Regional Hydrostratigraphic Relations

Productive aquifers occur in all the late Wisconsin depositional systems and in older deposits as well; however, the sequence of Saginaw Lobe deposits constitutes the most extensive, productive and clearly integrated group of glacigenic aquifers in northern Allen county and thus is widely utilized for domestic, industrial, agricultural and public water supplies. These aquifers are collectively referred to as the Huntertown aquifer system

(Fleming, 1992a, 1994). In contrast, the predominantly clay-rich Erie Lobe sequences above the aquifer system comprise a significant regional confining unit. Local confining unit characteristics vary according to (1) type of depositional sequence, which determines the bulk vertical permeability and heterogeneity of the confining unit; and to (2) variations in surface terrain, which determine the nature of groundwater-surface water interaction and thus affect the distribution of recharge and discharge for the subjacent aquifer system. Similarly, the buried surface of the Trafalgar Formation effectively is the base of the Huntertown aquifer system, with the sequence of loamy diamictons below also constituting a regional-scale confining unit. The Huntertown aquifer system and its overlying and underlying eastern-source confining units thus are the main elements of a well-defined shallow hydrogeologic system.

At the scale of the whole system, each of these elements could be considered a regional hydrostratigraphic unit. On the other hand, a variety of local interacting hydrofacies and surface terrain elements is present that greatly affects flow into and out of the system as well as internal properties, such as transmissivity and potential well yields. The distributions of these heterogeneities tend to be closely correlated with certain glacial terrains (Fig. 2; Fleming, this volume). For example, where well-confined by thick clayey diamictons of the Wabash Moraine, the Huntertown aquifer system appears to be characterized by strong horizontal flow and little or no interaction with shallower aquifers or surface water. In contrast, where it is poorly confined in internally drained uplands within the Eel River valley and Huntertown interlobate area, the top of the system is susceptible to rapid, direct recharge from the surface. Similarly, productivity of the aquifer system varies systematically according to the distribution of facies that resulted from the interaction of the Saginaw Lobe and its meltwaters with its depositional basin. Understanding the architecture of the terrains thus is useful for interpreting the distribution of internal aquifer system properties as well as the recharge-discharge function and geochemical characteristics in different parts of the system.

HYDROGEOLOGIC SYSTEMS

The Huntertown Aquifer System

The Huntertown aquifer system can be thought of as having two geologically and hydraulically distinct subsystems. The northwestern part of the system is dominated by relatively large, coarse, subaerially deposited outwash bodies. This depositional sequence generally consists of a widespread basal outwash apron that is succeeded upward by a variable thickness of diamictons (chiefly basal till). These, in turn, are capped by a highly heterogeneous supraglacial assemblage composed mainly of interbedded diamictons and granular units, but locally consists entirely of thick, ice-contact fan sediments. In contrast, the southeastern part of the aquifer system corresponds to a predominantly glaciolacustrine facies that constitutes the distal part of the depositional system. Associated sequences appear to constitute a system of small, coalesced deltas that were locally overrun by coarser glaciofluvial sediments and locally by ice. The transition between these two subsystems is a facies change that ranges from relatively sharp to as much as several kilometers wide, and it generally parallels the St. Joseph River Valley (Fig. 2). The valley also serves as the major groundwater discharge area, separating the system into two hydraulically distinct parts that coincide almost precisely with the geologic division.

Aquifer System Characteristics.—
Three regional facies associations have been identified within the Saginaw Lobe depositional system in northern Allen County (Fig. 3 of Fleming, this volume): (1) a basal outwash association consisting of sandy, proglacial outwash locally capped by basal till; (2) a distal glaciolacustrine association consisting mainly of numerous local deltaic sequences; and (3) an upper supraglacial complex best developed in the northernmost parts of the system and typically containing a heterogeneous assemblage of diamictons and small to large granular units. From an aquifer system perspective, these associations are further subdivided into various *hydrogeologic facies tracts* defined by the principal type(s) of aquifers and their relations to other elements of the aquifer system (Fig. 3). Each facies tract is characterized by a different set of aquifers, internal confining-unit geometry and resulting range of hydrogeologic properties and interconnections among different aquifers (Table 1). The basal outwash apron, ice-contact fans and the largest deltaic sand bodies are the major aquifers of the Huntertown aquifer system. Smaller ice-contact stratified bodies and lacustrine sands also are utilized at places, but they tend to be finer grained, less well sorted and(or) smaller in dimension, and thus less productive.

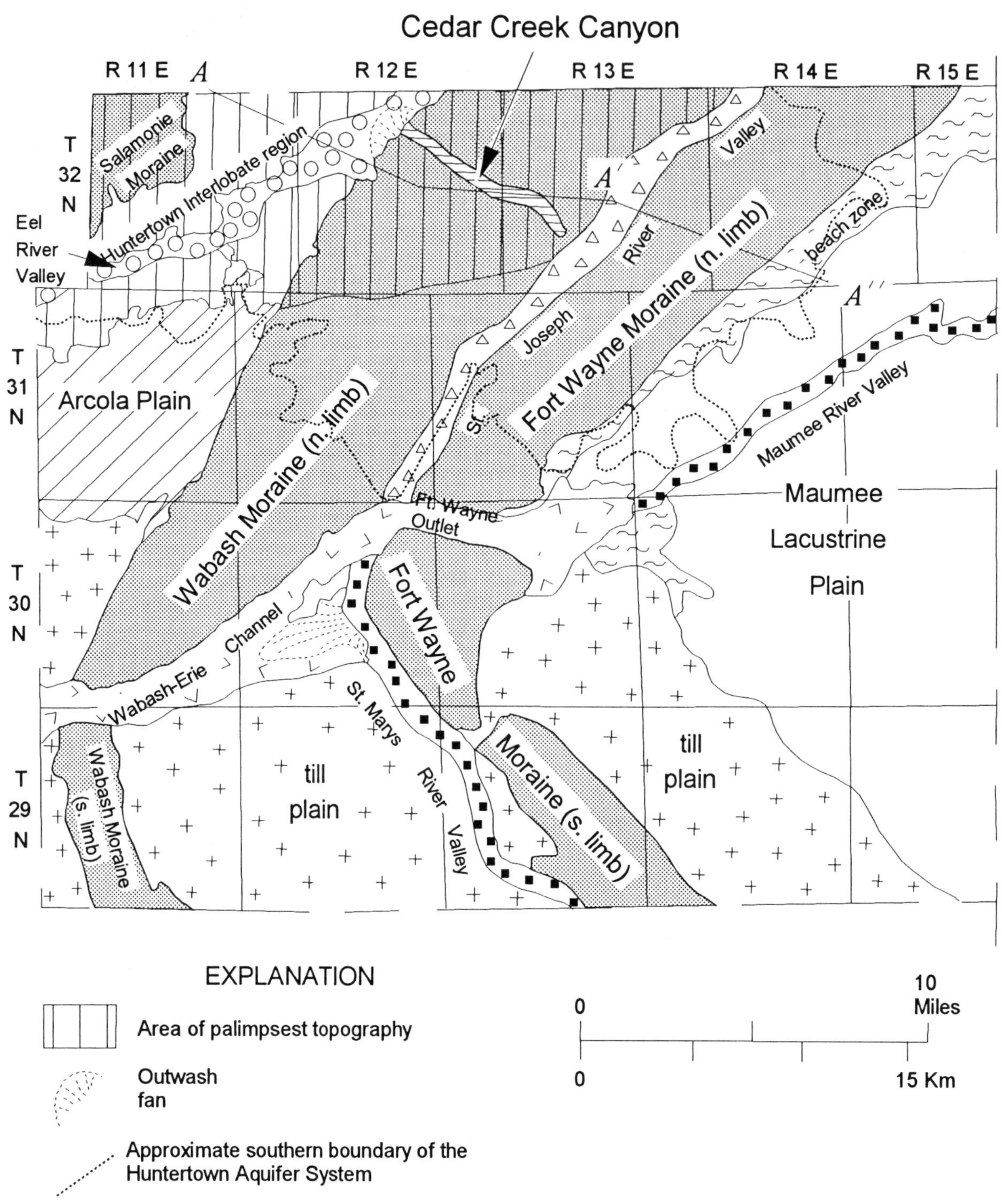

FIG. 2.— Map of Allen County showing distribution of glacial terrains. The Huntertown aquifer system underlies the area north of the dashed line. Clay-rich diamictons of the Erie Lobe that form the regional confining unit atop the aquifer system occur as three main types of sequences: (1) a thick, inner constructional sequence composed mainly of basal tills that underlies the Fort Wayne and Wabash Moraines, as well as till plains inboard of their southern limbs; (2) an outer sequence outboard of the Wabash Moraine, which is discontinuous and partly palimpsest over collapsed supraglacial sequences of the Saginaw Lobe in the Huntertown interlobate region; and (3) chiefly glaciolacustrine diamictons and lake mud that underlie the Maumee Lacustrine Plain and adjacent areas.

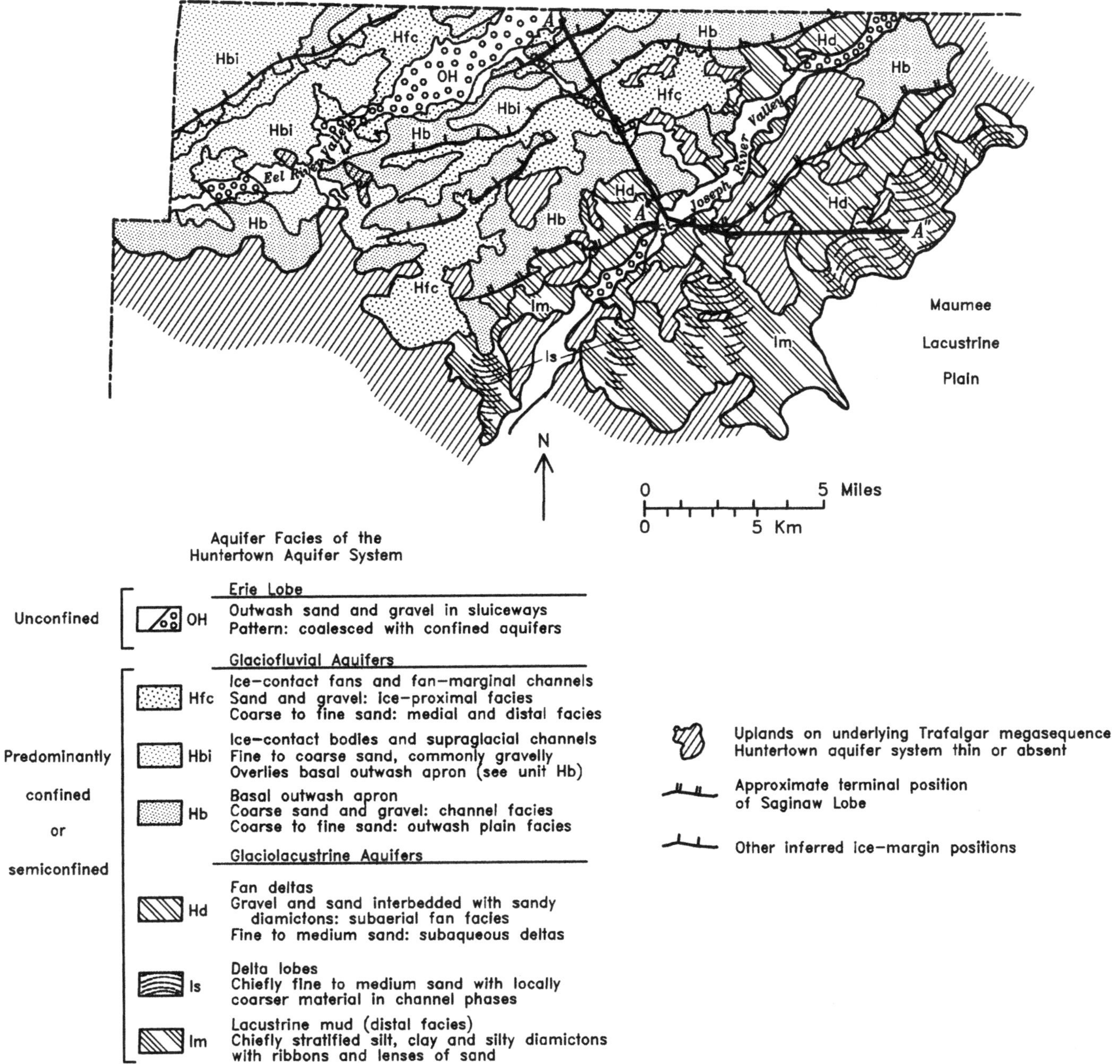

FIG. 3.—Map of northern Allen County showing distributions of facies tracts of the Huntertown aquifer system.

The basal outwash apron is continuous throughout most of the area northwest of the St. Joseph River other than atop large prominences on the subjacent Trafalgar surface. It is represented in three kinds of facies tracts—*Hb*, *Hbi* and *Hfc* (Fig. 3). Unit *Hb* represents one end member in which the part of the sequence above the basal outwash apron consists chiefly of sandy basal till and overlying supraglacial diamictons. The diamicton sequence effectively isolates the underlying outwash from other granular units that may be present higher in the system. On the other hand, the upper part of the sequence in unit *Hbi* contains many ice-contact stratified sand and gravel bodies. This facies tract generally corresponds to areas having a well-developed supraglacial complex superposed on the basal outwash association. Ice-contact bodies are locally in lateral hydraulic connection to one

51

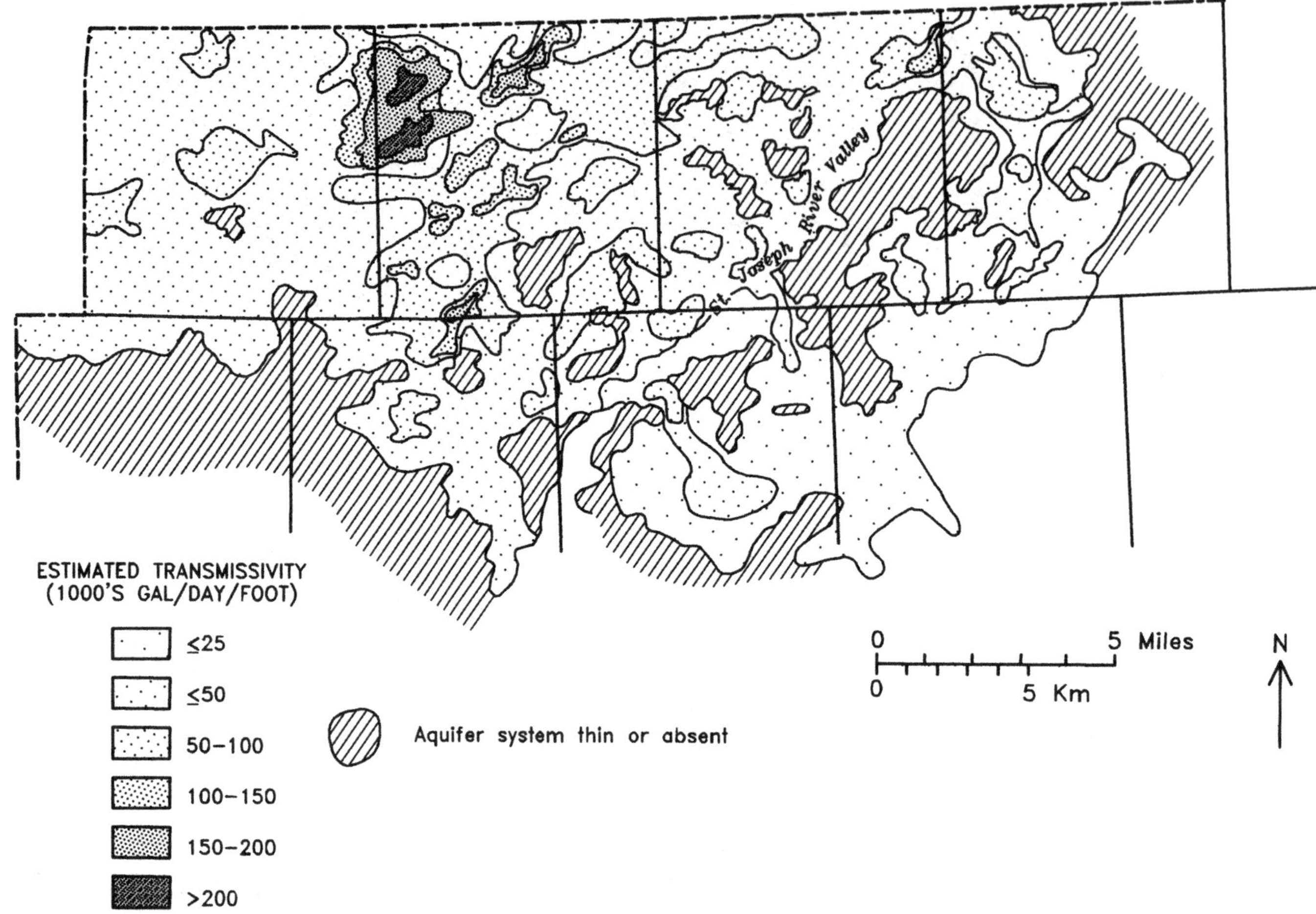

FIG. 4.— Map of northern Allen County showing estimated transmissivity ranges of the Huntertown aquifer system based on specific capacity data, scattered pumping tests and distributions of aquifer facies.

another and are occasionally in contact with the top of the basal outwash. Some are sufficiently large and transmissive to act as significant local aquifers. The intervening diamictons tend to exhibit abrupt discontinuities at places and probably act only as localized internal confining units. Channel-like bodies in the supraglacial complex may serve as conduits to recharge deeper parts of the system because they commonly exhibit higher water levels than the basal outwash.

Facies tracts identified by unit *Hfc* represent the other extreme, where virtually the entire sequence locally consists of sand and gravel believed to have been deposited mainly as large ice-contact fans during recession of the ice margin. In places, thin to moderately thick diamictons separate the fans from the basal outwash below, but elsewhere the fans have coalesced with the basal outwash or truncated it altogether. In the latter instances, a few thin, discontinuous debris flows in the upper part of the fans may be the only fine-grained

sediments within a 20- or 30-m section of outwash. The fans probably serve as locations for recharge to the basal outwash, and their presence greatly increases transmissivity and storage of the aquifer system (Fig. 4).

Southeast of the St. Joseph River, a variety of glaciolacustrine aquifers predominates. These aquifers tend to be smaller, finer grained and less interconnected than those in the northwestern part of the aquifer system. Glaciolacustrine aquifers comprise three types of facies tracts (Fig. 3) distinguished mainly by size and coarseness of the predominant aquifers. Unit *lm* represents the mode, or typical condition, within this part of the system and consists chiefly of lacustrine mud with scattered bodies of fine and very fine sand. The sand units typically are small and lenticular or ribbon-like in geometry, although a few somewhat larger lobate bodies are present as well. These bodies are believed to represent distal deltaic elements that were deposited at considerable distance from the meltwater streams or ice that furnished the sediment to the lake basin. Most of these appear to be enclosed in lake mud; direct interconnection among sand bodies appears to be limited, and groundwater probably circulates very slowly within this facies tract.

In contrast, unit *ls* represents large, relatively high-energy delta lobes deposited in greater proximity to major sediment sources and meltwater outlets. It is thus characterized by a greater abundance of sand bodies and is the most productive facies tract in this part of the aquifer system. Although relatively fine-grained, the sand typically is very well sorted, and some bodies are 10 m or more thick. They commonly are capped by and contain scattered intercalations of lake mud and various diamictons derived from bergs or nearby blocks of stagnant ice; they also contain rare, small lenticular zones of gravelly sand that locally enhance permeability. Unit *Hd* is typified by small fanlike bodies of coarse, occasionally gravelly sand that interfinger laterally with finer deltaic sand and with muddy sediments. This facies appears to represent the subaerial component of fan-deltas deposited in front of one or more Saginaw Lobe margins adjacent to the lake basin. Most of the granular units are relatively thin (2 to 10 m) and locally intercalated with sandy diamictons (debris flows) that limit their productivity.

At places, all these glaciolacustrine facies contain intercalations of gravelly loam and granule sand with abundant fragments of pyritic black shale, gypsiferous limestone and other lithologies suggestive of an *eastern* source. These presumably were derived either from masses of stagnant ice that bounded some of the local lake basins, or perhaps as distal, subaqueous flows from a receding Huron-Erie Lobe margin some distance to the east. The distinction between northern versus eastern source areas can have considerable bearing on the interpretation of ambient geochemistry, particularly the presence and form of sulfur (Fleming and Yarling, 1994; Michael Yarling, Indiana Dept. of Environmental Management, pers. commun., 1996).

Prominences on the Trafalgar surface (Fig. 3) produce discontinuities in the aquifer system. These paleouplands commonly are cored by thick diamict sequences, and aquifers associated with the Huntertown aquifer system are commonly thin or absent over them. They typically interrupt the hydrogeologic continuity of the system and create localized areas of poor groundwater availability.

Major surficial glaciofluvial sequences of the Erie Lobe, represented by unit "OH" on Figure 3, are locally incised into various facies tracts of the Huntertown aquifer system. Thick, unconfined, composite granular units have formed where the surface outwash intersects some of the larger aquifers. These situations likely enhance the transmissivity (Fig. 4) and flow into and out of the subjacent aquifer system, and they could, from that standpoint, be construed as an additional "hydrogeologic facies." However, it is difficult to justify generalizing these units into a single "facies tract" because of the large number of possible aquifer configurations that result from superposition of Erie Lobe outwash on different Saginaw Lobe facies tracts. Instead, the hydrogeologic characteristics of these situations are specific to the glacial terrains in which they occur and are best represented as hydrogeologic settings (Fleming, this volume).

Distribution of Hydraulic Properties.—
Detailed information concerning the hydraulic properties of the Huntertown aquifer system is sparse. Data from pumping tests are available from several well fields, but the water levels reported at all but a small number of the tested wells were recorded only in the pumping well, which greatly limits their interpretive value. On the other hand, specific-capacity data are available from hundreds of water well records that can readily be identified with specific aquifers and hydrogeologic facies tracts. Although information from such records typically is imprecise and sometimes inaccurate, the relatively large number of available records affords at least some statistical assurance of deriving reasonably representative *ranges* of hydraulic properties. Therefore, specific-capacity data from about 350 wells in the Huntertown aquifer system were input into with the computer program TGUESS (Bradbury and Rothschild, 1985) to derive hydraulic conductivity and transmissivity values for different types of aquifers (Fig. 5, Table 2). Unfortunately, the data available for aquifer facies *lm* and *Hbi* were insufficient for a meaningful comparison to other types; so, the values for these units were grouped with units *ls* and *Hfc*, respectively.

The resulting hydraulic conductivity (K) ranges for the different aquifer facies are similar. In view of the gross sedimentological differences that characterize these facies, this result might seem to cast doubt on the specific-capacity data from which the values were derived. Recognize, however, that these ranges are attached to *facies*, which are themselves interpretive models characterized by overlapping ranges of sedimentary bodies and structures, many of which exhibit substantial spatial

TABLE 1.— Characteristics of hydrogeologic facies tracts of the Huntertown aquifer system.

Aquifer Facies Tract	Aquifers	Depositional Environment	Sequence	Thickness(m)	Lithology	T range (1000sgpd/ft)
Hb	Basal outwash apron	Proglacial outwash plain	Stacked channel and bar facies in upward coarsening sequences; capped by basal till and overlying supra-glacial diamictons	1-20	Coarse sand with gravelly zones	15 - 75
Hbi	Basal outwash apron	Proglacial outwash plain	Similar to facies *Hb* but upper part of sequence contains a variety of small to large channel fills and ice-contact stratified units, which have locally coalesced with each other and the basal outwash	1-20	Coarse sand with gravelly zones	15 - 100[b]
	Ice-contact stratified and small channel fills	Stagnant, debris-covered ice		1-15	Ranges widely but commonly gravelly sand	10 - 40
Hfc	Ice-contact fans	Prograding fan lobes in front of ice margin	Upward-coarsening fan lobe facies with embedded fining-upward channel facies, locally capped by or intercalated with sandy and silty diamictons and commonly coalesced with basal outwash apron	10-30[a]	Medium to coarse gravelly sand	25 - 200[c]
	Basal outwash apron	Proglacial outwash plain		1-10[a]	Coarse sand with gravelly zones	25 - 200[c]
lm	Lacustrine sand	Small distributaries in distal deltas	Many minute fining-upward packages of sandy silt and silty clay within an over-upward-coarsening sequence; sand units form small embedded fining-upward inter-vals indicative of channel abandonment	<1-5	Very fine to medium sand	<1 - 10
ls	Lacustrine sand	Proximal deltas and drowned paleovalleys	Agglomeration of fining-upward packages within an overall fining-upward sandy sequence; commonly contains lenses of and grades laterally and up into dense sandy silt with rafted or resedimented diamictons	3-15+	Fine and medium sand; rare lenticular gravelly facies	5 - 40[d]
Hd	Subaerial and subaqueous fans	Proximal fan-deltas	Deltaic sand and silt of units ls or lm progressing upward into heterogeneous assemblage of ice-proximal fan facies that are locally gravelly and intercalated with silty and sandy diamictons near the top	1-10	Fine to coarse sand, some gravelly sand	5 - 50[e]

a. The upper thickness range typically represents superposition of fans directly on basal outwash. In these situations there is generally no practical distinction between the two types of aquifers. The upper thickness range of the basal outwash is somewhat arbitrary and represents areas where localized confining units separate it from overlying fan aquifers.

b. The higher values occur where large ice-contact stratified units have coalesced with basal outwash.

c. Highest values are associated with thickest fans or coalescing of fans and basal outwash into one thick aquifer.

d. Values greater than about 10,000 gpd/ft are restricted to the coarsest, thickest sand bodies within the most proximal parts of deltas.

e. Values greater than about 10,000 gpd/ft mainly occur in thick, localized channel facies in subaerial fans.

Table 2. Summary of hydraulic conductivity values for the Huntertown aquifer system.

Aquifer	Typical Lithology	Number of Wells	Geometric Mean of K (cm/sec)	Standard Deviation of Log Values in Log Cycles
Basal outwash (Hb)	Medium to coarse sand, some gravelly sand at places	159	5.65×10^{-2}	0.97
Ice-contact fans (Hfc)	Fine to coarse gravelly sand; some bodies contain intercalations of till-like sediment	64	4.94×10^{-2}	1.45
Fan-deltas (Hd)	Medium sand with scattered gravelly zones and interbeds of fine sand and till-like sediment	48	3.27×10^{-2}	1.44
Sub-aqueous deltas (ls)	Fine to medium sand, well sorted, with silt intercalations and rare gravelly zones	79	3.62×10^{-2}	1.03

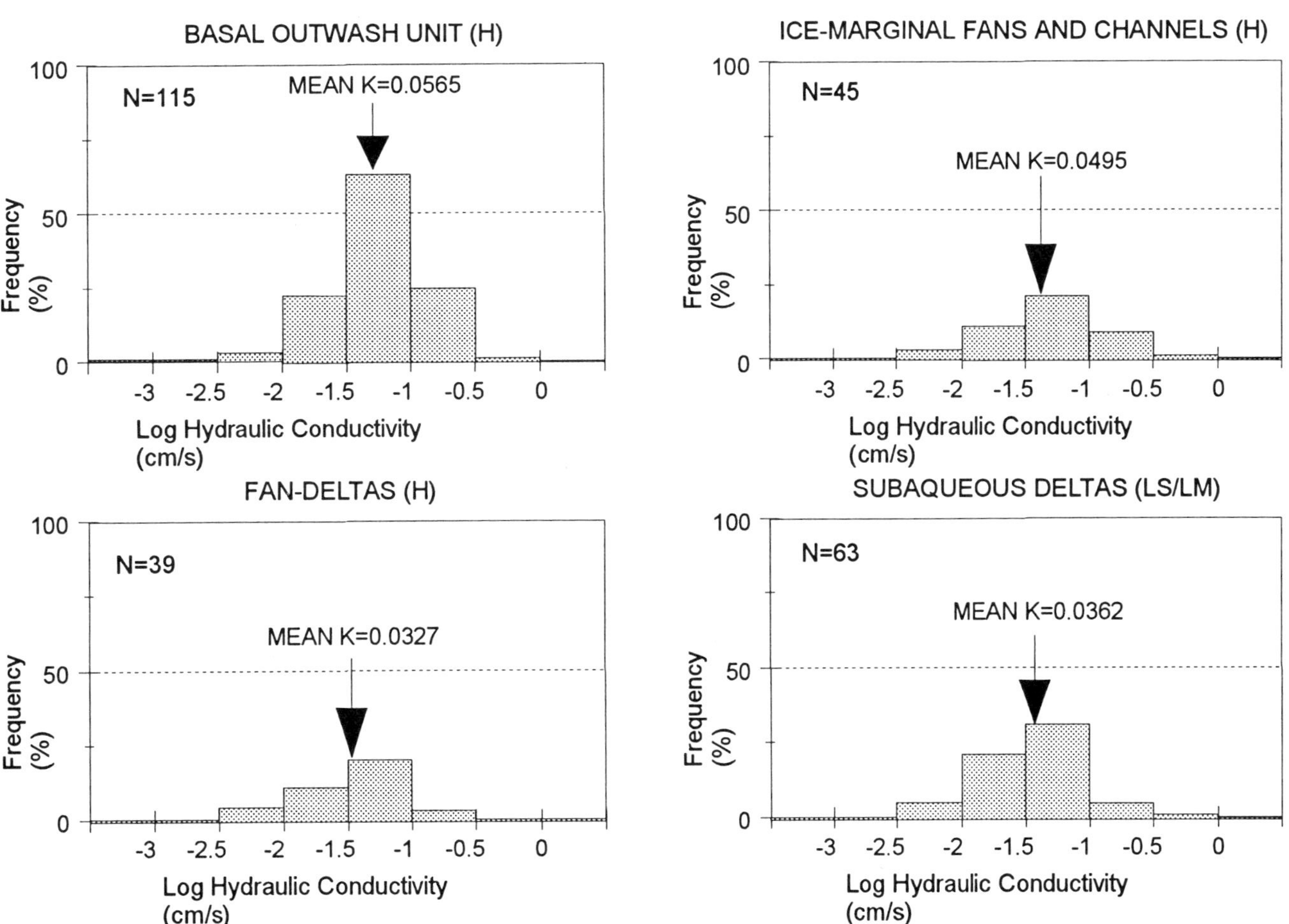

FIG. 5.— Frequency-distribution graphs of hydraulic conductivity for four aquifer facies of the Huntertown aquifer system.

variation. It may or may not be significant that the relatively small number of available pumping test results tends to support the values derived from specific-capacity data. This is true both in a broad sense (by aquifer facies) and for comparisons of both types of data taken from very localized areas in specific aquifers.

Mean K values are on the order of 10^{-1} to 10^2 cm/s, a typical range for well-sorted sands (Freeze and Cherry, 1979). The relative rankings of mean K values for the different aquifer facies also are compatible with their respective sedimentological characteristics. The highest mean values are associated with the basal outwash and ice-contact fan facies—extensive aquifers characterized by thick sequences of well-sorted, medium to coarse, gravelly sand. The lower mean values of delta and fan-delta facies reflect the finer texture typical of sands deposited in glaciolacustrine environments.

The statistical distributions of hydraulic properties for the different aquifers may be of greater importance than their absolute values. The K values (Fig. 5) are log-normally distributed, a condition widely observed in heterogeneous geologic formations elsewhere, including many unconsolidated deposits (e.g., Davis, 1969; Warren and Price, 1961; Bennion and Griffiths, 1966). Similarly, standard deviations are on the order of 1.0 to 1.5 log cycles (Table 2), well within the range typical of most geologic environments (Freeze, 1975). These values are a quantitative representation of the relative degree of internal heterogeneity associated with each type of hydrogeologic facies tract, which should, in turn, generally reflect the relative diversity of depositional processes. Substantially lower standard deviations are associated with the basal outwash and lacustrine deltas—aquifers deposited in settings characterized by relatively uniform depositional processes and less transient environments. In contrast, the greater standard deviations associated with ice-contact fans and fan-deltas appear to reflect the greater sedimentary heterogeneity of ice-proximal environments. These differences are manifested graphically by variations in the frequency distribution curves (Fig. 5).

A comparison of aquifer facies and transmissivity ranges across the entire aquifer system (Figs. 3 and 4) highlights several features that tend to support geologically derived inferences about aquifer characteristics. Particularly evident is the sharp, regional division between generally higher transmissivity values in the northwest and uniformly lower values in the southeastern part of the aquifer system. This division coincides with the regional facies change in the depositional system from predominantly subaerial, outwash-dominated sequences of the northwest to finer-grained and more geographically restricted glaciolacustrine aquifers in the southeast. Likewise, nodes of high transmissivity are associated with some of the largest ice-contact fans; such nodes commonly indicate places where the fans have coalesced with the basal outwash, producing very thick vertical sequences composed entirely of sand and gravel. Transmissivity values for the northwestern part of the system also show a minor systematic decrease toward the southeast that is too slight to appear at the scale of Figure 4. As most of the specific-capacity data are from wells developed in the basal outwash, it seems reasonable to assume that this trend reflects a decrease in grain size and(or) thickness in the distal parts of this aquifer. The most extreme transmissivity values are in the Eel River Valley, where unconfined Erie Lobe outwash is directly superposed atop thick unbroken sections of coarse Saginaw Lobe outwash.

Confining Units

Erie Lobe Sequences.—
The predominantly fine-grained Erie Lobe sequences form an extensive, low-permeability, regional confining unit atop the Huntertown aquifer system. The integrity of the confining unit is affected by its sedimentary characteristics, variations in thickness, presence and nature of discontinuities and characteristics of the landscape that influence infiltration rates and other interactions between surface water and groundwater. All these qualities exhibit a distinct relationship to the three types of fine-grained facies assemblages of the Lagro megasequence (Fig. 2; Fleming, this volume):
1. a discontinuous outer assemblage of thin, resedimented clay-loam diamicts that interfinger with silty to sandy channel and kettle fills. This assemblage lies outboard of the Wabash Moraine in northwestern Allen County and is partly palimpsest over hummocky Saginaw Lobe topography.
2. an inner constructional sequence of thick, repetitively sheared and stacked silty clay diamictons that makes up the Wabash and Fort Wayne Moraines; and
3. a glaciolacustrine sequence of various waterlain diamictons, lake muds and subaqueous meltwater deposits underlying the Maumee Lacustrine Plain and immediately adjacent parts of neighboring terrains.

Data concerning the hydraulic characteristics of fine-grained Erie Lobe diamicts in Allen County are limited to several triaxial permeameter tests and to one field site. Triaxial tests on large undisturbed blocks of massive diamicts yielded hydraulic conductivity values between 10^{-7} to 10^{-8} cm/ (Fleming, 1994). Tritium analyses from wells nested at different depths in a thick diamicton sequence in the southern limb of the Wabash Moraine indicated the presence of "modern" (post-bomb testing) water at depths up to about 9 m (Ferguson, 1992); wells at and above this depth also appear to respond relatively rapidly to precipitation events. These results are consistent with the presence of an "active" groundwater zone associated with fractures and weathering in the upper part of the sequence and with a thin (≤ 3 m) cap of supraglacial sediments deposited upon final melting of the ice. In contrast, water at greater depths is not tritiated and is more mineralized; nitrogen isotope ratios and other data indicate a vertical seepage velocity of less than 1 ft (0.3 m) per year at depth below this site (Ferguson and others, 1992).

Jointing appears to be a common feature of some Erie Lobe diamictons. Basal tills exhibit particularly well-developed, near-vertical fractures that show abundant evidence of water movement (Fleming, 1994). In large exposures these fractures commonly extend to depths of about 6 to 8 m and appear to be oriented in statistically distinct sets that are arrayed conjugately about local ice-flow direction. Known exposures showing this kind of fracture pattern are limited to the inner constructional sequence within the Wabash and Fort Wayne Moraines, where significant shear stress from basal sliding and ice thrusting affected the sediments during and after deposition. Scattered, oxidized fractures also have been observed in resedimented diamicts exposed in small excavations in hummocks associated with the outer sequence. Most of these fractures are inclined parallel to the modern surface slope and appear to be related to stress relief during deposition of the debris flow and(or) during erosional unloading along the modern slope. Such exposures were too small to reveal the dimensions of individual fractures or whether any systematic fracture pattern exists in the outer sequence. No fractures have been observed in the glaciolacustrine sequence, although they may well be present. Based on current knowledge, fractures are expected to be most pervasive, and thus of greatest hydrogeological significance, in sequences composed chiefly of till or overridden diamicts, whereas their role in other types of sequences is less certain.

It is difficult to extrapolate results from a single site to other types of sequences, but generalizations are possible based on overall sedimentological characteristics of the Erie Lobe depositional system. The overall character of the confining unit suggests that it may be less effective where thinner than about 8 m and(or) where it is composed mainly of supraglacial sequences. The outer sequence is the most heterogeneous of the three Erie Lobe assemblages in all respects, and it probably possesses the most uncertain qualities as a confining unit. As the thinnest and most texturally variable of the three, it is characterized by numerous discontinuities associated with irregularities on the underlying Saginaw Lobe supraglacial complex. These localities also tend to exhibit well-developed internal drainage, which is in turn likely to promote recharge. In contrast, the Wabash and Fort Wayne Moraines are composed of thick sequences of very clayey diamictons that are locally up to 30 m thick. Many of the diamictons are basal tills that vary little in thickness or texture, and the sequence as a whole exhibits relatively less heterogeneity. Leakage through this sequence is inferred to occur at exceedingly slow rates. The properties of the glaciolacustrine sequence are intermediate between the other two. Although typically quite fine-grained and moderately thick, it exhibits considerable local heterogeneity, both in thickness and texture. The thickness of fine-grained glaciolacustrine facies is as little as 1 to 3 m over parts of the northern lake plain, and fine-grained facies locally are absent along wave-stripped beach zones. Internal variability is chiefly in the form of (1) small subaqueous fans and other more permeable units interbedded with finer-grained facies, and (2) relatively thinner depositional units that exhibit many local facies changes. Bulk permeability of this sequence is inferred to range widely; however, flow likely is upward at most places since the unit occupies a regional groundwater discharge area (Fleming, 1994).

Trafalgar Surface.—
In addition to its impact on younger depositional systems, the Trafalgar surface constitutes a significant hydrostratigraphic feature within the modern hydrogeologic regime. Extensive areas of the buried surface are underlain by thick (20 to 50 m) sequences of strongly overconsolidated, poorly permeable basal till and other fine-grained diamictons that contrast sharply with the much sandier and more transmissive Saginaw Lobe sequences above. The characteristically sharp hydraulic contrast along this interface causes it to act as a major regional flow boundary that separates the relatively active and distinctive shallow flow system of the Huntertown aquifer system above from deeper and less clearly

integrated systems below.

Localized discontinuities in this flow boundary are represented by small to moderately large, tabular and channel-like sand and gravel bodies within the Trafalgar megasequence that subcrop at places along the buried surface (Fleming, 1994). The nature of hydraulic interaction across these discontinuities depends on the relative hydraulic potential between the truncated bodies and the shallow system above, which is in large part a function of the particular setting within the buried Trafalgar landscape. Discontinuities situated at relatively elevated locations on the buried surface, for example, appear to serve as recharge areas for deeper systems, whereas those present within regional lows on the surface generally are characterized by discharge from deeper units into the shallow system. Such discontinuities clearly are significant, if not the dominant, features at a local scale, but they become considerably less important at the broad scale of the entire aquifer system. In that context the predominantly fine-grained sequences underlying the buried Trafalgar surface collectively act as a regional confining unit and lower flow boundary for the Huntertown aquifer system.

Ground-Water Flow Systems

Potentiometric Surface of the Aquifer System.—
The Huntertown aquifer system is characterized by a well-defined flow system with relatively distinct recharge and discharge areas (Fig. 6). Groundwater enters the northwestern corner of Allen County and flows southeast toward the Eel River Valley under a gradient of about 10 ft/mile (2 m/km), generally paralleling the regional change in elevation of the modern land surface. Over a broad region in the vicinity of the Eel River Valley (Fig. 2) and adjoining areas, however, the potentiometric surface becomes virtually flat and contains several small mounds, which suggests the presence of a significant vertical flow component. A downward gradient is indicated here by the relationship between the water table, represented by peatlands and other depressional wetlands, and the potentiometric surface of the aquifer system, whose elevation appears to be as much as 5 m lower at certain times.

This region of the flattened potentiometric surface coincides with the course of the eastern continental drainage divide. The landscape exhibits well-developed internal drainage, and the aquifer system is unconfined below the Eel River Valley and locally very poorly confined beneath adjacent areas. All these features are highly characteristic of a regional recharge area. To the southeast, the potentiometric surface resumes its southeasterly gradient, which continues essentially unchanged beneath the Wabash Moraine until the St. Joseph River Valley. Several small recharge areas may exist where overlying diamicts are thin or pierced by channel-like sand units, but for the most part little recharge apparently is taking place beneath the thick clayey sequences of the moraine.

The St. Joseph River is the principal discharge area for the aquifer system. The regional configuration of the potentiometric surface (Fig. 6) is essentially oriented about the valley, and the course of the river coincides with a prominent valley in the potentiometric surface. Elevation of the potentiometric surface is just below, equal to or slightly greater than river level, indicating that groundwater from the aquifer system is discharging upward into the river or into Erie Lobe outwash below the channel. The river receives some 14 million gallons (53 million liters) per day of groundwater discharge, mainly from the Huntertown aquifer system, along a 10 mile (16 km) reach between two gauging stations in northern Allen County (Arvin, 1989).

The part of the aquifer system southeast of the river appears to be hydraulically distinct. Potentiometric surface elevation is greatest beneath crestal areas of the Fort Wayne Moraine. Water levels at different elevations exhibit greater variability in this part of the aquifer system owing to the smaller and more poorly interconnected aquifers. Nevertheless, the potentiometric surface clearly slopes northwestward from the moraine crest into the St. Joseph River Valley as well as southeastward toward the Maumee Lacustrine Plain. No geologically distinct recharge area is associated with the potentiometric high in this part of the system, which is generally well confined. Local recharge occurs to the upper parts of several deltaic aquifers where they are poorly confined beneath the large beach ridges that bound the lake plain. However, the surface of the lake plain is a regional discharge area, evidenced by flowing wells and other indications of strong upward gradients.

Geochemical Characteristics.—
Geochemical and isotopic properties of groundwater within the Huntertown aquifer system (Fleming and Yarling, 1994) are compatible with the regional flow pattern inferred from geological and water-level characteristics and with the distributions and

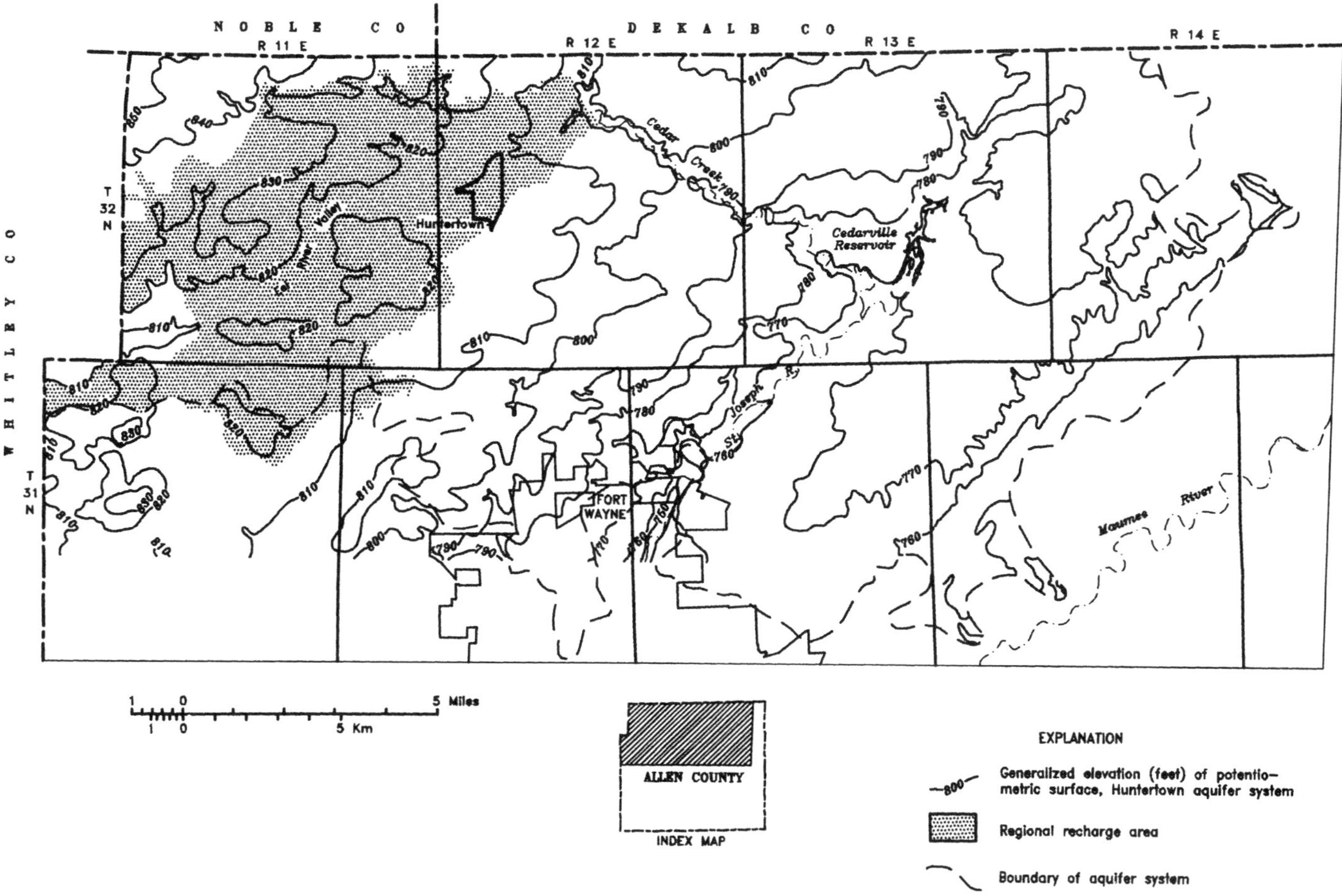

FIG. 6.— Map of northern Allen County showing potentiometric surface of the Huntertown aquifer system. Most of the water levels northwest of the St. Joseph River are from the basal outwash aquifer, whereas those to the southeast are dominantly from various glaciolacustrine sand units. The shaded area in northwestern Allen County corresponds to the regional recharge area, a region characterized by thick sand and gravel, unconfined or poorly confined conditions, internally drained landscapes and a marked flattening of the piezometric surface. Extension of potentiometric contours beyond the boundary of the aquifer system represents water levels from other shallow aquifers associated with different depositional systems; many of these are not known to be laterally interconnected with aquifers of the Huntertown system. A prominent flow divide parallels the aquifer system boundary in the southwest part of the map area. Contour interval 10 ft (approx. 3 m).

characteristics of hydrofacies. Analyses of groundwater from the inferred recharge area in northwestern Allen County indicated the presence of tritium at many places. Concentrations up to 15 to 30 T.U. were found in shallow aquifers close to peatlands in the Eel River Valley and in areas of abundant large enclosed depressions, whereas somewhat lower tritium values were typical at greater depths in this area. In the better confined parts of the aquifer system beneath the Wabash and Fort Wayne Moraines, however, tritium concentrations were mostly below the detection limit of 0.8 T.U.

Isotopic values for oxygen and deuterium from wells throughout the Huntertown aquifer system were within a narrow range that is consistent with modern temperate precipitation in northern Indiana (Drever, 1988). This result is comparable to isotopic values in groundwater from the thick, clayey Erie Lobe sequence farther south in Allen County (Ferguson, 1992; Ferguson, et al., 1992), which suggests that the aquifer system has been completely flushed by post-Pleistocene precipitation. Groundwater from many of the same wells that exhibited elevated tritium also showed a minor but systematic depletion of heavy isotopes (^{2}H and ^{18}O) relative to other parts of the aquifer system. The meaning of this pattern is unclear but conceivably could relate to the fractionation of these isotopes by vegetation, decomposition processes or precipitation of carbonate minerals within depressional wetlands.

Other geochemical patterns also suggest the presence of relatively young groundwater below the inferred recharge area. Wells there demonstrate slight to strong enrichment in ^{13}C relative to wells elsewhere. Although analyses of the carbon cycle of this system are incomplete, this characteristic probably corresponds to an increase in dissolved organic carbon, which in turn may be attributable to organic acids derived from biodegradation of peat. Concentrations of total dissolved solids (TDS) are relatively low in groundwater below the recharge area (commonly <400 mg/L) but show a slow, progressive increase down-gradient to the southeast. The down-gradient increase in TDS also coincides with a change to sulfate as the co-dominant anion (along with bicarbonate). Concentrations of iron and several trace metals also appear to be less near the recharge area than in areas further down-gradient.

Strongly elevated levels of sulfate and sulfide (500 to 2,000 ppm) occur locally in the southeastern part of the aquifer system. Analyses indicate that the bulk of the sulfur is derived from the oxidation of pyrite and to a lesser extent from gypsum (Michael Yarling, Indiana Dept. of Environmental Management, pers. commun., 1996). Both minerals are abundant in Huron-Erie Lake sediments, which contain numerous fragments of pyritic shale and gypsiferous limestone, but are relatively less common in Saginaw Lobe deposits. Hence, instances of elevated total sulfur probably indicate areas of significant Huron-Erie Lobe input to the glaciolacustrine sequences in the southeastern section of the system. However, they can also be related to the generally finer texture and less hydraulic continuity typical of aquifers in this part of the system. These aquifers are chiefly small lacustrine bodies, many of which are enclosed in lake mud and diamictons, hence, groundwater should have considerably longer residence time in these hydrofacies than in the much coarser and strongly interconnected outwash aquifers characteristic of the northwest part of the aquifer system. The significantly less mineralized nature of groundwater in the northwest may well be attributable to high transmissivities of these aquifers and correspondingly short residence times.

Water Table and Local Flow Systems in Confining Units.—
The water table typically lies at depths of 1 to 3 m in the regional confining unit. It marks the top of a relatively active shallow flow system that circulates through fractures, small granular units and other discontinuities within the upper, weathered zone of the fine-grained Erie Lobe sequences. The configuration of the water table system is expected to closely mimic the overlying surface topography. In places this shallow system is the top of the true zone of saturation, particularly in low-lying areas and where fine-grained Erie Lobe sequences are relatively thin and overlie saturated sand and gravel. However, where the confining sequences are very thick, such as in some moraines, the water-table system may be perched on thick sections of very poorly permeable, unfractured till at depth.

The water-table system probably does not interact with the flow system in the subjacent aquifer system except where the confining unit is sufficiently thin to be penetrated by fractures or other discontinuities, or where it is broken by depressions, such as in the outer Erie Lobe sequence in the northwestern part of the county. In these situations local flow systems in the confining unit probably contribute significant recharge to the aquifer system. In most places, however, groundwater discharging from the shallow system maintains the base flow of numerous streams that flow over the confining unit and contributes to depressional bogs and other wetlands in hummocky areas. The system is inferred to receive widespread recharge via fractures and other macropores as well as upland depressions that collect runoff and gradually release it to the subsurface.

DISCUSSION

Nature and Significance of Facies Changes

One of the most prominent aspects of the Huntertown aquifer system is its division into two geologically distinct subsystems--one northwest of the St. Joseph River, composed chiefly of subaerially deposited outwash aquifers, and another to the southeast, consisting chiefly of glaciolacustrine sediments. In practical terms this results in widespread, coarse granular units containing abundant groundwater supplies in northwestern Allen County, versus finer, more areally restricted sands typified by somewhat more sporadic groundwater availability to the southeast--a distinction recognized by many local water supply contractors. Superficially, these subsystems appear so different that one might easily conclude that they represent completely separate depositional systems. Although relatively sharp at places, the boundary between the two subsystems is, in fact, a major facies change largely within a single depositional system. This rather

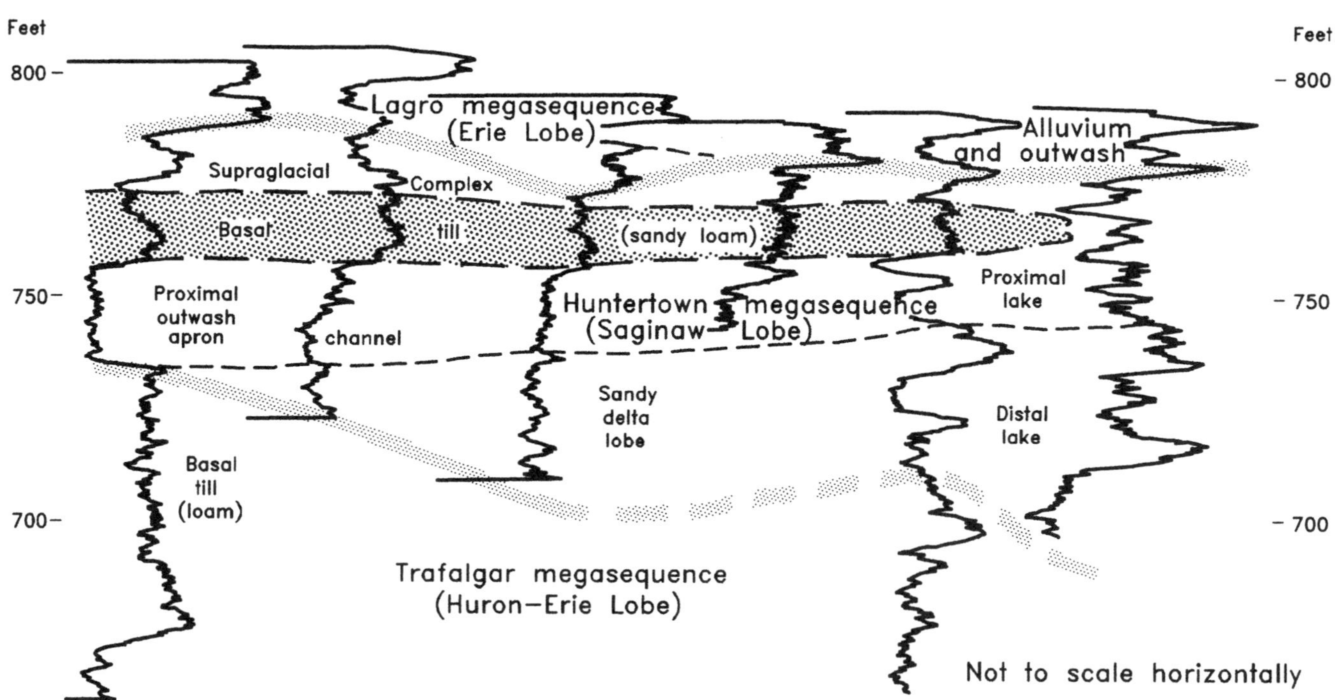

FIG. 7.— Gamma-ray log section along the north side of the St. Joseph River Valley illustrating regional facies change in the Saginaw Lobe depositional system between the basal outwash and glaciolacustrine associations, as well as aspects of eastern-source sequences above and below. Heavy stippled lines represent regionally extensive bounding surfaces between major depositional sequences. Most Saginaw Lobe sequences exhibit an overall upward increase in grain size and complexity, suggesting that they were deposited during ice advance, whereas individual sediment packages (i.e., glaciolacustrine facies, basal outwash) become somewhat finer-grained in a down-ice (NE) direction. Overlying Erie Lobe sequences consist of outwash (and alluvium) to the northeast and thin, meltwater-washed and reworked diamictons to the southwest. Logs are from general area of point A' in figure 3. Total length of section is about 4 km.

marked division suggests that some systematic control acted on the depositional system to produce the observed facies distribution.

The basal outwash association and facies-equivalent parts of the glaciolacustrine sequences to the southeast appear to represent a continuum in which a consistent, similar pattern is recognizable (Fig. 7):

1. Many vertical sequences exhibit a marked coarsening-upward pattern suggestive of relatively rapid ice advance, particularly in the northwest. This is reflected in the generally sandy proglacial outwash by an abrupt upward progression into gravelly sand and, in some instances, coarse gravel, just below the characteristically sharp contact with the overlying diamicts. The gravelly sands are interbedded with thin sandy diamicts (debris flows) and are interpreted as proximal fans. The characteristically sandy, compact and massive overlying diamicts contain numerous faceted clasts and exhibit a uniform, blocky

pattern on gamma-ray logs; they are interpreted as basal tills. In the more proximal glaciolacustrine sequences, a similar pattern reflects an overall increase in depositional energy, expressed mainly by an upward increase in the size and number of sand units; gravelly channel facies and thin diamictons may cap the most ice-proximal of these sequences.

2. Meltwater facies become progressively finer grained in a southeasterly direction, reflecting greater distance from the ice front at any given time as well as a greater influence of glaciolacustrine deposition in the distal parts of the depositional system. This is manifested especially well in the glaciolacustrine association, which progresses from a succession of small, upward-coarsening packages of sand and silty sand in proximal positions to a thin zone of silty-clay rhythmites in the most distal sequences.

3. Downhole samples and logs along and near the St. Joseph River Valley show well-sorted, medium to coarse

sand overlying and presumably prograding over finer grained glaciolacustrine sequences. The sand is slightly finer than typical basal outwash but otherwise similar and locally capped by thin till. This relation implies that much of the glaciolacustrine sediment was furnished by the same meltwater that deposited the basal outwash and that the former is simply the distal equivalent of the latter.

4.The greatest heterogeneity is associated with the glaciolacustrine sequences in general, but especially those overridden by ice and(or) located along the regional facies change in the aquifer system.

The above sedimentological characteristics suggest that hydraulic conductivity values, and perhaps lateral continuity, also may increase upward in individual aquifers within proglacial parts of this system. However, no data are available from different zones within a given vertical sequence to test this inference.

Effect of Depositional Surface Form on Facies Distribution

The gross distribution of hydrofacies in a large part of this system, as well as the particular sedimentological patterns noted above, clearly reflects the prograding of increasingly proximal facies into a complex of proglacial lakes during ice advance. Regional distribution of these facies, as well as the behavior of the depositional system of which they are part, ultimately is determined by overall form and internal morphology of the depositional basin, which in this case is represented by the buried surface of the Trafalgar Formation (Fig. 8) or, more simply, the "Trafalgar surface." This important bounding surface usually is identified readily by various criteria, most notably its extreme hardness. Between 100 and 400 blows usually are required to drive a split spoon through 1 ft (0.3 m) of the overconsolidated diamicts; the surface is thus readily identifiable by an abrupt increase in blow counts reported in the logs of geotechnical borings. In addition, this horizon is typically and uniquely reported as "hardpan" on many water well records.

Several different regions and features are defined by the gross morphology of the Trafalgar surface in northern Allen County, each of which produced a somewhat different effect on the succeeding depositional system (Fig. 9):

1. The surface in northwest Allen County constitutes a dissected, south- to southeast-facing *regional slope* that helped direct the flow of northern-source ice and meltwater into north-central Allen County. Consequently,

the basal outwash association is well developed along this regional slope and in associated south- to southeast-trending lowlands and paleovalleys on the Trafalgar surface that conducted meltwater away from the Saginaw Lobe as it advanced into northwest Allen County. Most of the outwash probably aggraded relatively rapidly in a braided, proglacial plain during rapid ice-margin advance, resulting in the characteristic strongly upward-coarsening sequences abruptly truncated by basal till.

2. A prominent series of elongate ridges and smaller irregular hills forms a lobate, *morainal highland* that is broadly concentric about the central and eastern part of the county. The highland appears to have formed a significant drainage divide that helped impound earliest phases of ancestral Lake Erie within the much lower-lying area to the east (Fleming, 1992b), whereas relief associated with the irregular northern edge of the highland in north-central Allen County may have helped stall further southward advance of thin Saginaw Lobe ice. Not surprisingly, the major regional facies change that divides the aquifer system coincides with the irregular boundary between the northern edge of this buried highland and the regional slope that lies to the north. The more elevated parts of this highland are often devoid of northern-source sediments, whereas internal basins typically contain proximal glaciolacustrine sequences.

3. Several broad southeast-trending *paleovalleys* are superposed on the regional slope in northwestern Allen County and appear to terminate within a series of irregular basins and embayments that interfinger with the morainal highland. Some of the thickest and coarsest facies of the basal outwash aquifer commonly are found within these paleovalleys, whereas some of the more sand-rich glaciolacustrine sequences occur at or outboard of their apparent mouths.

4. A broad, low-lying *plain* of low relief is situated inboard of the morainal highland in eastern Allen County. The most distal glaciolacustrine sequences are concentrated along this region of the depositional surface.

Distribution and Significance of Low-Permeability Bodies

The Huntertown aquifer system fits the classic definition of "a heterogeneous body of permeable and poorly permeable materials that functions regionally as a water-yielding unit; it consists of two or more aquifers separated at least locally by confining units that impede ground-water movement but do not affect the overall

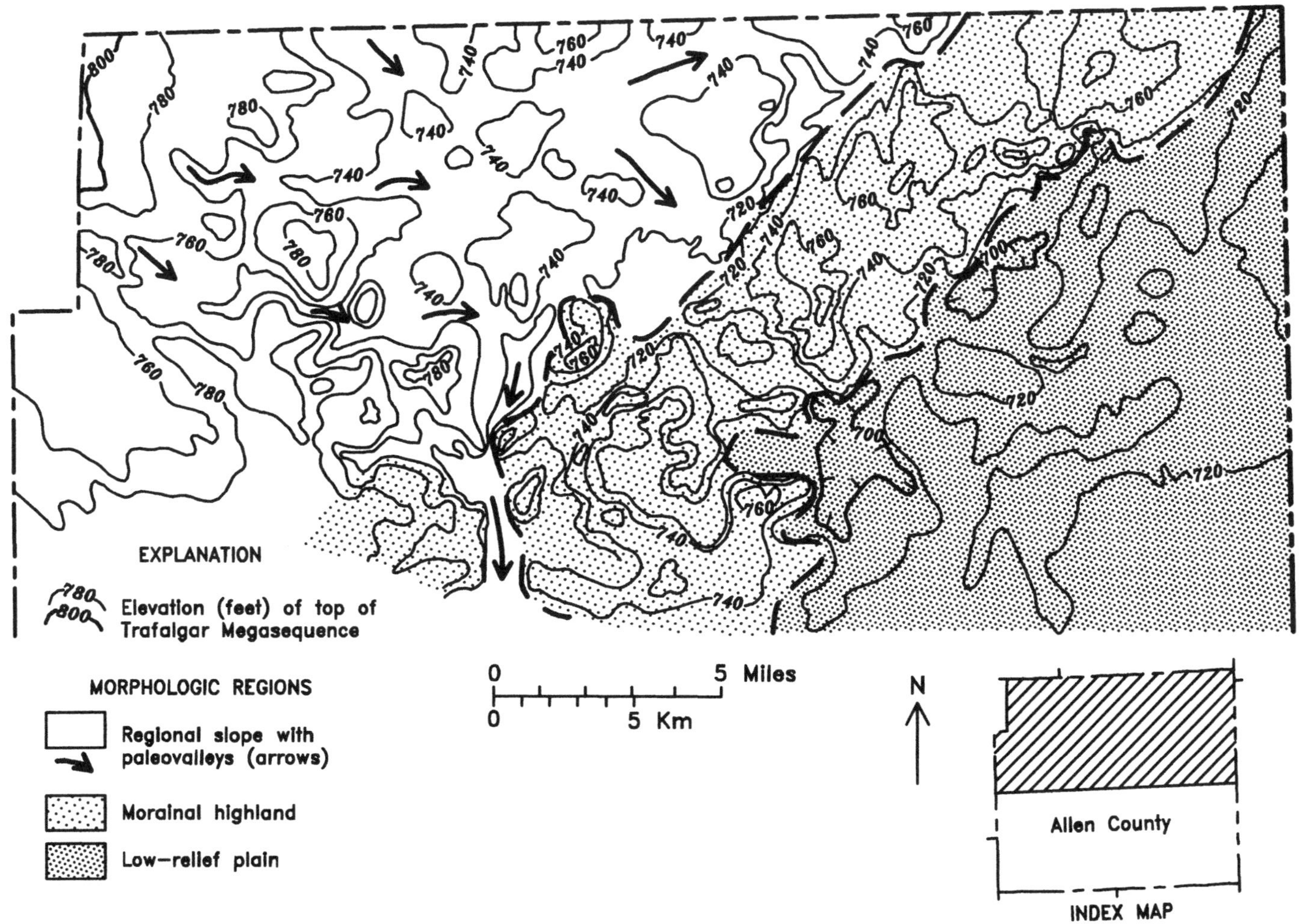

FIG. 8.— Map of northern Allen County showing topography of the buried surface of the Trafalgar megasequence and its major morphologic regions. Contour interval 20 ft.

hydraulic continuity of the system" (Poland and others, 1972). Thus, although it can be thought of primarily as a major water-bearing unit sandwiched between two regional confining units, the Huntertown aquifer system also contains a variety of internal low-permeability units that are of great importance to the hydraulic function of the system, especially on a localized basis. These units range from relatively uniform, sheet-like bodies of basal till that may have considerable continuity over some areas to texturally varied and irregularly shaped diamictons associated with the supraglacial complex (Fig. 10). In terms of hydraulic continuity, they range from relatively thin, isolated lenses of diamicton, which probably do not greatly affect flow within their thick, granular host sequences to extensive fine-grained sequences, which likely are the dominant controls on flow into and out of their enclosed small sandy aquifers.

In all these cases, understanding the distribution, nature and geometry of low-permeability bodies within the aquifer system is necessary for such undertakings as large-capacity water supply development and contaminant-impact assessment. For example, the upper surfaces of these internal fine-grained units not only can and do intercept "heavy" contaminants that tend to sink through granular materials, they also may direct the migration of separate contaminant phases along the slope of the top of the unit in entirely different directions from local or regional groundwater flow. Knowledge about environments of deposititon can provide valuable insight into the probable continuity and surface slope of these units and thus assist in evaluating of potential migration pathways, a key element in the emerging field of risk assessment (Nelson and Fleming, 1996). Similarly, the presence of internal low-permeability units can provide

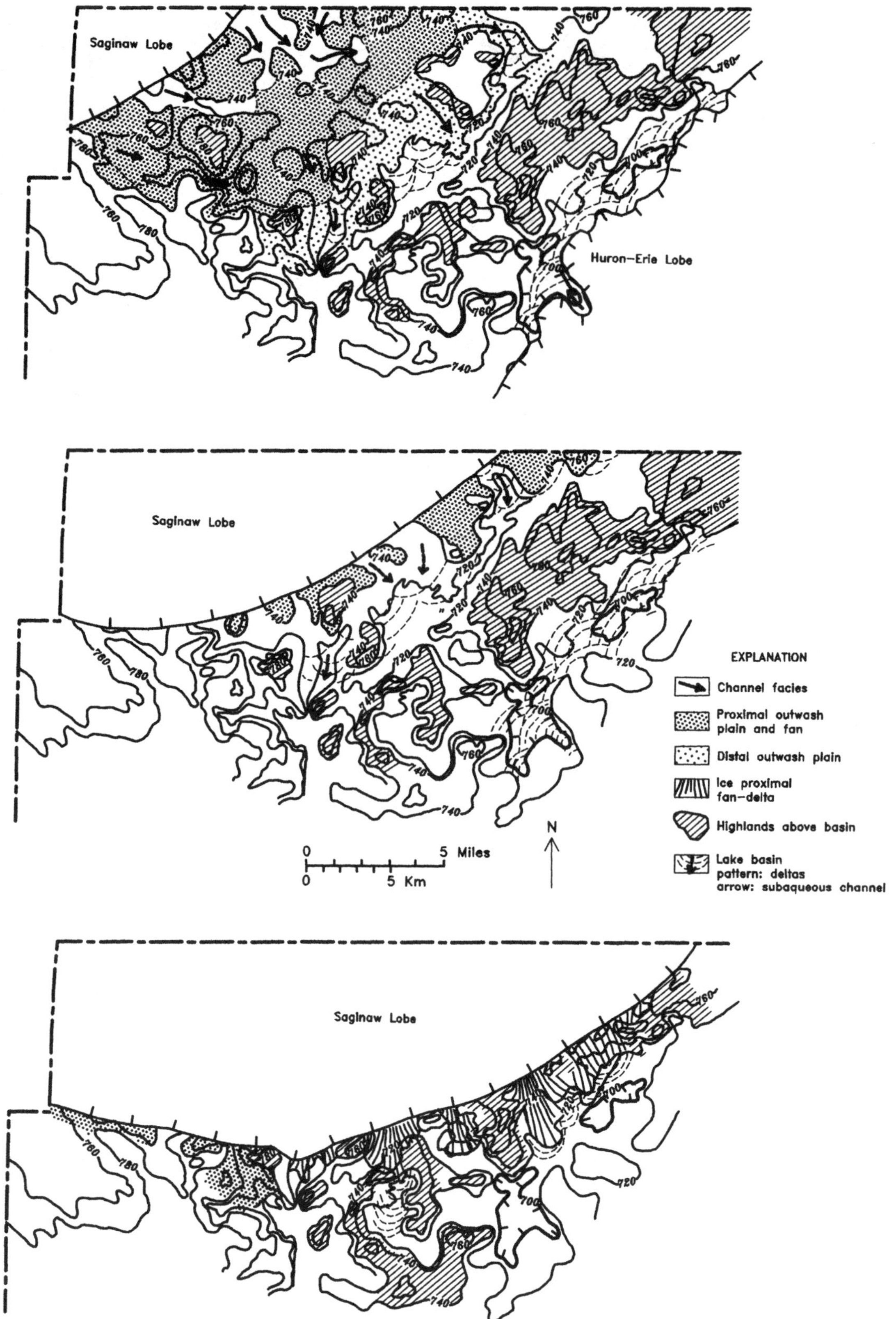

FIG. 9.— Evolution of depositional environments and facies associated with advance of the Saginaw Lobe into Allen County. Depositional events are superposed on the map of the Trafalgar surface shown in figure 8 to illustrate their relationship to the morphology of the depositional surface. Contour interval 20 ft.

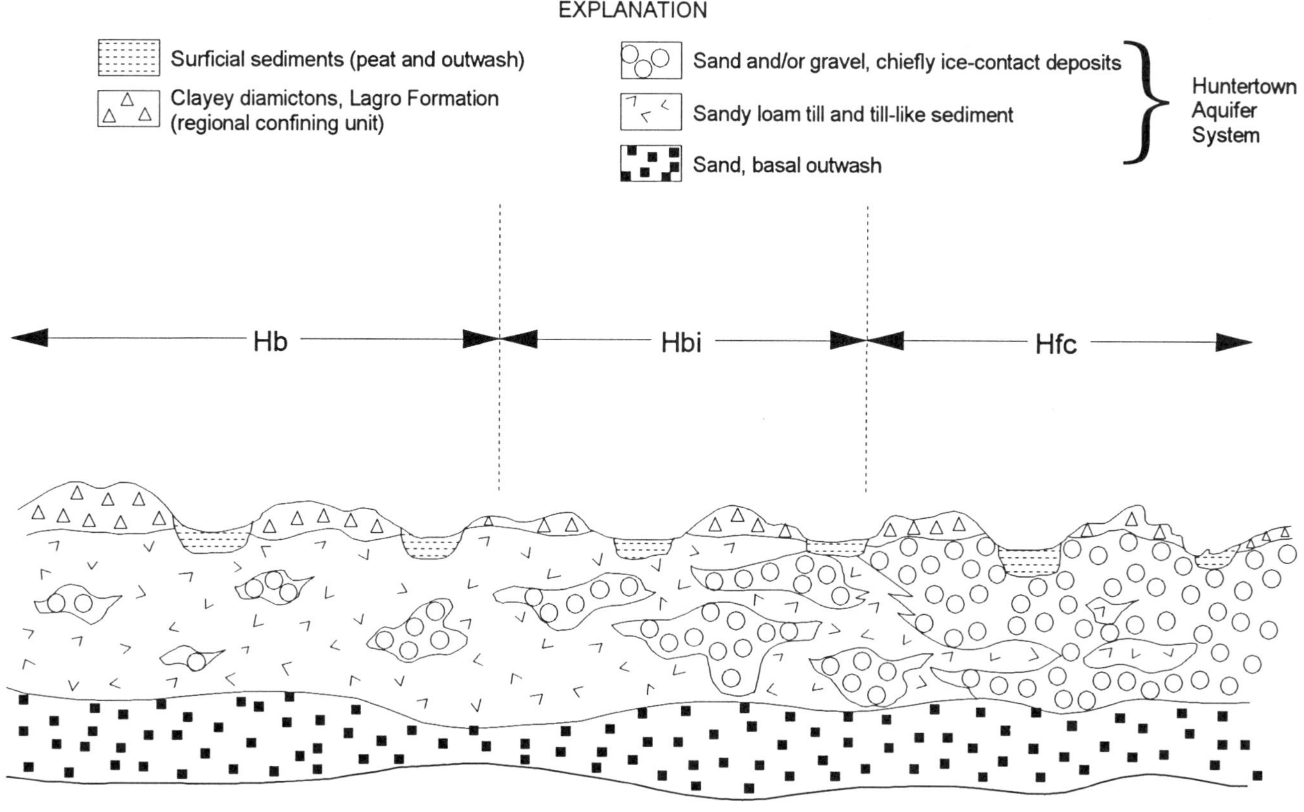

FIG. 10.— Schematic section showing vertical sequences and aquifer-confining unit geometries associated with aquifer facies *Hb, Hbi, and Hfc* (Fig. 3) in the Huntertown interlobate area. Erie Lobe diamictons that act as the regional confining unit above the entire aquifer system are generally thin and discontinuous in this terrain. Consequently, recharge reaches the top of the aquifer system relatively readily, and characteristics of internal confining units within the system assume considerably greater importance in terms of both productivity of individual aquifers degree of natural protection they afford from surface-derived contamination.

additional protection for water supplies, especially high-capacity installations that generate large or steep zones of influence; however, they can also limit well field productivity as well, by restricting recharge to the production zone from higher zones in the aquifer system.

Several statements can be made about the character of internal confining units in different parts of the Huntertown aquifer system according to their facies associations and, to some extent, the regional geometry of the depositional surface below. Basal till commonly forms sheets just above the basal outwash aquifer in the northwestern part of the system. The till is relatively uniform in its physical properties, commonly a moderately overconsolidated sandy loam with few sand bodies. Analyses of numerous geophysical logs and water well records suggest that, despite being pierced by large granular units in areas of thick supraglacial deposition, the

till has considerable continuity, being thickest (as much as 12 m) in the northwest and thinning progressively to the southeast, where it blends into glaciolacustrine facies. The surface of the till appears to slope gently to the south-southeast, parallel to the regional slope on the underlying Trafalgar surface. In contrast, the diamictons associated with supraglacial sequences generally are much more heterogeneous, appearing to have limited or uncertain continuity, widely varied surface slopes, textures ranging from silty clay to loamy sand and a plethora of intercalated granular units and channeling by meltwater. Those diamictons that cap the largest ice-contact fans, however, appear to dip down-fan to the southeast, although their continuity is limited. Fine-grained bodies associated with sandy deltaic sequences deposited at some distance from active ice probably are tabular and may exhibit little surface slope or channeling by meltwater, whereas those deposited in ice-proximal or ice-contact positions may be considerably more varied in configuration. Consequently, the basal till, as well as some of the more distal glaciolacustrine units, likely are the only internal confining units that exert a regional effect on flow within the aquifer system, probably causing it to act as a layered system at many places. In contrast, the other types of confining units, which are associated with much more heterogeneous assemblages, exhibit considerably less certain continuity, physical properties, thickness and other qualities. In short, they should have limited integrity as confining units, except perhaps on a highly localized basis, and their effects on the hydraulics of the whole aquifer system likely are less pronounced.

SUMMARY AND CONCLUSIONS

Regional aquifer systems associated with large glacial depositional systems often are typified by complex boundary conditions and various internal heterogeneities that frequently control the locations of key productive zones as well as the distribution of preferred pathways for contaminant migration. In many instances, efforts to develop, protect and remediate groundwater resources at specific sites or within jurisdictional localities are hindered by the lack of information concerning the architecture of the entire depositional system and attendant variability within individual parts of the system. These activities often can be performed more efficiently and effectively in the context of a regional conceptual model that documents the nature and spatial distribution of heterogeneities in an aquifer system. The model thus provides a basis for forecasting the impact of these activities on specific parts of the hydrogeologic system.

One such approach involves extensive downhole sampling and geophysical logging of domestic wells and selected environmental sites to establish relationships among the local sedimentary sequences and associated landscape elements that collectively constitute basin-scale, genetically based glacial-depositional systems. The glacial terrain models that result from this process emphasize the nature of variation *within* map units as well as relationships between entire depositional sequences. This approach has been used effectively in northern Allen County to derive and depict large-scale facies trends and heterogeneity in a glacigenic aquifer system and to show how the origin and distribution of these heterogeneities can be traced back to depositional environments of the aquifer system and its confining (bounding) units.

Glacial terrains in northern Allen County represent the transitional southern terminus of the interlobate region of northeastern Indiana. Individual terrains reflect different kinds of interaction and superposition of two large, latest Wisconsin depositional systems and their relationship to the depositional basin as represented by the surface of a third late Wisconsin depositional system. These depositional systems collectively constitute a hydrogeologic system that contains the major source of groundwater for most of northern Allen County. The principal aquifers in this system consist of a group of extensive proglacial outwash apron, outwash fan and glaciolacustrine facies of the Saginaw Lobe that collectively form the Huntertown aquifer system. Within this aquifer system, hydraulic properties of distinct facies tracts vary systematically according to the sedimentary environments that existed in the depositional system. In particular, the greatest heterogeneity is statistically associated with aquifers deposited in ice-proximal environments that shifted rapidly or experienced fluctuating depositional energy, whereas less heterogenetly characterizes outwash plain and glaciolacustrine aquifer facies, which formed in relatively more stable environments. The characteristics of internal confining units and their hydraulic significance to the entire aquifer system also exhibit distinct, and generally predictable, relationships to the types of depositional sequences in which they occur. The distribution of all these hydrofacies ultimately is determined by overall form and internal morphology of the depositional basin, which is represented by the buried surface developed on distinctively overconsolidated diamictons of the Huron-Erie Lobe. Because of its largely fine-grained composition, the depositional surface also constitutes the lower flow boundary of the Huntertown aquifer system.

At most places, the aquifer system is capped by a suite of clayey diamictons deposited by the Erie Lobe. These fine-grained sediments generally form a regional confining unit for the Huntertown aquifer system, but distinct differences in the origin, sedimentologic properties and landscape associated with the capping sequences result in significant geographic differences in hydraulic integrity of the confining unit. Thin, discontinuous diamicts associated with the collapsed landscape of the Huntertown interlobate region are palimpsest over hummocky Saginaw Lobe topography and exhibit well-developed internal drainage. This results in abundant recharge to the semiconfined aquifer system below, primarily via runoff into depressional wetlands that interact extensively with the flow system in the aquifers immediately below. This terrain forms the regional recharge area which is characterized by a flat or mounded potentiometric surface, strong downward flow gradients and young, tritiated groundwater with relatively low TDS. In contrast, thick constructional sequences of repetitively sheared and stacked, massive basal tills that core the Wabash and Fort Wayne Moraines appear to possess extremely low bulk permeabilities and thus drastically limit flow into and out of the subjacent aquifer system. This is manifested by a nearly total absence of mounding or other effects on the regional slope of the potentiometric surface beneath these sequences, despite the fact that moraines form broad, prominent uplands standing as high as 35 m above adjacent areas. In yet another variation, the confining unit, which lies below the Maumee Lacustrine Plain, consists chiefly of fine, glaciolacustrine sequences, whose hydrogeologic characteristics appear to range between the two other types of confining sequences noted above. This terrain, a regional discharge area for the Huntertown aquifer system and for bedrock aquifers, is characterized by significant upwelling of strongly mineralized waters across the confining unit. The continuity of the regional confining unit is punctuated by three extensive Erie Lobe glaciofluvial sequences that are at places incised into the aquifer system and act as localized hydraulic interconnections through which recharge and discharge occur.

The Huntertown aquifer system and its overlying and underlying eastern-source confining units are the main elements of a well-defined shallow hydrogeologic system. At the scale of the entire system, each of these elements could properly be regarded as a regional hydrostratigraphic unit. On the other hand, a variety of local hydrofacies and surface terrain elements is present, whose interactions greatly affect not only flow into and out of the system but also internal properties, such as transmissivity and potential well yields. Development of the glacial terrain model for northern Allen County provides the regional context necessary for interpreting, depicting and forecasting the nature and occurrence of heterogeneities that characterize different parts of this aquifer system.

ACKNOWLEDGMENTS

Many of the results presented herein were derived from a hydrogeologic study of Allen County funded in large part by the Allen County Board of Commissioners and coordinated by the Allen County Department of Planning Services. I wish to acknowledge Kimberly Bowman and Dennis Gordon of the latter agency; their interest and initiative in understanding and protecting the county's groundwater resources directly led to this study. N. K. Bleuer of the IGS introduced me to the concept and practice of glacial terrain analysis, and his earlier studies of Allen County formed the cornerstone of this research. I am also grateful to Victoria Ferguson of Indiana University, who provided information from her studies of till hydrogeology, and to Michael Yarling of the Indiana Department of Environmental Management, who shared geochemical and isotopic data from the Huntertown aquifer system. Discussions with both of these individuals stimulated many of the ideas presented herein. Finally, I thank the many water well contractors for providing access to well construction sites for sampling and logging.

REFERENCES

ANDERSON, M.P., 1989, Hydrogeologic facies models to delineate large-scale trends in glacial and glaciofluvial sediment: Geological Society of America Bulletin, v.101, p. 501-511.

ANDERSON, M.P., 1991, Aquifer heterogeneity—A geological perspective, *in* Bachu, S. ed., Parameter Identification and Estimation for Aquifer and Reservoir Characterization: Proceedings of the 5th Annual Canadian/American Conference on Hydrogeology, Banff, Alberta, Canada, National Water Well Association, p. 3-22.

ARVIN, D.V., 1989, Statistical summary of streamflow data for Indiana: U.S. Geological Survey Open-File Report 89-62.

ASHLEY, G.M., SHAW, J., and SMITH, N.D., 1985, Glacial sedimentary environments: Society of Economic Paleontologists and Mineralogists Short Course 16, 246 p.

BENNION, D.W., and GRIFFITHS, J.C., 1966, A stochastic model for predicting variations in reservoir rock properties: Transaction of American Institute of Mining and Metalurgical Engineers, v. 232, no. 2 p. 9-16.

BLEUER, N.K., 1985, The nature of some glacial and manmade sedimentary sequences and their downhole logging by natural gamma ray, *in* West, T.L., ed., Building On and With Sedimentary Bedrock: Proceedings of the 36th Annual Highway Geology Symposium, p. 204-219.

BRADBURY, K.R., and ROTHSCHILD, E.R., 1985, A computerized technique for estimating the hydraulic conductivity of aquifers from specific capacity data: Ground Water, v. 23, no.2, p. 240-246.

DAVIS, S.N., 1969, Porosity and permeability of natural materials, *in* De Weist, R.J.M., ed., Flow Through Porous Media: New York, Academic Press, p. 54-89.

DREVER, J.I., 1988, The geochemistry of natural waters: Englewood Cliffs, N.J., Prentice Hall 388 p.

EYLES, N., and MIALL, A.D., 1984, Glacial facies, in Walker, R.G., ed., Facies Models: Geological Association of Canada Reprint Series 1, p. 15-38.

EYLES, N., CLARK, B.M., KAYE, B.G., HOWARD, K.W.F., and EYLES, C.H., 1985, The application of basin analysis techniques to glaciated terrains— An example from the Lake Ontario Basin, Canada: Geoscience Canada, v. 12, no. 1, p. 22-32.

FERGUSON, V.R., 1992, Hydrogeology and hydrogeochemistry of fine-grained glacial till, northeastern Indiana: Indiana University, Bloomington, Unpublished M.S. thesis, 75 p.

FERGUSON, V.R., FLEMING, A.H., KROTHE, N.C., and STEEN, W. J., 1992, Hydrogeology and hydrogeochemistry of fine grained glacial till, northeastern Indiana (abs.): Geological Society of America Abstract with Program, v. 24, no. 1, p. 302.

FLEMING, A.H., 1992a, The hydrogeologic framework of Allen County, Indiana: Hydrostratigraphic atlas emphasizing subsurface sequence stratigraphy and ground-water contamination potential: Indiana Geological Survey Open File Report 92-14, 55 p.

FLEMING, A.H., 1992b, Late Wisconsin evolution of the Maumee Lacustrine Basin in northeastern Indiana (abs.): Proceedings Geological Society America Ammia Meeting, v. 24, no. 7, p. 347.

FLEMING, A.H., 1994, The hydrogeology of Allen County, Indiana—A geologic and ground-water atlas: Indiana Geological Survey Special Report 57, 111 p.

FLEMING A.H., and YARLING M., 1994, Facies distributions, recharge-discharge relations, and aquifer sensitivity in a glacial aquifer system, northeastern Indiana (abs.): Geological Society of America, Abstract with Program, v. 26, no. 5, p. 15.

FREEZE, R.A., 1975, A stochastic conceptual model of one-dimensional groundwater flow in non-uniform homogeneous media: Water Resources Research, v.11, p. 725-741.

FREEZE, R.A., and CHERRY, J.A., 1979, Groundwater: Englewood Cliffs, N.J., Prentice-Hall 604 p.

LEVELL, B.K., BRAAKMAN, J.H., and RUTTEN, K.W., 1988, Oil-bearing sediments of Gondwana glaciation in Oman: American Association of Petroleum Geologists Bulletin, v. 72, no. 7, 775-796.

POLAND, J.F., LOFGREN, E.E., and RILEY, F.S., 1972, Glossary of selected terms useful in studies of aquifer systems and land subsidence due to fluid withdrawals: U.S. Geological Survey Water-Supply Paper 2025, 9 p.

SHAW, J., and ASHLEY, G.M., 1988, Glacial facies models-continental terrestrial environments: Geological Society of America Short Course Notes, 121 p.

STRAW, W.T., PASSERO, R.N., and KEHEW, A.E., 1993, Conceptual hydrogeologic glacial facies models—Implications for aquifers and agrichemicals in southwest Michigan, in Eckstein, Y., and Zaporozec, A., eds., Hydrogeologic Investigation, Evaluation, and Ground-Water Modeling: Proceedings of the Second USA/CIS Joint Conference on Environmental Hydrology and Hydrogeology, Water Environment Federation, Alexandria, Va. p. 35-68.

WARREN, J.E., and PRICE, H.S., 1961, Flow in heterogeneous porous media: Society Petroleum Engineering Journal, v.1, p. 153-169.

WAYNE, W.J., 1963, Pleistocene formations in Indiana: Indiana Geological Survey Bulletin 25, 85 p.

WEBB, E.K., and ANDERSON, M.P., 1991, Using a sedimentary depositional model to simulate heterogeneity in glaciofluvial sediments, in Bachu, S. ed., Parameter Identification and Estimation for Aquifer and Reservoir Characterization: Proceedings of the 5th Annual Canadian/American Conference on Hydrogeology, Banff, Alberta, Canada, National Water Well Association, p. 3-22.

USING THREE-DIMENSIONAL GEOLOGIC MODELS TO MAP GLACIAL AQUIFER SYSTEMS: AN EXAMPLE FROM NEW JERSEY

SCOTT D. STANFORD[1] AND GAIL M. ASHLEY[2]

1New Jersey Geological Survey, Trenton, New Jersey 08625
2Department of Geological Sciences, Rutgers University, New Brunswick, New Jersey 08903

ABSTRACT: Glacial aquifer systems are sediment bodies and their bounding surfaces. While the sedimentologic and geometric character of these systems can be predicted from general depositional models, the actual configuration of sediment bodies and contacts in a given aquifer system is determined by the sequence of glacial geologic events at that location. Geologic analysis integrates depositional models with local data to define the (1) bedrock surface containing or underlying the aquifer system; (2) geometry of the sediment units comprising the aquifer system; (3) contacts between the units; and (4) texture, sorting and stratification within the units. The bedrock surface may be a fluvial valley unmodified by glacial erosion, a basin created entirely by glacial erosion or a glacially modified fluvial valley. The geometry of the sediment units may be tabular, basin-filling, beaded, ridged or blanketing; and they may be bounded by fluvial erosional, glacial erosional, onlapped, gradational, interfingered or deformed contacts. Texture, sorting and stratification within the sediment bodies may be nearly homogeneous and isotropic or heterogeneous and anisotropic.

INTRODUCTION

Glacial deposits are characterized by extreme variations in texture and sorting over short distances. They have complex sediment-body geometries and bounding surfaces that reflect a wide variety of depositional processes by ice, water and wind. The resulting mosaic of sediments creates complex permeability relationships and groundwater flow paths. Geologic mapping combined with stratigraphic and sedimentologic study are important first steps in the process of modeling these aquifer systems because a geologic approach provides the only means of predicting sediment-body geometries, permeability properties and the nature of bounding surfaces. Understanding the hetereogeneity of the sediments allows realistic delineation of aquifers, aquitards, and aquicludes. In this paper we describe the basic geologic elements of glacial aquifer systems and their importance in hydrogeologic modeling.

Most previous studies on this subject have focused on mathematical modeling of permeability variations using geostatistical or stochastic methods (Freeze, 1975; Delhomme, 1979; Dagan, 1982; Kitanidis and Vomvoris, 1983; Hoeksema and Kitanidis, 1984, 1985; Sudicky, 1986; Silliman and Wright, 1988; Gomez-Hernandez and Gorelick, 1989). Several studies emphasize that understanding depositional environments and geologic architecture is an important element in hydrogeologic investigations (Kempton and Cartwright, 1984; Fraser and Bleuer, 1987; Stephenson et al., 1988; Anderson, 1989; Webb, 1994). We believe that in most investigations of regional aquifers, where study areas generally are at 1:24,000 or larger

scales, and where money and time constraints limit the ability to acquire the detailed data required for calibration of mathematical models, a simpler modeling strategy is more effective. This strategy uses the geologic concepts described below to map permeability units and to assign reasonable permeability values to them. Because variations in permeability within units are significantly less than the permeability contrast between units, a workable simulation can be achieved without detailed knowledge of small-scale permeability distribution. This approach also provides a starting point for finer-scale studies of landfill and hazardous waste sites, where small-scale permeability distributions significantly affect contaminant transport. Conversely, the results of detailed, data-rich studies of small areas can be extrapolated to larger areas in an informed way using these geologic concepts.

We first describe the basic architectural components of glacial aquifer systems and then outline how these components can be assembled to create a site-specific hydrogeological model. The last section illustrates this process with an example from a glacial valley-fill aquifer in New Jersey.

COMPONENTS OF GLACIAL AQUIFER SYSTEMS

Considered as simple geometric entities, glacial aquifer systems are composed of bodies of sediment of irregular shape bounded by surfaces. Each sediment body has physical properties (texture, sorting, bedding) that distinguish it from other bodies. The bounding surfaces include the bedrock surface that underlies or borders the aquifer system and the contacts between the sediment bodies within the aquifer system. Groundwater moves through this physical environment in response to equipotential fields that reflect, in part, the geometry and permeability of the sediment bodies and bedrock.

Glaciers are unique sedimentation agents for five reasons. First, within the limits of glaciation, they deposit sediment on an entire subaerial landscape of irregular topography, not just in base level-controlled sedimentary basins. Second, they entrain and transport sediment of all grain sizes equally, providing a constant supply of various textures, depending on the source materials, at the site of deposition. Third, they generate large volumes of meltwater that vary greatly in discharge and location, both in time and space. Fourth, they continually encounter and create new topography and, thus, new sedimentary environments, as their margins move. Fifth, ice is the substrate and supporting material for some depositional basins, creating high-standing inverted sediment bodies upon melting.

For these reasons, glacial aquifer systems differ from other sedimentary aquifer systems in several important ways: (1) physical properties of the sediment bodies vary over a large range within short distances; (2) contacts between sediment bodies commonly have irregular or complex topography; (3) the bedrock bounding surface likewise commonly has irregular or complex topography; and (4) the scale of these geologic components is larger than is generally the case with marine or fluvial sediments.

In order to manage this complexity in a hydrogeologic investigation, the basic architectural components of glacial aquifer systems must be identified and mapped. The following discussion outlines these components.

Bedrock Surface

The bedrock surface that underlies or borders glacial deposits can be, at one extreme, a preglacial fluvial valley essentially unmodified by glacial erosion or, at the other extreme, a basin produced entirely by glacial erosion. More commonly it is a fluvial valley modified to varying degrees by glacial erosion (Fig. 1).

Unmodified fluvial valleys (Fig. 1A) most likely occur at the glacial limit or beyond, where there was little to no glacial erosion. Fluvial valleys also can be preserved farther up-glacier if they were filled with stratified deposits before burial by ice, thus escaping later subglacial erosion. They have branching drainage patterns and slope forms that reflect fluvial erosion of the rock substrate. The valley floors descend continuously in the direction of preglacial fluvial drainage and are not significantly overdeepened.

Bedrock surfaces created entirely by glacial erosion are produced under thick, active ice some distance upglacier from the margin (Fig. 1B). Multiple erosional events can remove most traces of preglacial fluvial topography, particularly in low-relief areas. Elongate scour troughs form along strike belts of weak rock that are subparallel to glacier flow. Streamlined uplands eroded into more resistant rock border the troughs. Irregular crescentic or elliptical pluck-out basins that form in belts of weak or finely jointed rock are subperpendicular to glacier flow. These basins are bounded by plucked, cliffed rock surfaces on their upglacier sides and by abraded, sloped rock surfaces on their downglacier sides. Intricate knob-and-basin or ridge-and-swale topography are local details superimposed on these larger basin forms.

Glacially modified fluvial valleys (Fig. 1C) dominate in the zone between deep glacial erosion under thick ice and the noneroding margin. This zone includes most of the glaciated United States. They include elements of both fluvial valleys and glacially eroded basins. Fluvial valley slopes are variously modified by plucking and abrasion, and valley bottoms locally are overdeepened by plucking or scour. The location and type of erosional modification are determined by the interaction of ice flow, rock structure, lithology, subglacial thermal regime and preglacial valley orientation.

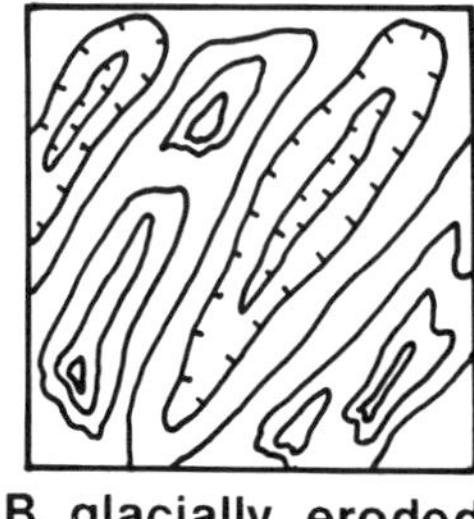

A. fluvial valley   **B. glacially eroded basin**   **C. glacially modified fluvial valley**

FIG. 1.—Contour maps of the three basic types of bedrock surface underlying and containing glacial aquifer systems.

The configuration of the bedrock surface is hydrogeologically important, particularly in establishing model boundary conditions. Groundwater in buried fluvial valleys can drain from the aquifer system along the preglacial valley thalweg, without entering bedrock. Groundwater in glacially eroded basins can leave the aquifer system only through the bedrock or by discharge at the land surface. Groundwater in glacially modified fluvial valleys encounters segments of both thalweg drains and closed basins. The intersection of the bedrock surface with both high- and low-permeability bedrock units determines the location of potential subsurface drains and no-flow boundaries in the aquifer system.

Sediment Body Geometry

The sediment bodies comprising glacial aquifer systems have a wide range of potential geometries because of the variety of depositional basins and constructional landforms in the glacial environment. The basic forms are shown on Figure 2.

Tabular forms, where the upper surface has a continuous gradient and the horizontal dimensions greatly exceed the vertical dimension, are characteristic of glaciofluvial sand and gravel deposits laid down in valleys that slope away from the glacier margin (Fig. 2A). The gradient, thickness and extent of these units depend on the form and slope of the valley they occupy and on the discharge, texture and duration of the glacial deposition (Church and Gilbert, 1975; Boothroyd and Ashley, 1975). In areas of low relief, glaciofluvial deposits may form regionally extensive tabular sheets covering hundreds of square kilometers. Deltas also have tabular geometries, although they are generally thicker and smaller in extent than an equivalent volume of fluvial sediment. The dimensions of deltaic bodies depend on water level, shape of the depositional basin and location of the meltwater point sources supplying the delta.

Basin-filling forms, where the upper surface is nearly flat and the lower surface is the unmodified predepositional basin surface, are characteristic of fine-grained glacial lake-bottom deposits, particularly those deposited from turbid underflows (Fig. 2B). These density-driven flows progressively accumulate in the lowest parts of the lake basin (Gustavson, 1975). In some smaller lake basins, large deltas also may prograde and fill the entire basin, with an upper surface grading to the lake spillway (Koteff and Pessl, 1981). Again, horizontal dimensions greatly exceed the vertical dimension, but these sediment bodies generally are thicker than tabular bodies because base levels are higher.

Beaded and ridged forms are positive landforms created by deposition of till or stratified deposits in subaerial or sublacustrine ice-contact environments or in subglacial environments (Figs. 2C, 2D). The particular shape and size of their forms are determined by the geometry of the ice margins, ice-walled basins or subglacial cavities against or within which they were deposited. Examples of these forms include (1) moraines, which may be simple ridged forms or complex combinations of ridged and beaded forms, with intervening closed basins and swales; (2) eskers, which are generally ridged when they represent subglacial tunnel deposits, and beaded where they represent successive tunnel-mouth deposits (Banerjee and McDonald, 1975; Warren and Ashley, 1994); and (3) lacustrine fans, which are deposited on the lake floor at the mouths of subglacial or englacial channels (Rust and Romanelli, 1975; Gustavson and Boothroyd, 1987) and typically have beaded or, more rarely, ridged form. Figure 3 is an example of a ridged lacustrine-fan or esker form exposed by lowering of the lake in which it was deposited. Some deltas, particularly small deltas deposited in deep lakes, have a ridged or beaded form with a flat summit marking the lake level.

Blanketing forms, where a layer of roughly constant thickness is fitted to the form of the substrate surface, is characteristic of till sheets and, to a lesser extent, eolian deposits (Fig. 2E). Note, however, that many till sheets may

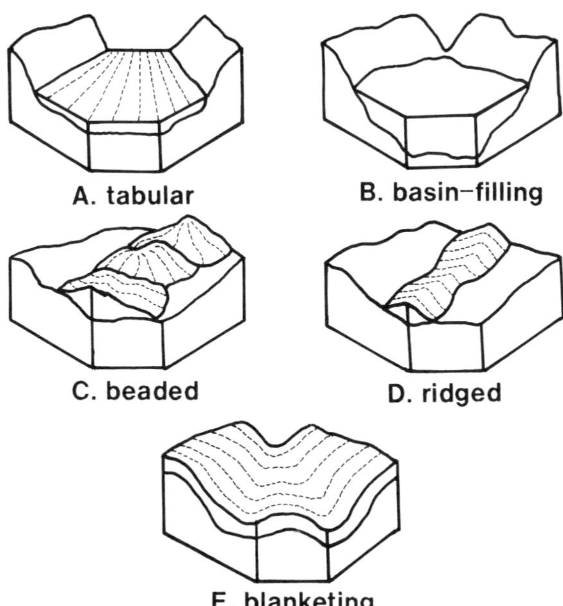

Fig. 2.—Basic geometric forms of sediment bodies in glacial environments.

FIG. 3.—Lacustrine fan and esker deposit exposed by lowered lake level, Iceland. Note ridged geometry. Deposits of this type commonly are buried by fine-grained lake-bottom deposits, forming confined aquifers. Photo by J. C. Boothroyd.

vary considerably in thickness and, in areas of high substrate relief or low substrate erodibility, are patchy.

These are all simple and familiar forms, but they become conceptually important when they are stacked vertically and juxtaposed horizontally in an actual glacial aquifer system. For example, a relatively simple single ice advance into and out of a lowland might result in a basal tabular glaciofluvial sand, deposited before ice-damming created a proglacial lake, followed by a lake clay with basin-filling geometry. Then follows a blanketing or ridged till sheet representing ice advance and retreat, with the possible formation of a sublacustrine moraine, in turn followed by beaded or ridged lacustrine fans or eskers deposited during retreat through the lake. This, in turn, is followed by a second basin-filling lake clay and topped by a second glaciofluvial sand deposited after lake drainage. Accurate mapping and modeling of the multiple aquifers and confining units in this package depend on recognition of the various geometries of the stacked sediment units. Multiple advances and retreats add more complexity. Similar stacking of coarse and fine sediments also can occur in fluvial environments beyond the glacial border, where coarse glaciofluvial deposits are stacked on and inset into finer interglacial alluvium (Fraser, 1994).

Contacts

The boundaries of sediment bodies are defined by contacts which, like the other elements of glacial depositional environments, cover a wide range (Fig. 4). Fluvial erosional contacts occur at the base of fluvial units (Fig. 4A). They are sharp and have channelform topography. The channel

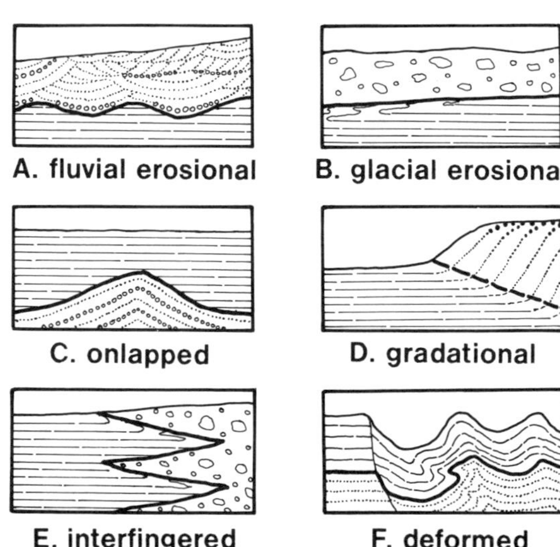

FIG. 4.—Basic types of contacts in glacial environments.

forms become more pronounced as grain size and channel slope increase (Church and Gilbert, 1975). Scour surfaces are common within fluvial deposits, where they may occur with gravel lags or gravel beds of locally high permeability, but these contacts have regional hydrogeologic significance where fluvial deposits are laid down on lower permeability material, such as till or fine-grained lacustrine and alluvial deposits.

Glacial erosional contacts (Figs. 4B, 5) occur at the base of till sheets. They also are sharp but generally more planar than fluvial erosional contacts, although they may be complicated locally by thrust faulting, recumbent folding and other soft-sediment compressional deformation structures. They occur on both regional and local scales and may be formed by multiple glaciations, multiple lobe advances during a single glaciation, regional readvances of a single lobe during a single glaciation or local ice-margin fluctuations.

Onlapped contacts (Figs. 4C, 5) are created when a deposit passively buries an earlier substrate material, for example, when a lake clay accumulates atop a lacustrine fan or a delta progrades over a moraine. They are sharp and have the topography of the surface of the underlying substrate. Onlap surfaces are common within lacustrine and fluvial deposits but have regional hydrogeologic significance where they mark a major permeability contrast, for example, a lake-bottom clay onlapping a lacustrine-fan gravel.

Gradational contacts occur where one sediment body passes into another by a gradual textural change over some distance (Fig. 4D). An example is the transition along the lower edges of delta fronts from sandy, lower foreset beds through silt and sand bottomset beds into silt and clay lake-bottom rhythmites (Church and Gilbert, 1975). Another example, in the vertical dimension, is stacked proglacial fluvial sediments deposited from multiple advances within a single glaciation, without a significant hiatus in deposition (Fraser, 1994). In this situation vertical coarsening and fining mark successive advances and retreats farther upstream in the basin. Gradational contacts are of hydrogeologic importance where they separate materials of significantly different permeability, such as delta sand and lake-bottom clay.

Interfingered contacts (Fig. 4E) occur where the boundary between the sediment bodies is marked by a vertical alternation of the two materials over some horizontal distance (Fig. 4E). This type of contact may be formed either where a glacier repeatedly advances and retreats over a short distance into a sedimentary basin, resulting in till interfingered, by alternating deposition or thrust faulting, with stratified deposits; or at the margin of a moraine, where flowtills from the moraine slope may interfinger with stratified deposits accumulating against the moraine (Hartshorn, 1958). Note that this contact is scale dependent—when the scale of the interfingering becomes mappable at the scale of an investigation, the contact then becomes a set of erosional and onlapped contacts in vertical succession. Finally, any of these contacts can be affected by later fold-and-fault deformation, either from ice push, melting of ice blocks or slope failure (Fig. 4F). These deformed areas are localized along the ice-contact part of deposits or in areas of once extensive stagnant ice.

Fɪɢ. 5.—Till deposited by a readvance in a glacial-lake delta, Wayne, New Jersey, U.S.A. Base of till is a glacial erosional contact on delta foreset beds; delta bottomset beds are in onlap contact on the top surface of the till.

Internal Variability

A fourth basic element of glacial aquifer systems is the variability of texture, sorting and stratification within the sediment bodies and the effect this internal variablity has on their hydraulic conductivity. Figure 6 summarizes the variability in texture and sorting for the six major sediment types in glacial environments. Till has a wide range of matrix texture, depending on the texture of the bedrock or surficial materials from which it is derived (Dreimanis and Vagners, 1971; Karrow, 1975). However, it is generally poorly sorted over all textures. Locally, till contains lenses of well-sorted stratified sediment. These generally are too small and isolated to be hydrogeologically significant at regional aquifer scale, and are rarely mappable but are important at site scale. Lacustrine-fan and ice-contact sediments also have a wide textural range, reflecting variable flow conditions in those environments. Although individual beds may be well-sorted, the overall deposit is generally moderately to poorly sorted as a result of sharp textural changes over short distances (Rust and Romanelli, 1975; Banerjee and McDonald, 1975). Proximal deltaic and fluvial sediments are coarse textured but better sorted than lacustrine-fan and ice-contact sediments because current velocities are less variable. Distal deltaic and fluvial sediments are finer and better sorted than proximal deposits (Gustavson et al., 1975; Shaw, 1975; Boothroyd and Ashley, 1975). Eolian sediment is very well-sorted and has a narrow textural range of fine sand and silt, reflecting the narrow range of grain sizes that are transported by wind (Lea, 1990; Ashley and Hamilton, 1993). Lake-bottom sediment is fine grained and well sorted, also reflecting the relatively narrow range of grain sizes that can be transported long distances onto lake bottoms by turbidity flows and in suspension (Ashley, 1975).

The effects of these physical properties on hydraulic conductivity (K) of the sediments are summarized in Figure 7. K anisotropy increases as the degree of alternating fine- and coarse-textured stratification increases in stratified deposits, for example, in proximal rhythmically layered lake-bottom sediment and in ice-contact and lacustrine-fan sediment. Fractured till, which generally has a cohesive silt- and clay-rich matrix, also behaves anisotropically (Grisak et al., 1976). Isotropic K distribution is favored where little textural variation occurs between strata, as in distal deltaic and fluvial sediments or in nonstratified sediments, such as some eolian deposits and nonfractured till. Homogeneous K distributions can be expected where spatial variation in the physical character of the deposit is slight; for example, in eolian and lake-bottom sediments, which maintain nearly uniform texture, stratification and sorting over their entire extent. Heterogeneous K distributions can be expected where texture, stratification and sorting change over short distances, for example, in ice-contact and lacustrine-fan environments and in till sheets containing variably textured subglacial, englacial and superglacial facies.

From a practical perspective, much of the internal variability in hydrogeological properties of these glacial sediment units cannot be measured with specificity because of the large number of aquifer or laboratory tests required. This difficulty has led to development and use of statistical and probabilistic techniques to model K. We believe it is important to identify and map sediment bodies having major K contrasts, and their bounding surfaces, using geologic methods before applying mathematical techniques.

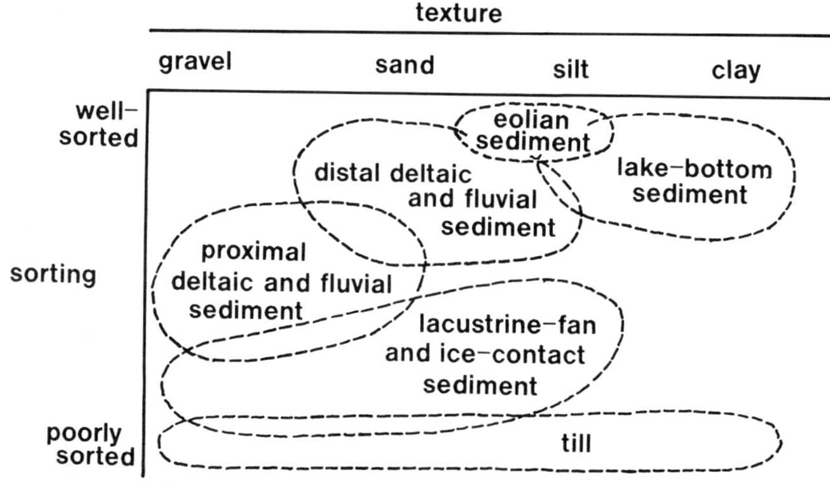

FIG. 6.—Variability in texture and sorting for the major types of glacial sediment.

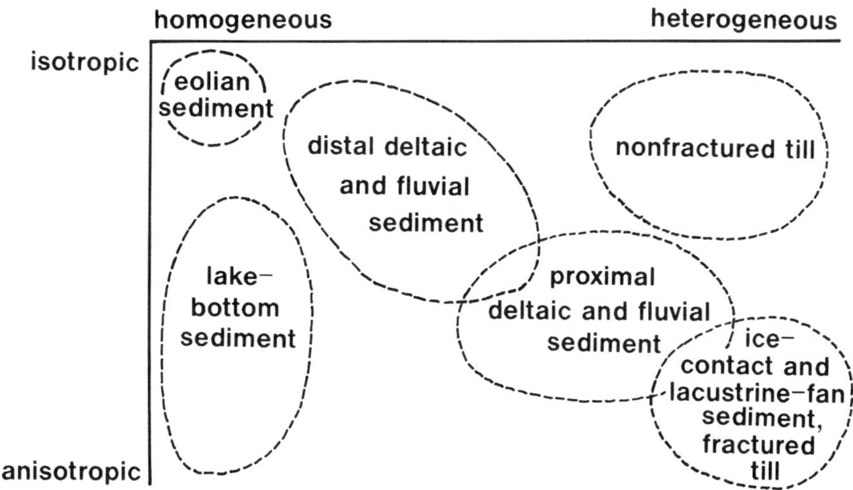

FIG. 7.—Variability of hydraulic conductivity for the major types of glacial sediment. Fields indicate hydraulic conductivities considered as point distributions within the sediment bodies. Some of these distributions would change with a coarser sampling resolution. The distribution shown for till assumes a variable matrix texture and clast content, reflecting the presence of basal, englacial, and superglacial facies. Tills with uniform matrix texture and low gravel and boulder content have a more homogeneous conductivity distribution.

The variability of K within sediment bodies of similar texture generally ranges over one to three orders of magnitude, whereas the contrast between sediment bodies (such as between sand and gravel stratified deposits and silt and clay lake-bottom deposits) generally is four or more orders of magnitude (Davis, 1969; Stephenson et al., 1988). Once the basic geologic architecture of the aquifer system, and the resulting major K contrasts, are successfully modeled, then the internal K variability of individual sediment bodies can be investigated if necessary. By this method, the apparently chaotic or random K variability of the aquifer system, when considered as a bulk sediment package, will be analyzed into component pieces of more uniform K.

This approach is illustrated by the distribution of K values in glacial aquifers in New Jersey (Fig. 8). Sand and gravel stratified deposits, whether deposited in deltaic, fluvial or lacustrine-fan environments, have K values nearly all in the 10^{-1}- and 10^{-2}-cm/s range. Silt and clay lake-bottom deposits, although data are sparse, are in the 10^{-6}- to 10^{-8}-cm/s range. This indicates a two or three order of magnitude range of values within a given sediment unit, and a four to seven order of magnitude difference between different sediment units.

CONSTRUCTING GLACIAL GEOLOGIC MODELS

The actual set of sediment bodies, contacts and bedrock surfaces present in a particular glacial aquifer system depends on the temporal and spatial sequence of erosional and depositional events in that location. This sequence, in turn, depends on the location, through time, of ice margin positions, meltwater drainage routes and depositional basins. This dependence on local geography and local geologic history makes each system unique. This uniqueness, in turn, means that investigating the sedimentary architecture of a system requires integrating depositional models and local geologic data.

The internal variability of the sediment bodies is, however, less dependent on geographic and historic events. It can therefore be more successfully predicted from general models (Eyles, 1983; Ashley et al., 1985). For example, glacial-lake deltas everywhere show a standard bottomset-foreset-topset bed architecture, but the dimensions of any given delta depend on local geologic factors, such as spillway elevation, basin shape, ice-margin configuration and feeder-channel location.

Some aspects of internal variability depend on local circumstances. These include till matrix texture, which depends on lithology of the source materials from which the till was eroded. Because till matrix texture is a major control on till permeability and on the susceptibility of till to fracturing, the ability to predict this texture is hydrogeologically significant (Morris and Johnson, 1967; Stephenson et al., 1988). Even at a single location, matrix texture, lateral continuity, thickness and degree of fracturing may vary significantly in vertically stacked tills. This may be due to variable subglacial thermal conditions, changing ice-flow directions of successive advances or different substrate materials encountered by successive

SCOTT D. STANFORD AND GAIL M. ASHLEY

advances in the same flow direction (for example, Willman and Frye, 1970). For example, glaciers traversing sandy-textured substrate will deposit sandy till; those traversing fine-textured substrate will deposit silty-clayey till. Successive advances along the same flowline may overrun sandy glaciofluvial deposits during one advance, produc-

ing a sandy till. A second advance may overrun clayey glaciolacustrine deposits, laid down during the previous recession, producing a clayey till. Knowing the direction, timing and extent of advances is thus essential for mapping the distribution of till in aquifer systems and predicting its internal composition.

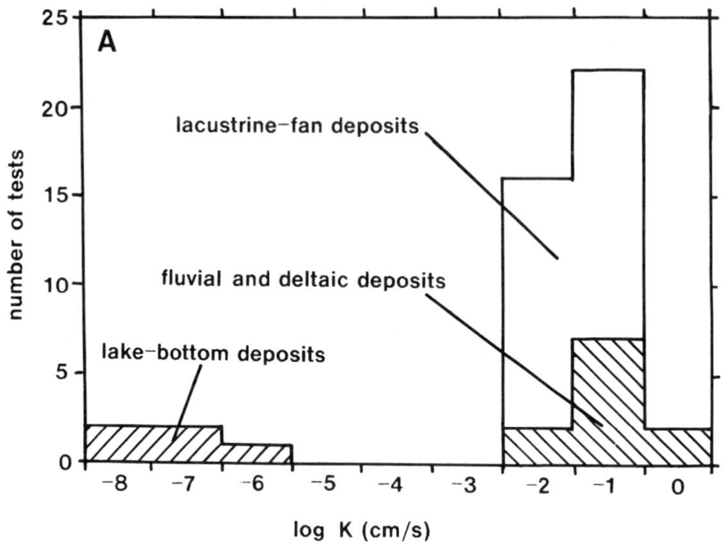

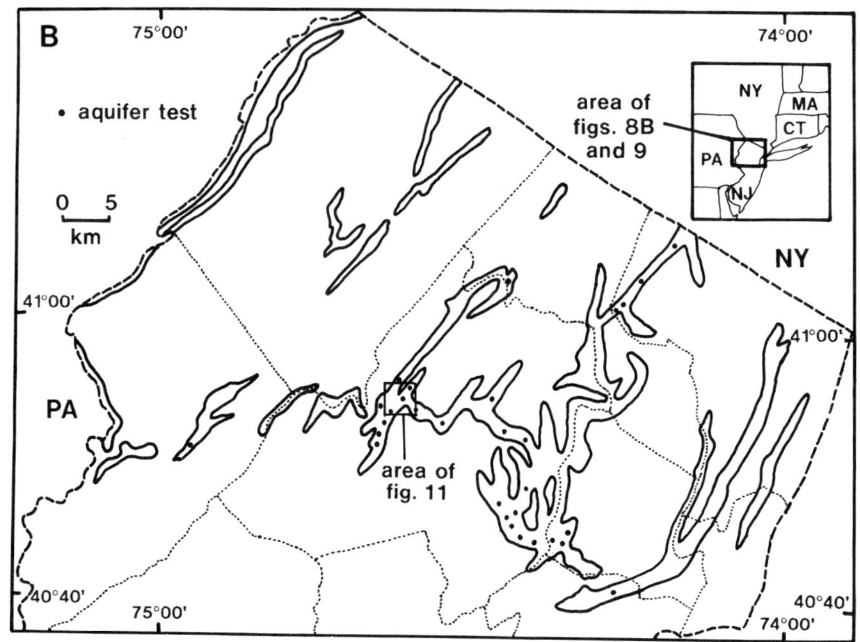

FIG. 8.—(A) Distribution of hydraulic conductivity in glacial valley-fill aquifers in New Jersey, and (B) location of the aquifers (heavy outline) and aquifer tests (dots). Hydraulic conductivities for lacustrine-fan, fluvial and deltaic deposits are from aquifer pump tests; values for lake-bottom deposits are from laboratory tests on undisturbed samples recovered in the field and from modelling results. Extent of aquifers is from Stanford and others (1990) and Hoffman and Stone (1991). Hydraulic conductivity data are from Gill and Vecchioli (1965), Anderson (1968), Vecchioli and Miller (1973), Meisler (1976), Hutchinson (1981), Canace et al. (1983), Hill (1985), Sirois (1986), Hill et al. (1992) and Nicholson et al. (1996). Dotted lines are county boundaries.

The same considerations are true, to a lesser degree, for stratified deposits. Extensive gravel deposits should not be expected where source materials do not yield gravel-sized clasts; thick clays are unlikely where source materials are not fine-grained.

Glacial Geologic Mapping

In any given glacial aquifer system, the geologic elements outlined above are rarely exposed except in a fragmentary way. Identification of these elements thus depends on accurate, detailed geologic mapping, both within the area of the aquifer and regionally, to define the temporal and spatial extent of the glacial events that produced the sediment bodies and contacts. In most glacial aquifer systems the relatively fine scale of the depositional environments requires mapping at detailed scales (in the United States, generally at 1:24,000). This scale is larger than most existing U.S. state or federal glacial mapping.

The goal of geologic mapping for hydrogeologic investigation is to delineate permeability units. Many glacial geologic maps use morphologic or chronostratigraphic criteria to define map units, and each map unit may include mixed lithologies. This procedure generally is unsatisfactory for aquifer studies because it does not differentiate permeability units. For aquifer studies, map units must be lithostratigraphic. Morphology and age, while important as tools to mapping, are subordinate to lithology. Likewise, mapping materials alone is unsatisfactory because it omits the stratigraphic information needed to infer the sequence of depositional and erosional events that determine sediment-body and contact geometry. These requirements impose a rigor and detail that may be difficult to achieve, particularly in complex ice-contact environments that traditionally have been mapped as single morphostratigraphic units, such as "moraines" or "ice-contact stratified drift." These settings can be successfully analyzed into their component lithologic units if field data are sufficiently detailed.

A lithostratigraphic mapping investigation integrates geomorphic analysis, outcrop observations and subsurface data to produce surficial maps, cross sections, isopach maps of sediment bodies and contour maps of contacts and surfaces. Integration is achieved by recognizing the type of contacts and classes of sediment bodies present, and using these components to infer the distribution of materials where data is absent.

A four-step mapping procedure usually is most effective. First, geomorphic analysis, using airphotos, produces a preliminary outcrop map of contacts and sediment bod-

ies. Generally, these features have distinctive geomorphic signatures. Erosional features of potential stratigraphic significance, such as lake spillways, shorelines, meltwater channels and ice-sculpted forms, also are mapped. Second, sediment bodies and contacts are checked in the field and described where exposed. Where exposures are absent, hand or power excavation is used to verify materials. Third, subsurface data are acquired, including logs of wells and borings, geophysical surveys and project test drilling. Finally, the map is compiled. During compilation, the geomorphic analysis is evaluated together with the outcrop and subsurface sedimentologic and stratigraphic data to produce a three-dimensional lithostratigraphic model.

Like hydrogeologic models, geologic models can be calibrated as the model parameters are adjusted to match observed head data. Fixed control is provided by the surface map, subsurface data points, and the geologic limits on sediment body geometry and internal variability.

AN EXAMPLE FROM NEW JERSEY

Glacial aquifers provide about 40 percent of the groundwater supply of northern New Jersey (New Jersey Department of Environmental Protection, 1987). Glacial deposits in New Jersey, as in much of the northeastern United States outside the Great Lakes basin, include stratified glaciolacustrine and glaciofluvial deposits filling glacially modified fluvial valleys, separated by broad, till-veneered bedrock uplands (Fig. 9). Glacial lakes formed in valleys that sloped toward ice margins, and so were dammed, or in valleys that were dammed by glacial deposits. Glaciofluvial sediment was deposited in valleys that sloped away from ice margins, and so were not dammed.

Glacial Valley-Fill Aquifers in New Jersey

In New Jersey the stratified valley-fill deposits are productive aquifers where they exceed about 30 m in thickness (Fig. 8). Four main classes of valley-fill aquifer are found in New Jersey. Valley fills in *unfilled lake basins* (Fig. 10A) include sediment deposited in glacial lakes that were either too large or too deep or received too little sediment to have filled completely. Landforms in these basins include deltas and lacustrine fans alternating with low-lying lake plains. The fans typically occur as north-trending beaded ridges that mark successive positions of subglacial or englacial tunnel channels. In places they aggrade upward to lake level and are the nucleus of deltas. In the subsurface, silt and clay lake-bottom deposits cover permeable

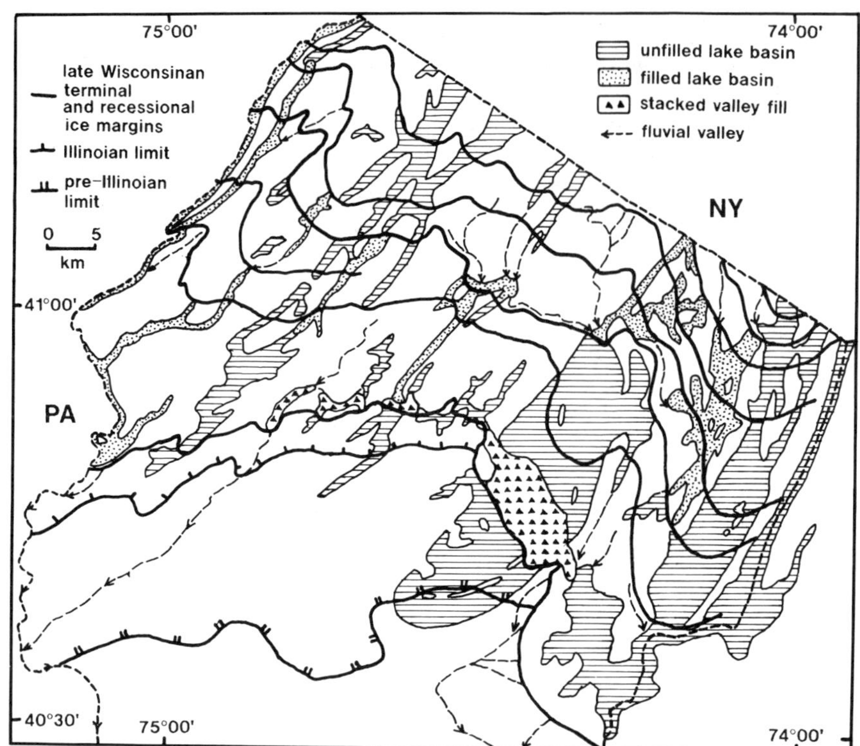

FIG. 9.—Principal filled and unfilled glacial-lake basins, glaciofluvial drainage and stacked valley fills in New Jersey. Heavy lines are terminal and recessional ice-margin positions. Sources: Stone et al. (1989), Stanford and others (1990) and Witte (1991). Location as in Fig. 8.

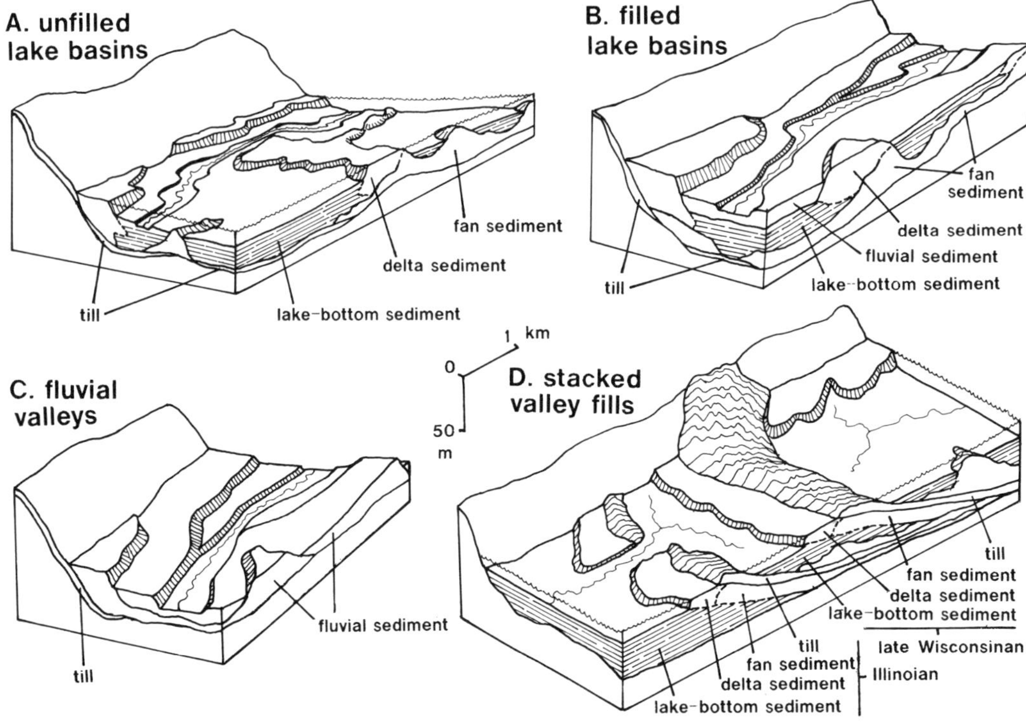

FIG. 10.—Four types of glacial valley-fill aquifer in New Jersey. Dashed contacts are gradational. Lacustrine-fan sand and gravel deposits are productive confined aquifers where they are overlain by fine-grained lake-bottom sediment or till (diagrams A, B, D). Deltas and outcrop areas of fans are potential recharge areas for these aquifers. Deltaic and fluvial sand and gravel are unconfined aquifers where saturated thickness is sufficient (diagrams A, B, C, D). Lacustrine valley fills may have both confined and unconfined aquifers (A, B, D); fluvial valley fills are generally unconfined (C). Fluvial deposits form shingled terraces (B, C) representing deposition from successive recessional ice margins.

fan deposits in places, forming productive confined aquifers. These aquifers are recharged, in part, from outcropping fans and deltas that are vertically and laterally connected to the buried fan segments.

Filled lake basins were sufficiently small, shallow, or received enough sediment to have filled completely (Fig. 10B). Delta, fan and lake-bottom sediments were deposited as in unfilled lake basins, but the lake either filled with sediment or drained while meltwater was still discharging into the basin. Thus, an upper fluvial deposit, laid down after the lake filled or drained, caps the lake-bottom sediments and, in places, is inset within earlier deltaic deposits (Von Schondorf, 1987). This stratigraphy creates a two-aquifer system: a lower confined lacustrine-fan aquifer, overlain by confining silty lake-bottom deposits, in turn overlain by an unconfined fluvial or fluvial-deltaic aquifer. Again, outcropping lacustrine fans and deltas that break the continuity of the lake-bottom sediments are potential recharge conduits.

Fluvial valley fills were deposited in valleys that sloped away from the ice margin and that contained no glacial lakes (Fig. 10C). These valley fills are predominantly sand and gravel glaciofluvial deposits, although some glaciolacustrine deposits may be present in the subsurface, filling basins in the bedrock surface. Fine-grained confining materials generally are not present in fluvial valley fills because silt and clay are retained in suspension and flushed from the valley. However, loess and fine-grained postglacial alluvial deposits locally may cap the coarser glacial deposits. If sufficiently thick, fluvial valley fills are unconfined aquifers, particularly if they are in hydraulic connection with lakes or streams. They may be locally confined by fine-grained alluvium.

Stacked valley fills occur along the terminal moraine of the most recent advance, of late Wisconsinan age (oxygen isotope stage 2, about 20 ka). In valleys and lowlands along the moraine, this ice overlapped but did not completely erode older valley-fill sediment of Illinoian age (oxygen isotope stage 6, about 150 ka). Glaciofluvial and glaciolacustrine deposits and till of late Wisconsinan age were deposited on top of similar deposits of Illinoian age, forming multiple aquifers and confining units (Fig. 10D).

The Rockaway Valley-Fill Aquifer

A geologic and hydrogeologic investigation of a part of one of these stacked valley fills, the Rockaway valley-fill aquifer (Fig. 9), illustrates how geologic data can contribute to hydrogeologic modelling (Stanford, 1989; Schaefer et al., 1993; Canace et al., 1994; Nicholson et al., 1996). This aquifer system occupies a preglacial fluvial valley system that is floored, in places, by carbonate bedrock and bordered by gneiss and quartzite. The valley system has been slightly modified by glacial scour in places. Part of this valley system was north draining, and so was dammed by the ice margin during both the Illinoian and late Wisconsinan glaciations to form a glacial lake. Other valley segments were south or east draining. During retreat of the late Wisconsinan ice margin, these were dammed by terminal moraine deposits, forming another set of glacial lakes. These lakes filled with sediment, and their outlets were erosionally lowered, allowing deposition of glaciofluvial sediment atop the lacustrine deposits.

These events created a valley fill with a variety of sediment body geometries and contacts (Fig. 11). Initial Illinoian advance into the rising glacial lake deposited a discontinuous layer of till on the preglacial bedrock substrate, which includes thick weathered carbonate residuum on the valley floor and colluvium derived from weathered gneiss and quartzite along the valley sides. As the Illinoian ice margin retreated, lacustrine fans, with beaded to ridged geometry, were deposited on the lake floor at the mouths of subglacial or englacial channels. The fans then were covered, with onlapping contact, by fine-grained lake-bottom sediment with a basin-filling geometry. The lake was deep (the spillway elevation is as much as 75 m above the bedrock floor of the valley) and so did not fill completely with sediment. As the retreating ice margin uncovered the eastward-draining Rockaway valley, the lake drained. An extended period of stream erosion followed, modifying the Illinoian surface. However, this erosion surface is not hydrogeologically significant because it separates two units of lake-bottom deposits, and so does not mark a permeability contrast.

During late Wisconsinan advance, the lake reformed and rose to the level of its Illinoian spillway, when eastward drainage was blocked by the advancing ice front. Lake-bottom deposits began to fill in on top of the eroded Illinoian surface in front of the advance. These deposits grade upvalley into lacustrine-fan sediments laid down at the advancing margin. With continued advance to its maximum position, ice overode these fans, modifying their beaded or ridged forms with a smooth glacial erosional contact and depositing till on the eroded surface. At the terminal position, lacustrine fans aggraded nearly to lake level, and a delta prograded out into the lake basin, forming a basin-filling deposit with a lower gradational contact into lake-bottom sediments. As the sandy delta foresets prograded into the lake basin, gravelly delta topset beds aggraded over the foresets to produce a tabular fluvial unit at the surface, in sharp contact with the underlying foreset sand. While the delta prograded into the lake basin, till was deposited subglacially and by superglacial sediment flows along the ice margin as it receded from its maximum position. This deposition formed the terminal moraine, with multiple ridge-and-bead topography.

While the ice margin stood at the terminal position, erosion by ice and subglacial meltwater stripped the Illinoian and advance-phase late Wisconsinan deposits from the valley bottom north of the moraine. The till sheet of the moraine thus marks an erosional discontinuity that breaks the valley-fill stratigraphy to the south from that to the north. The till, however, has a silty sandy matrix derived primarily from granular-weathering gneiss and so is not a low-permeability barrier.

As the ice margin retreated from the moraine, a second set of glacial lakes formed, dammed to the south by the moraine. These lakes each filled with the same stratigraphy as the valley south of the moraine—a basal ridged-and-beaded lacustrine fan sand and gravel, overlain by

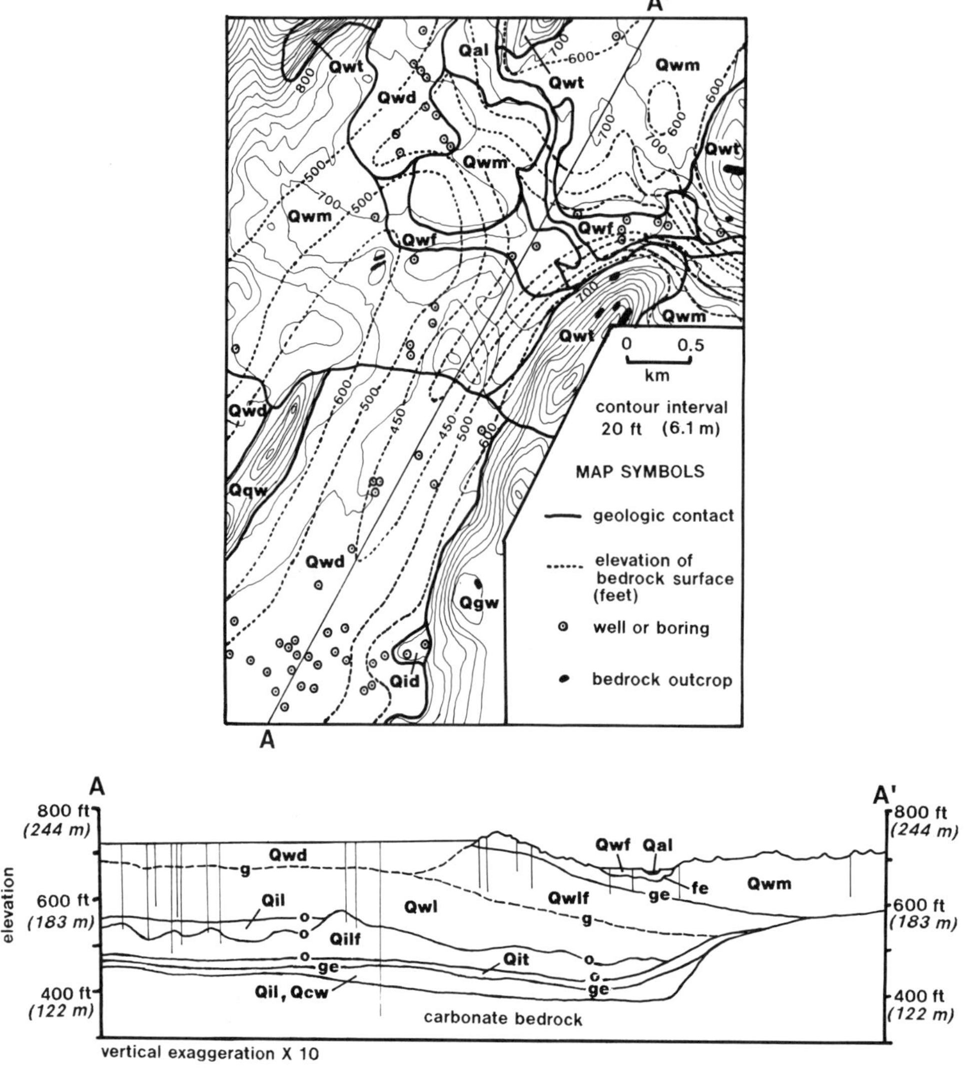

Fig. 11.—Map and section of a part of the Rockaway valley-fill aquifer, showing types of sediment bodies and contacts. Location shown on Fig. 8. Map units: Qal = alluvium; Qwt, Qit = till; Qwm = till of the terminal moraine; Qwd, Qid = deltaic sand and gravel; Qwf = fluvial sand and gravel; Qwl, Qil = lake-bottom silt, clay, fine sand; Qwlf, Qilf = lacustrine-fan sand and gravel; Qgw = weathered gneiss; Qqw = weathered quartzite, Qcw = weathered carbonate rock. Glacial units labeled with "w" are of late Wisconsinan age; those with an "i" are of Illinoian age. Units Qwd and Qwf show tabular to basin-filling geometry; Qwl and Qil are basin-filling; Qwlf and Qilf are beaded and ridged (the upper surface of Qwlf has been modified by glacial erosion); Qwm and Qit are blanketing (with beaded and ridged forms on the surface of Qwm). Contacts on section AA' are labeled according to type: fe = fluvial erosional, g = gradational, ge = glacial erosional, o = onlapped.

basin-filling lake-bottom silts in onlapping contact, capped by tabular to basin-filling deltaic-fluvial sand and gravel in gradational contact—except that the sequence is entirely of late Wisconsinan age.

Hydrogeologic Modelling of the Rockaway Valley Fill

This valley fill is modeled hydrogeologically as a four-layer aquifer system (Fig. 12) (Nicholson et al., 1996). The upper fluvial-deltaic sand and gravel and, locally, sandy late Wisconsinan till, is an unconfined upper aquifer. The lake-bottom silt, fine sand, clay and, in places, late Wisconsinan till, comprise an upper confining unit. The lacustrine-fan sand and gravel is a lower semiconfined aquifer, and the underlying Illinoian till and silty carbonate-rock residuum are a lower confining unit. Vertical connections between the aquifers or confining units represent either of two situations. The first is where lacustrine fans aggraded to lake level and connect with deltas, thus cutting out the upper confining unit. The second case is areas of the lake bottom away from subglacial tunnel channels where fans were not deposited, thus permitting deposition of lake-bottom sediments directly on carbonate residuum and till and connecting the upper and lower confining units. Alternatively, erosion by subglacial channels or by ice (particularly north of the moraine) has removed the basal carbonate residuum and till in places, thus breaking the lower confining unit. At several sites along the valley sides, the upper fluvial-deltaic units may be in direct contact with the eroded carbonate bedrock surface.

No-flow boundaries form the southern borders of the upper and lower aquifers, representing their termination against lake-bottom clay and the rising bedrock surface at the distal end of the lake basin. Flux boundaries form the northern border of the two aquifer units, representing their continuation northward in the Rockaway valley.

Recharge enters the aquifer by precipitation and surface water leakance on the surface of the upper aquifer, which covers nearly the entire valley floor. In places where Illinoian fans and deltas crop out near the Illinoian terminal position in the southern part of the lake basin, recharge is directly into the lower aquifer. Recharge was estimated as total base flow for all streams draining the model area (minus the through-going base flow contributed from upland tributaries that do not drain the valley fill), plus well pumpage, minus wastewater discharge. Recharge was discretized using streambed conductance along stream cells, and as precipitation (minus evapotranspiration) on outcropping aquifer or confining unit cells.

The working model was built by discretizing isopach maps of the various sediment units (Fig. 13). These maps were built from the geologic control assembled during mapping. This control includes the surface map and cross sections, and the test-boring, water-well and geophysical data. Hydraulic conductivities were assigned to the aquifers based on aquifer-test and well-performance data; these values then were adjusted to achieve model calibration. Hydraulic conductivities of the confining units were assigned based on laboratory test data and model calibration results. During model calibration, layer configurations were modified between control points to account for observed head distributions. For example, part of the moraine (unit Qwm, Fig. 11), which is composed of silty sand till of moderate permeability, was reassigned to the upper aquifer layer from its initial classification in the upper confining unit because head distributions indicated substantial ground water flow through the moraine.

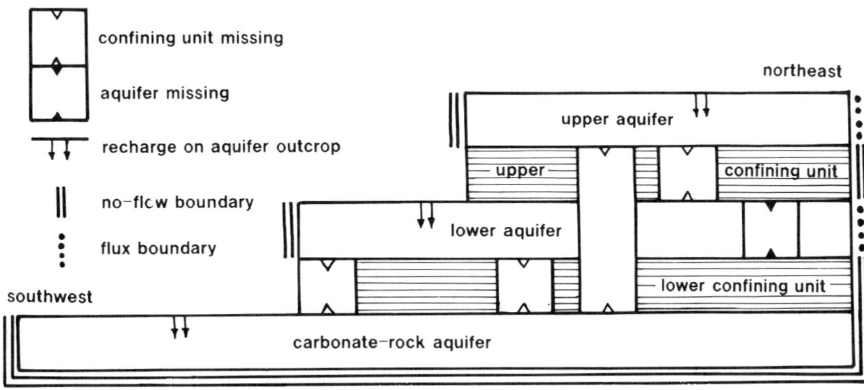

FIG. 12.—Schematic diagram of the hydrogeologic model for part of the Rockaway valley-fill aquifer, from Nicholson et al. (1996).

CONCLUSIONS

Glacial aquifer systems are texturally varied and have abrupt permeability contrasts, intricate sediment-body ge-ometries and bounding surfaces. This complexity is best managed through the application of geologic mapping and geomorphic, stratigraphic and sedimentologic analysis at the outset of the aquifer study. The nature of glacial aqui-

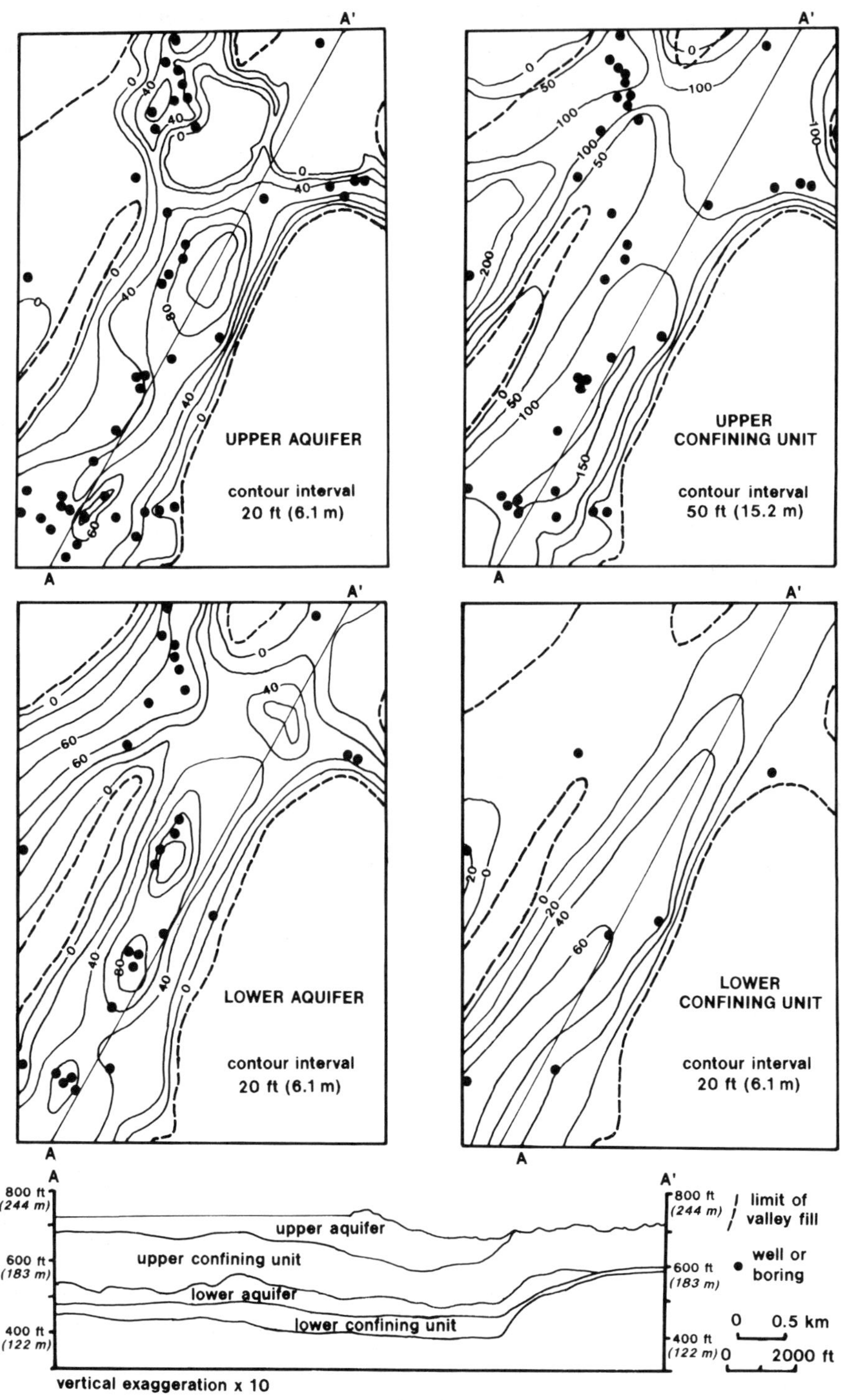

FIG. 13.—Isopachs of model layers for the area of Fig. 11. The upper aquifer includes units Qwd, Qwf, and Qwlf (including thin overlying till, Qwm). The upper confining unit includes Qwl, Qil, and thick till (Qwm). The lower aquifer includes Qilf and Qwlf (where it is overlain by thick till). The lower confining unit includes Qit, Qil and Qcw. Note larger contour interval for upper confining unit.

fer systems generally requires that this work be done at 1:24,000 or similar scales.

Geologic analysis identifies the sedimentary units and their geometry, the nature of their contacts and their internal properties. These elements all vary within predictible ranges. However, their vertical stacking and horizontal distribution in a given aquifer system are unique because these characteristics are determined by the specific set of geologic events at that location. These events, and the unique sedimentary architecture they produce, are deciphered by using depositional models to analyze data from onsite geologic investigation. The completed geologic analysis, presented as isopach and contour maps, provides the data framework within which hydrogeologic models can be constructed and hydrogeologic parameters assigned.

ACKNOWLEDGMENTS

We thank G.S. Fraser, S. Brown and an anonymous reviewer for their helpful comments on the manuscript. Discussions with J.L. Hoffman, R. Canace, B.D. Stone, R.W. Witte and L. Nicholson during data collection helped refine our geologic and hydrogeologic interpretations. This work was partially funded by the New Jersey Water Supply Bond of 1981.

REFERENCES

ANDERSON, H.R., 1968, Geology and ground-water resources of the Rahway area, New Jersey: New Jersey Department of Conservation and Economic Development, Division of Water Policy and Supply Special Report 27, 72 p.

ANDERSON, M.P., 1989, Hydrogeologic facies models to delineate large-scale spatial trends in glacial and glaciofluvial sediments: Geological Society of America Bulletin, v. 101, p. 501-511.

ASHLEY, G.M., 1975, Rhythmic sedimentation in glacial Lake Hitchcock, Massachusetts and Connecticut, *in* Jopling, A.V., and McDonald, B.C., eds., Glaciofluvial and Glaciolacustrine Sedimentation: Society of Economic Paleontologists and Mineralogists Special Publication 23, p. 304-320.

ASHLEY, G.M., AND HAMILTON, T.D., 1993, Fluvial response to Late Quaternary climatic fluctuations, central Kobuk valley, northwestern Alaska: Journal of Sedimentary Petrology, v. 63, no. 5, p. 814-827.

ASHLEY, G.M., SHAW, J., AND SMITH, N.D., 1985, Glacial sedimentary environments: Society of Economic Paleontologists and Mineralogists, Short Course 16, 246 p.

BANERJEE, I., AND MCDONALD, B.C., 1975, Nature of esker sedimentation, *in* Jopling, A.V., and McDonald, B.C., eds., Glaciofluvial and Glaciolacustrine Sedimentation: Society of Economic Paleontologists and Mineralogists Special Publication 23, p. 132-154.

BOOTHROYD, J.C., AND ASHLEY, G.M., 1975, Processes, bar morphology, and sedimentary structures on braided outwash fans, northeastern Gulf of Alaska, *in* Jopling, A.V., and McDonald, B.C., eds., Glaciofluvial and Glaciolacustrine Sedimentation: Society of Economic Paleontologists and Mineralogists Special Publication 23, p. 193-222.

CANACE, R., HUTCHINSON, W.R., SAUNDERS, W.R., AND ANDRES, K.G., 1983, Results of the 1980-81 drought emergency ground water investigation in Morris and Passaic counties, New Jersey: New Jersey Geological Survey Open-File Report 83-3, 132 p.

CANACE, R., STANFORD, S.D., AND HALL, D.W., 1994, Hydrogeologic framework of the middle and lower Rockaway river basin, Morris County, New Jersey: New Jersey Geological Survey Report GSR-33, 68 p.

CHURCH, M., AND GILBERT, R., 1975, Proglacial fluvial and lacustrine environments, *in* Jopling, A.V., and McDonald, B.C., eds., Glaciofluvial and Glaciolacustrine Sedimentation: Society of Economic Paleontologists and Mineralogists Special Publication 23, p. 22-100.

DAGAN, G., 1982, Stochastic modeling of groundwater flow by unconditional and conditional probabilities, part 1—Conditional simulation and the direct problem: Water Resources Research, v. 18, no. 4, p. 813-833.

DAVIS, S.N., 1969, Porosity and permeability of natural materials, *in* DeWiest, R.J.M., ed., Flow through Porous Media: New York, Academic Press, p. 54-89.

DELHOMME, J.P., 1979, Spatial variability and uncertainity in ground-water flow parameters—A geostatistical approach: Water Resources Research, v. 15, no. 2, p. 269-280.

DREIMANIS, A., AND VAGNERS, U.J., 1971, Bimodal distribution of rock and mineral fragments in basal till, *in* Goldthwait, R.P., ed., Till—A Symposium: Columbus, Ohio State University Press, p. 237-250.

EYLES, N., ed., 1983, Glacial Geology—An Introduction for Engineers and Earth Scientists: New York, Pergamon Press, 409 p.

FRASER, G.S., 1994, Sequences and sequence boundaries in glacial sluiceways beyond glacial margins, *in* Dalrymple, R., Zaitlin, B., and Boyd, R., eds., Incised-Valley Systems—Origin and Sedimentary Sequences: Society of Economic Paleontologists and Mineralogists Special Publication 51, p. 337-351.

FRASER, G.S., AND BLEUER, N.K., 1987, Use of facies models as predictive tools to locate and characterize aquifers in glacial terrains, *in* Proceedings of the NWWA Focus Conference on Midwestern Ground Water Issues: Dublin, Ohio, National Water Well Association, p. 123-143.

FREEZE, R.A., 1975, A stochastic-conceptual analysis of one-dimensional groundwater flow in nonuniform homogeneous media: Water Resources Research, v. 11, no. 5, p. 725-741.

GILL, H.E., AND VECCHIOLL, J., 1965, Availability of ground water in Morris County, New Jersey: New Jersey Division of Water Policy and Supply Special Report 25, 56 p.

GOMEZ-HERNANDEZ, J.J., AND GORELICK, S.M., 1989, Effective ground-water model parameter values—Influence of spatial variability of hydraulic conductivity, leakance, and recharge: Water Resources Research, v. 25, no. 3, p. 405-419.

GRISAK, G.E., CHERRY, J.A., VONHOF, J.A., AND BLUEMLE, J.P., 1976, Hydrogeologic and hydrochemical properties of fractured till in the interior plains region, *in* Legget, R.F., ed., Glacial Till—An Interdisciplinary Study: Royal Society of Canada Special Publication 12, p. 304-333.

GUSTAVSON, T.C., 1975, Sedimentation and physical limnology in proglacial Malaspina Lake, southeastern Alaska, *in* Jopling, A.V., and McDonald, B.C., eds., Glaciofluvial and Glaciolacustrine Sedimentation: Society of Economic Paleontologists and Mineralogists Special Publication 23, p. 249-263.

GUSTAVSON, T.C., AND BOOTHROYD, J.C., 1987, A depositional model for outwash, sediment sources, and hydrologic characteristics, Malaspina glacier, Alaska—A modern analog of the southeastern margin of the Laurentide ice sheet: Geological Society of America Bulletin, v. 99, p. 292-302.

GUSTAVSON, T.C., ASHLEY, G.M., AND BOOTHROYD, J.C., 1975, Depositional sequences in glaciolacustrine deltas, *in* Jopling, A.V., and McDonald, B.C., eds., Glaciofluvial and Glaciolacustrine Sedimentation: Society of Economic Paleontologists and Mineralogists Special Publication 23, p. 264-280.

HARTSHORN, J.M., 1958, Flowtill in southeastern Massachusetts: Geological Society of America Bulletin, v. 69, p. 477-482.

HILL, M.C., 1985, An investigation of hydraulic conductivity estimation in a ground-water flow study of northern Long Valley, New Jersey: Unpub. Ph.D. dissertation, Princeton University, Princeton, N. J., 341 p.

HILL, M.C., LENNON, G.P., BROWN, G.A., HEBSON, C.S., AND RHEAUME, S.J., 1992, Geohydrology of, and simulation of groundwater flow in, the valley-fill deposits in the Ramapo river valley, New Jersey: U.S. Geological Survey Water Resources Investigation Report 90-4151, 92 p.

HOEKSMA, R.J., AND KITANIDIS, P.K., 1984, An application of the geostatistical approach to the inverse problem in two-dimensional groundwater modeling: Water Reources Research, v. 20, p. 1003-1020.

HOEKSMA, R.J., AND KITANIDIS, P.K., 1985, Analysis of the spatial structure of properties of selected aquifers: Water Resources Research, v. 21, no. 4, p. 563-572.

HOFFMAN, J.L., AND STONE, B.D., 1991, Geohydrologic framework, pumpage, ground-water levels and dewatering of the central Passaic stratified-drift aquifer, New Jersey (abs.): Geological Society of America Abstracts with Programs, v. 23, no. 1, p. 45.

HUTCHINSON, W.R., 1981, A computer simulation of the glacial/carbonate aquifer in the Pequest valley, Warren County, New Jersey: Unpub. M.S. thesis, Rutgers University, New Brunswick, N. J., 115 p.

KARROW, P.F., 1975, The texture, mineralogy, and petrography of North American tills, in Legget, R.F., ed., Glacial Till—An interdisciplinary study: Royal Society of Canada Special Publication 12, p. 83-98.

KEMPTON, J.P., AND CARTWRIGHT, K., 1984, Three-dimensional geologic mapping—A basis for hydrogeologic and land-use evaluations: Bulletin of the Association of Engineering Geologists, v. 21, no. 3, p. 317-335.

KITANIDIS, P.K., AND VOMVORIS, E.G., 1983, A geostatistical approach to the inverse problem in groundwater modeling (steady state) and one-dimensional simulations: Water Resources Research, v. 19, no. 3, p. 677-690.

KOTEFF, C., AND PESSL, F., JR., 1981, Systematic ice retreat in New England: U.S. Geological Survey Professional Paper 1179, 20 p.

LEA, P.D., 1990, Pleistocene periglacial eolian deposits in southwestern Alaska—Sedimentary facies and depositional processes: Journal of Sedimentary Petrology, v. 60, p. 582-591.

MEISLER, H., 1976, Computer simulation model of the Pleistocene valley-fill aquifer in southwestern Essex and southeastern Morris Counties, New Jersey: U.S. Geological Survey Water Reources Investigations 76-25, 76 p.

MORRIS, D.A., AND JOHNSON, A.I., 1967, Summary of hydrologic and physical properties of rock and soil materials as analyzed by the hydrologic laboratory of the U.S. Geological Survey, 1948-1960: U.S. Geological Survey Water-Supply Paper 1939-D, 39 p.

NEW JERSEY DEPARTMENT OF ENVIRONMENTAL PROTECTION, 1987, 1987 New Jersey water withdrawal report: New Jersey Department of Environmental Protection, Division of Water Resources, Bureau of Water Allocation, 34 p.

NICHOLSON, R.S., MCAULEY, S.D., BARRINGER, J.A., AND GORDON, A.D., 1996, Hydrogeology of, and ground-water flow in, a valley-fill and carbonate-rock aquifer system near Long Valley in the New Jersey highlands: U.S. Geological Survey Water Resources Investigations Report 93-4157, 159 p.

RUST, B.R., AND ROMANELLI, R., 1975, Late Quaternary subaqueous outwash deposits near Ottawa, Canada, in Jopling, A.V., and McDonald, B.C., eds., Glaciofluvial and Glaciolacustrine Sedimentation: Society of Economic Paleontologists and Mineralogists Special Publication 23, p. 177-192.

SCHAEFER, F.L., HARTE, P.T., SMITH, J.A., AND KURTZ, B.A., 1993, Hydrologic conditions in the upper Rockaway river basin, New Jersey, 1984-1986: U.S. Geological Survey Water Resources Investigations Report 91-4196, 103 p.

SHAW, J., 1975, Sedimentary successions in Pleistocene ice-marginal lakes, in Jopling, A.V., and McDonald, B.C., eds., Glaciofluvial and Glaciolacustrine Sedimentation: Society of Economic Paleontologists and Mineralogists Special Publication 23, p. 281-303.

SILLIMAN, S.E., AND WRIGHT, A.L., 1988, Stochastic analysis of paths of high hydraulic conductivity in porous media: Water Resources Research, v. 24, no. 11, p. 1901-1910.

SIROIS, B.J., 1986, Application of a modular three-dimensional finite difference ground-water flow model to a glacial valley fill stream-aquifer system in the Rockaway drainage basin, New Jersey: Unpub. M.S. thesis, Lehigh University, Bethlehem, Pa., 138 p.

STANFORD, S.D., 1989, Surficial geology of the Dover quadrangle, New Jersey: New Jersey Geological Survey Geologic Map Series 89-2, scale 1:24,000.

STANFORD, S.D., WITTE, R.W., AND HARPER, D.P., 1990, Hydrogeologic character and thickness of the glacial sediment of New Jersey: New Jersey Geological Survey Open-File Map 3, scale 1:100,000.

STEPHENSON, D.A., FLEMING, A.H., AND MICKELSON, D.M., 1988, Glacial deposits, in Back, W., Rosenshein, J.S., and Seaber, P.R., eds., Hydrogeology: Geological Society of America, The Geology of North America, v. O-2, p. 301-314.

STONE, B.D., STANFORD, S.D., AND WITTE, R.W., 1989, Preliminary surficial geologic map of northern New Jersey (abs.): Geological Society of America Abstracts with Programs, v. 21, no. 2, p. 69.

SUDICKY, E., 1986, A natural gradient experiment on solute transport in a sand aquifer—Spatial variability of hydraulic conductivity and its role in the dispersion process: Water Resources Research, v. 22, no. 13, p. 2069-2082.

VECCHIOLI, J., AND MILLER, E.G., 1973, Water resources of the New Jersey part of the Ramapo river basin: U.S. Geological Survey Water Supply Paper 1974, 77 p.

VON SCHONDORF, A., 1987, Sedimentary facies and paleohydraulics of an ice-contact glacial outwash plain, Germany Flats, New Jersey: Unpub. M.S. thesis, Rutgers University, New Brunswick, N.J., 122 p.

WARREN, W.P., AND ASHLEY, G.M., 1994, Origins of the ice-contact stratified ridges (eskers) of Ireland: Journal of Sedimentary Research, v. A64, no. 3, p. 433-449.

WEBB, E.K., 1994, Simulating the three-dimensional distribution of sediment units in braided-stream deposits: Journal of Sedimentary Research, v. B64, no. 2, p. 219-231.

WILLMAN, H.B., AND FRYE, J.C., 1970, Pleistocene stratigraphy of Illinois: Illinois State Geological Survey Bulletin 94, 204 p.

WITTE, R.W., 1991, Surficial geology of Kittatinny valley and vicinity in the southern part of Sussex County, northern New Jersey: New Jersey Geological Survey Open-File Map 7, scale 1:24,000.

CHARACTERIZING AQUIFER HETEROGENEITY WITHIN TERRIGENOUS CLASTIC DEPOSITIONAL SYSTEMS

WILLIAM E. GALLOWAY AND JOHN M. SHARP, JR.

Department of Geological Sciences, The University of Texas at Austin, Austin, Texas 78712-1101

ABSTRACT: Sedimentary facies, reflecting original depositional environment, define the trend, dimensions, connectivity and internal heterogeneity of transmissive zones within clastic aquifer systems. Three heterogeneity styles—layer cake, jigsaw puzzle and labyrinth—reflect increasing degree of complexity. Style is directly determined by the depositional origin of the aquifer. Heterogeneity occurs over a broad range of scales. Megascopic heterogeneity is determined by the external dimensions, trends and degree of interconnection of permeable units. Macroscopic heterogeneity, which occurs at the depositional facies scale, includes (1) compartmentalization due to flow barriers between specific facies within larger permeable units, (2) vertical and lateral permeability gradients created by patterns of grain size and sorting, and (3) stratification and low-permeability mud baffles, which create anisotropy. Mesoscopic heterogeneity reflects lithofacies, sedimentary structure and lamina-scale variability. All three scales of heterogeneity structure can be efficiently described, quantified, interpolated and predicted within the context of a well understood depositional system framework.

INTRODUCTION

Understanding of groundwater flow systems and their geological controls within active depositional systems can be applied to a variety of hydrologic and environmental studies. Although depositional environment is only one of several factors that determine the hydrogeology of sedimentary systems, reconstructions using paleogeographic or paleoenvironmental analyses provide a starting point for interpretation of paleohydrogeology and the effects of syngenetic and early diagenetic processes. The analysis of depositional systems provides a rational basis for inferring the physical characteristics of sediments and sedimentary rocks. It can be of considerable aid in guiding sample collection, interpretation of geophysical data, well development, siting of monitoring wells and remediation in shallow aquifers.

An important trend in applied sedimentology during the late twentieth century is the shift in focus from hydrocarbon and groundwater exploration to detailed reservoir or hydrostratigraphic characterization. The processing power of numerical flow and transport simulators requires sophisticated, quantitative description of the three-dimensional distributions of hydraulic parameters, such as porosity and permeability. In particular, new emphasis is placed on delineating the geometry of internal sedimentary facies and bedding architecture of hydrostratigraphic units and on accurately quantifying hydraulic properties. Genetic facies analysis has been applied successfully to this process. Outcrop, well log, core, ground-penetrating radar (GPR) and 3D geophysical data are being used to map, interpret and quantify hydrostratigraphic units. Production history (in effect, a long-term flow experiment), observed contaminant dispersal and tracer tests must be used to test, refine and calibrate the characterization. The goal is to predict fluid flow and solute transport and their effects within sediments and sedimentary rocks.

Within a single genetic depositional facies, heterogeneities are created by predictable patterns of bedding and spatial variability of textural parameters (e.g., Galloway and Hobday, 1996; Einsele, 1992). Bedding produces stratification that restricts cross-formational and cross-layer flow; it channels fluids into the more permeable layers. Early studies (e.g., Zeito, 1965) demonstrated the potential for internal permeability stratification and that the geometry and continuity of bedding correlative with the depositional environment of sand bodies. In addition, systematic lateral variation in permeability was indicated by studies of modern sand bodies (Pryor, 1973) and outcrop analogies (Barton, 1994; Davis, 1986; Ferris et al., 1993).

Driving the focus on aquifer characterization is the need for hydrogeologists, environmental scientists and regulatory agencies to accurately predict, delimit and devise remediation strategies for groundwater and soil contamination. The key questions are site specific; the answers require rather complete and sophisticated descriptions of the hydrogeologic properties of sediments and sedimentary rocks. A large body of information has been compiled on modern sand bodies and outcrop analogs. These data and case histories demonstrate that genetic interpretation allows prediction of a hierarchy of parameters that control fluid flow and transport in aquifers.

TYPES OF AQUIFER HETEROGENEITY

Alluvial strata are inherently inhomogeneous (or heterogeneous). Heterogeneity is created in any of various ways:

1. External boundaries of sand bodies create discontinuous permeable units that commonly have a well-defined orientation and trend (Fig. 1).
2. Individual permeable units are variably interconnected.
3. Within permeable units, porosity and permeability may exhibit lateral or vertical trends or spatial partitioning (Fig. 2).
4. Permeable units may be internally stratified (Fig. 2, zone I). Such stratification may be horizontal (subparallel) or systematically inclined relative to the top and bottom boundaries of the unit.
5. Variably continuous low-permeability layers (sometimes called "baffles") may occur within permeable units.
6. Permeability is commonly anisotropic. Typically, vertical permeability is substantially less than horizontal permeability in sedimentary aquifers. Alignment of beds or baffles commonly creates horizontal anisotropy (Fig. 2, zone I). Note that variable permeability layers (even where each is internally homogeneous) behave as an anisotropic medium on the scale of a pump test (Foreman and Sharp, 1981) so that our definitions of homogeneity and isotropy themselves can be considered scale dependent.

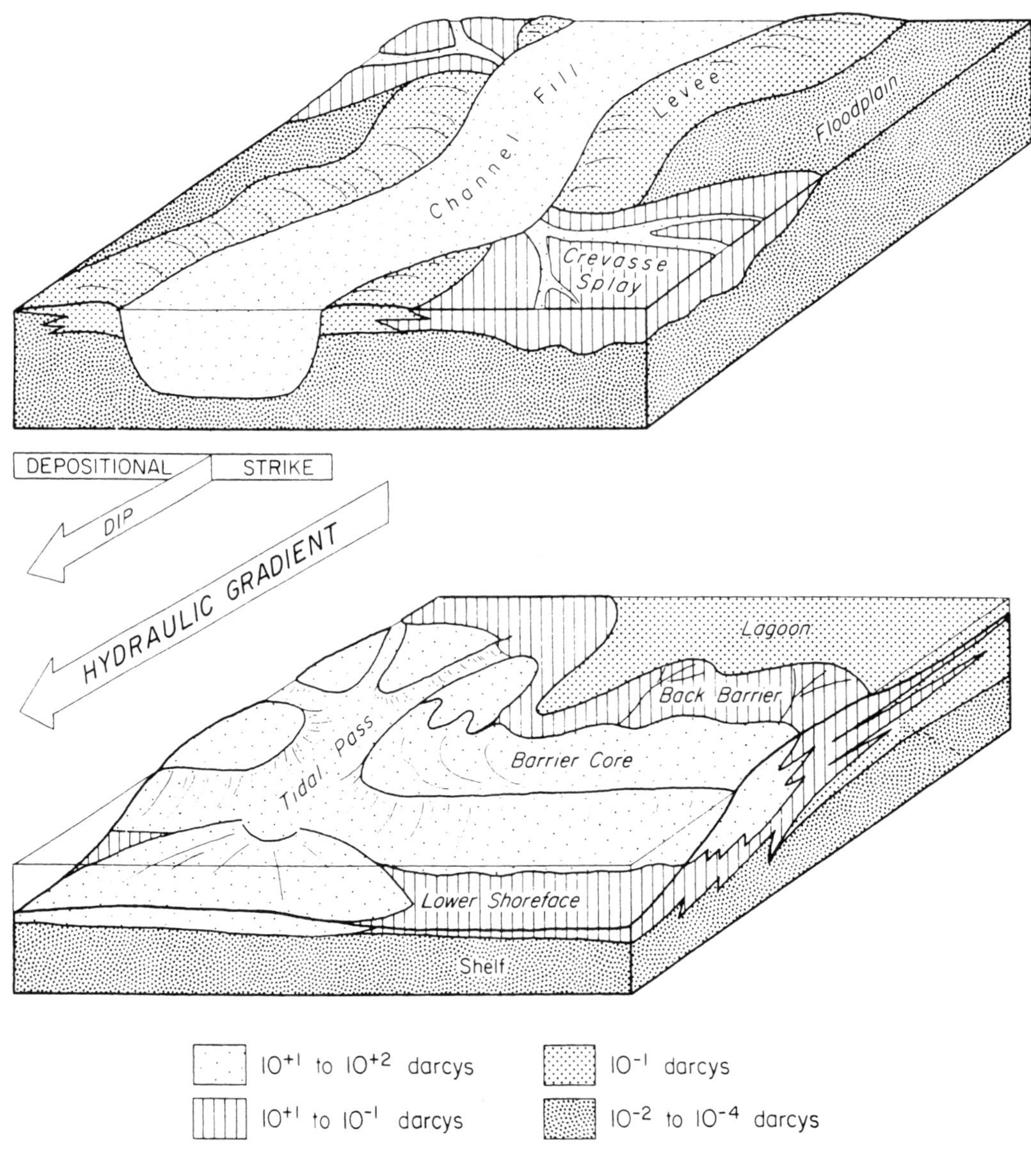

FIG. 1.—External geometry, trend relative to contemporary regional depositional and hydraulic gradients, and facies control of permeability pattern in fluvial and barrier island depositional systems (from Galloway et al., 1979).

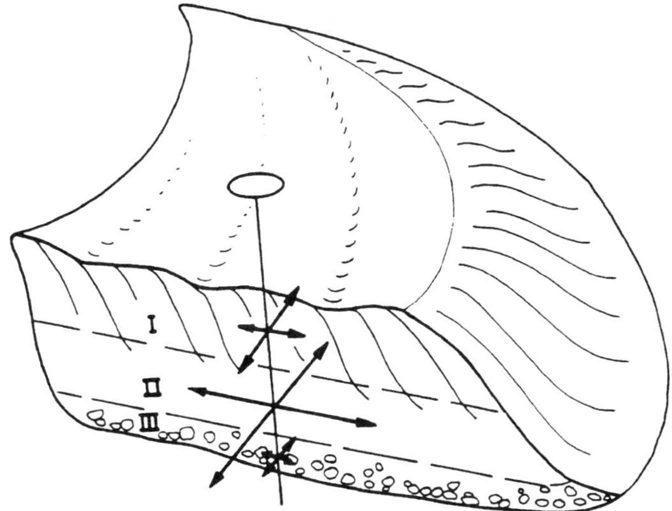

FIG. 2.—Permeability zonation and heterogeneities in a single point bar. Zone I includes relatively fine-grained lateral accretion beds, bounded by discontinuous clay drapes (baffles), of the upper point bar. Zone II consists of massive, cross-stratified sand of the middle to lower point bar. Zone III coincides with the poorly sorted heterogeneous basal lag of gravel, mud clasts and plant debris. Arrows indicate relatively permeability within each zone along and across the trend of the meander belt (modified from Galloway and Hobday, 1996, copyright, Springer-Verlag).

Like petroleum reservoirs, aquifers display three patterns or styles of facies and, consequently, hydrostratigraphic complexity (Fig. 3) (Weber and van Geuns, 1990):

1. "Layer cake" systems are characterized by lateral continuity, gradual lateral thickness change, relatively simple correlation and sheet to lobate geometries of permeable units. This is the system assumed for most conceptualizations and mathematical analyses.
2. "Jigsaw puzzle" systems are compartmentalized by complex, cross-cutting relationships. In such situations, correlation and delineation of general zones of high permeability may be relatively simple, but internal correlation is difficult.
3. "Labyrinth" systems consist of numerous partially to completely isolated permeable units. Both internal and external correlation are difficult. In its extreme form, dimensions of labyrinthine units are significantly less than average spacing of test or monitor wells. Quantification of sand-body distribution and interconnectivity, which is a necessity for input data to flow simulation models, may require statistical methods (Fogg, 1986; Stanley et al., 1990; Hirst et al., 1993).

For petroleum exploration and production, the well-data spacing necessary to determine facies geometry and permeability distribution varies according to the type of plumbing system. Weber and van Geuns (1990) have estimated adequate spacings of about 1 well/km for layer-cakes, 2 to 4 wells/km for jigsaw puzzles, and 10 to 30 wells/km for labyrinths. Although these spacings may be adequate for predicting long-term contaminant migration within an aquifer, they still may be far too coarse for site- and time-specific characterization of solute transport or groundwater remediation.

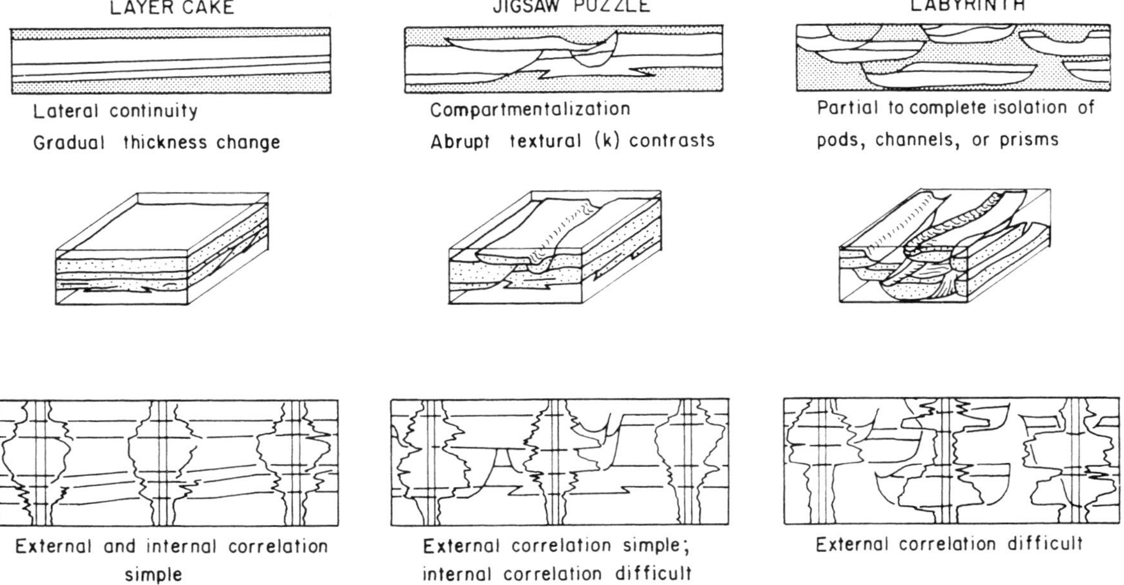

FIG. 3.—Three principal styles of facies and hydrostratigraphic complexity (from Weber and van Geuns, 1990).

SCALES OF AQUIFER HETEROGENEITY

Heterogeneities occur over a broad range of scales. Two factors are important in determining the scales of heterogeneity: (1) natural scaling of sedimentary processes and facies (Miall, 1988, 1991; Van Wagoner et al., 1990; Jordan and Pryor, 1992), and (2) scaling of the data-sampling systems, whether wells, direct samples or indirect geophysical evaluation. Scaling terminologies (Alpay, 1972; Schwartz, 1977; Haldorson, 1986; Krause et al., 1987) define different levels of heterogeneity:

Gigascopic heterogeneity is expressed at the scale of depositional systems and stratigraphic sequences on the order of 1+ Ma duration. This is the scale for resource estimation or regional aquifer studies.

Megascopic heterogeneity describes the external geometry and interrelationships of hydrostratigraphic units and is defined at the scale of the well field. Where sedimentary dimensions are relatively small or well spacing relatively large, megascopic heterogeneity may describe the interwell scale. At both gigascopic and megascopic scales, hyperpermeability fracture sets may be the dominating factors in controlling regional flow systems.

Macroscopic heterogeneity occurs at the sedimentary facies scale. This is the fundamental interwell variability commonly described by subsurface (well or borehole) data sets. Delineation and quantification of macroscopic heterogeneity are critical if flow is to be understood and accurately predicted. Macroscopic variability occurs near the limits of spatial resolution of the typical subsurface data base; consequently, interpretations must be guided or conditioned by predictive, quantified facies interpretation. Macroscopic features include:

1. Compartmentalization. Internally, sedimentary units consist of a mosaic of vertically or laterally amalgamated genetic facies, architectural elements and beds separated by bounding depositional surfaces or erosional diastems.

2. Permeability distribution. Each sedimentary facies has distinct vertical and lateral textural patterns that control the primary permeability and porosity distributions. Patterns may differ abruptly across facies boundaries and diastems. Specific successions and repetitions of sedimentary structures typify many sandstones. Bioturbation or diagenesis may create systematic trends of aquifer properties and alter the primary trends.

3. Stratification. Stratification reflects organized textural interbedding and is typified by sheet-like, tabular or lenticular units separated from adjacent units by perme-

ability contrasts or by baffles. Bedding is typically subhorizontal and parallel to the sand body, but it may also transect the sand body and cause anisotropy. Distinct stratification is created when bed surfaces are draped with mud. Such shale drapes are often termed "baffles." They restrict cross-bed or cross-formational flow; consequently, delineating their orientation, abundance and lateral continuity relative to the aquifer and to well locations is important (Zeito, 1965; Weber, 1986).

Mesoscopic heterogeneity reflects lithofacies, bedding and lamina-scale variation. It occurs within genetic facies or large depositional bodies, such as point bars or compound braid bars. Mesoscopic heterogeneity is a function of stratification. In general, wavy laminated and micaceous, ripple-laminated sands, and heterolithic laminites show the greatest degree of mesoscopic heterogeneity. Trough, swaley, and hummocky cross stratification cosets display moderate heterogeneity. Wind ripple, grain flow and planar lamination are least heterogeneous. At the mesoscopic scale, relative permeability also is related to sedimentary structure type (Dreyer et al., 1990; Hopkins et al., 1991, Lowry and Jacobson, 1993; Davies et al., 1993; Barton, 1994). At equivalent grain sizes, relative permeability is highest in massive, fluidized and grain-flow beds; intermediate in planar, tabular, and trough cross-stratified sets; and least in ripple-laminated beds, hummocky cross-bed sets and bioturbated beds. Muddy laminae and heterolithic cross-strata substantially reduce permeability. Within large cross-bed sets, permeability anisotropy is highly structured (Pryor, 1973; Weber, 1982; Hartkamp-Bakker, 1993). Toesets are relatively tight. Maximum permeability parallels foreset laminae.

Microscopic heterogeneity is present at the scale of individual grains and pores. It is primarily determined by sediment texture and diagenetic overprint. This is the scale most commonly studied in laboratory experiments. Trends in microscopic heterogeneity demonstrate close correlations among permeability, porosity, pore geometry and sediment texture (Beard and Weyl, 1973; Pryor, 1973; Fu et al., 1994). Porosity varies with sorting and sphericity and is theoretically independent of grain size. Permeability is highly sensitive and directly proportional to both grain size and sorting. Thus, the systematic textural trends and heterogeneities that characterize genetic facies provide qualitative predictions of porosity and permeability patterns and heterogeneity (e.g., Ferris et al., 1993; Barton, 1994; Chandler et al., 1989).

Microscopic and mesoscopic heterogeneities are at scales of millimeters to tens of meters. Therefore, parameters at this scale can be described geostatistically in most applications and are often treated stochastically in simulations. Several approaches (transition probabilities, autocorrelation, kriging, variograms and conditional simulation) have been applied or developed for statistical analysis and prediction of geologic information (e.g., Neuman, 1982; Hohn, 1988; Mackay and O'Connell, 1991).

Scaling of sedimentary units and the translation into scales of heterogeneity were illustrated by Jordan and Pryor (1992), who studied a Mississippi River meander belt. The megascale unit is the meander belt, which is encased by overbank muds and has dimensions of 25 km by 24 km and a volume of about 11 billion m^3. Individual point bars, which are partially isolated by abandoned channel plugs, form the largest macroscale units. Their dimensions are 8 km by 8 km with a volume averaging about 1.5 billion m^3. Internal macroscopic heterogeneity within point bars is created by individual lateral accretion units with average volumes of about 550,000 m^3. Bed sets within accretion units have volumes of less than 12,000 m^3 and create mesoscale heterogeneities. Within bed sets, various types of primary stratification create the smallest mesoscopic heterogeneity.

CONCLUDING STATEMENT

External geometry, internal compartmentalization, sedimentary structural and textural trends, and stratification all define the facies architecture of the aquifer. Geometry and compartmentalization both affect the three-dimensional flow field, which is most intense in zones of greatest transmissivity (Fogg, 1986; Phillips, 1991). Permeability trend and zonation control effective permeability distribution, modifying details of flow within aquifers. Sedimentary structures, bedding and shale baffles create mesoscopic heterogeneity. The abundance and nature of aquifer heterogeneities control macroscopic dispersion (Skibitzke and Robertson, 1963; Gillham and Cherry, 1982; Winograd and Thordarson, 1975; Gelhar and Axness, 1983).

Detecting, delineating and quantifying the three-dimensional distributions of porosity and permeability, defining patterns of anisotropy, and delineating the sedimentary features that create external and internal heterogeneity are the goals of hydrostratigraphic characterization. Our thesis is that this exacting task is most efficiently and accurately accomplished within the context of a genetic facies framework. Facies interpretation allows selection of appropriate analogs, guides interpolation of quantitative data, suggests requisite sample densities and boundary condi-

tions for statistical characterization, and conditions geologically realistic input parameters for flow simulations (Smith and Schwartz, 1981; Gillham and Cherry, 1982; Davis, 1986; Weber, 1986; Bryant and Flint, 1993; Martin, 1993; Slatt and Galloway, 1993). Simply stated, depositional facies are the fundamental units of such analyses. Although hydrogeologists have been more reticent than petroleum geologists in the application of depositional system analysis, many recognize that deterministic delineation of hydraulic conductivity and other properties is a prerequisite for accurate prediction (or retrodiction) of fluid flow and solute transport.

REFERENCES

Alpay, O.A., 1972, A practical approach to defining reservoir heterogeneity: Journal of Petroleum Technology, v, 24, p. 841-843.

Barton, M.D., 1994, Outcrop characterization of architecture and permeability structure in fluvial-deltaic sandstones, Cretaceous Ferron Sandstone, Utah: Unpub. Ph.D. dissertation, University of Texas, Austin, 262 p.

Beard, D.C., and Weyl, P.K., 1973, Influence of texture on porosity and permeability: American Association of Petroleum Geologists Bulletin, v. 57, p. 349-369

Bryant, I.D., and Flint, S.S., 1993, Quantitative clastic reservoir geological modeling: problems and perspectives, *in* Bryant, I.D., and Flint, S.S., eds., The Geological Modeling of Hydrocarbon Reservoirs and Outcrop Analogues: International Association of Sedimentologists Special Publication 15, p. 3-20.

Chandler, M.A., Kocurek, G., Goggin, D.J., and Lake, L.W., 1989, Effects of stratigraphic heterogeneity on permeability in eolian sandstone sequence, Page Sandstone, northern Arizona: American Association of Petroleum Geologists Bulletin, v. 73, p. 658-668.

Davies, D.K., Williams, B.P.J., and Vessell, R.K., 1993, Dimensions and quality of reservoirs in low and high sinuosity channel systems, Lower Cretaceous Travis Peak formation, east Texas, USA, *in* North, C.P., and Prosser, D.J., eds., Characterization of Fluvial and Aeolian Reservoirs: Geological Society of America Special Publication 73, p. 95-121.

Davis, A., 1986, Deterministic modeling of dispersion in heterogeneous porous media: Ground Water, v. 24, p. 609-615.

Dreyer, T., Scheie, A., and Walderhaug, O., 1990, Minipermeameter based study of permeability trends in channel sand bodies: American Association of Petroleum Geologists Bulletin, v. 74, p. 359-374.

Einsele, G., 1992, Sedimentary Basins—Evolution, Facies, and Sediment Budget: Berlin, Springer Verlag, 628 p.

Ferris, M.A., Kerans, C., and Sharp, J.M., Jr., 1993, Outcrop permeabilities within four facies of a single depositional parasequence, Upper San Andres Formation Guadalupian/Leonardian), Lawyer Canyon, Guadalupe Mountains, Otero County, New Mexico, *in* Love, D.W., and others, eds., Carlsbad Region, New Mexico and Texas: New Mexico Geological Society 44th Annual Field Conference, p. 205-210.

Fogg, G.E., 1986, Groundwater flow and sand body interconnectedness in a thick multiple-aquifer system: Water Resources Research, v. 22, p. 679-694.

FOREMAN, T.L., and SHARP, J.M., JR., 1981, Hydraulic properties of a major alluvial aquifer—An isotropic, inhomogeneous system: Journal of Hydrology, v. 53, p. 247-258.

FU, L., MILLIKEN, K.L., and SHARP, J.M., JR., 1994, Porosity and permeability variations in the fractured and liesegang-banded Breathitt Sandstone (Middle Pennsylvanian), eastern Kentucky—Diagenetic controls and implications for modeling dual porosity systems: Journal of Hydrology, v. 154, p. 351-381.

GALLOWAY, W.E., and HOBDAY, D.K., 1996, Terrigenous Clastic Depositional Systems, 2d ed.: Berlin, Springer Verlag, 489 p.

GELHAR, L.W., and AXNESS, C.L., 1983, Three-dimensional stochastic analysis of macro-dispersion in aquifers: Water Resources Research, v. 19, p. 161-180.

GILLHAM, R.W., and CHERRY, J.A., 1982, Contaminant migration in saturated unconsolidated geologic deposits, *in* Narasimhan, T.N., ed., Recent Trends in Hydrogeology: Geological Society of America Special Paper 189, p. 31-62.

HALDORSON, H.H., 1986, Simulator parameter assignment and the problem of scale in reservoir engineering, *in* Lake, L.W., Carroll, H.B., Jr., and Wesson, T.C., eds., Reservoir Characterization: London, Academic Press, p. 294-340.

HARTKAMP-BAKKER, C.A., 1993, Permeability heterogeneity in cross-bedded sandstones—Impact on water/oil displacement in fluvial reservoirs: Meppel, Netherlands, Krips repromeppel.

HIRST, J.P.P., BLACKSTOCK, C.R., and TYSON, S., 1993, Stochastic modeling of fluvial sandstone bodies, *in* Bryant, I.D., and Flint, S.S., eds., The Geological Modeling of Hydrocarbon Reservoirs and Outcrop Analogues: International Association of Sedimentologists Special Publication 15, p. 237-251.

HOHN, M.E., 1988, Geostatistics and Petroleum Geology: New York, Van Nostrand Reinhold.

HOPKINS, J.C., WOOD, J.M., and KRAUSE, F.F., 1991, Waterflood response of reservoirs in an estuarine valley fill—Upper Mannville G, U, and W pools, Little Bow Field, Alberta, Canada: American Association of Petroleum Geologists Bulletin, v. 75, p. 1064-1088.

JORDAN, D.W., and PRYOR, W.A., 1992, Hierarchical levels of heterogeneity in a Mississippi river meander belt and application to reservoir systems: American Association of Petroleum Geologists Bulletin, v. 76, p. 1601-1624.

KRAUSE, F.F., COLLINS, H.N., NELSON, S.D., MACHEMER, S.D., and FRENCH, P.R., 1987, Multiscale anatomy of a reservoir—Geological characterization of Pembine-Cardium pool, west-central Alberta, Canada: American Association of Petroleum Geologists Bulletin, v. 76, p. 1233-1260.

LOWRY, P., and JACOBSON, T., 1993, Sedimentological and reservoir characteristics of a fluvial-dominated delta-front sequence—Ferron Sandstone Member (Turonian), east-central, Utah, *in* Ashton, M., ed., Advances in Reservoir Geology: Geological Society of America Special Publication 69, p. 81-103.

MACKAY, R., and O'CONNELL, P.E., 1991, Statistical methods of characterizing hydrogeological parameters, *in* Downing, R.A., and Wilkinson, W.B., eds., Applied Groundwater Hydrology: Oxford, Clarendon Press, p. 217-242.

MARTIN, J.H., 1993, A review of braided fluvial hydrocarbon reservoirs, *in* Best, J.L., and Bristow, C. S., eds., Braided Rivers: Geological Society Special Publication 75, p. 333-367.

MIALL, A.D., 1988, reservoir heterogeneities in fluvial sandstones—Lessons from outcrop studies: American Association of Petroleum Geologists Bulletin, v. 72, p. 682-697.

MIALL, A.D., 1991, Hierarchies of architectural units in terrigenous clastic rocks and their relationship to sedimentation rate, *in* Miall, A.D., and Tyler, N., eds., The Three-Dimensional Facies Architecture of Terrigenous Clastic Sediments and Its Implications for Hydrocarbon Discovery and Recovery: Society of Sedimentary Geology, Concepts in Sedimentology and Paleontology 3, p. 6-12.

NEUMAN, S.P., 1982, Statistical characterization of aquifer heterogeneities, *in* Narasimhan, T.N., ed., Recent Trends in Hydrogeology: Geological Society of America Special Paper 189, p. 81-103.

PHILLIPS, O.M., 1991, Flow and Reactions in Permeable Rocks: Cambridge, Cambridge University Press.

PRYOR, W.A., 1973, Permeability-porosity patterns and variations in some Holocene sand bodies: American Association of Petroleum Geologists Bulletin, v. 57, p. 162-189.

SCHWARTZ, F.W., 1977, Macroscopic dispersion in porous media—The controlling factors: Water Resources Research, v. 13, p. 743-752.

SKIBITZKE, H.E., and ROBERTSON, G.M., 1963, Dispersion in groundwater flowing through heterogeneous materials: U.S. Geological Survey Professional Paper 386-B, p. 1-5.

SLATT, R.M., and GALLOWAY, W.E., 1993, Geological heterogeneities, *in* Morton-Thompson, D., and Woods, A.M., eds., Developmental Geology Reference Manual: American Association of Petroleum Geologists, p. 278-291.

SMITH, L., and SCHWARTZ, F.W., 1981, Mass transport, part 3—Role of hydraulic conductivity in prediction: Water Resources Research, v. 17, p. 1463-1479.

STANLEY, K.O., FORDE, D., RAESTAD, N., and STOCKBRIDGE, C.P., 1990, Stochastic modeling of reservoir sand bodies for input to reservoir simulation, Snorre Field, northern North Sea, Norway, *in* Boller, A.T., and others, eds., North Sea Oil and Gas reservoirs II: Norwegian Institute of Technology, Graham and Trotman, p. 95-101.

VAN WAGONER, J.C., MITCHUM, R.M., CAMPION, K.M., and RAHMANIAN, V.D., 1990, Siliciclastic sequence stratigraphy in well logs, cores, and outcrops: American Association of Petroleum Geologists, Methods in Exploration No. 7.

WEBER, K.J., 1982, Influence of common sedimentary structures on fluid flow in reservoir models: Journal of Petroleum Technology, v. 34, p. 665-672.

WEBER, K.J., 1986, How heterogeneity affects oil recovery, *in* Lake, L.W., and Carroll, H.B., Jr. eds., Reservoir Characterization: London, Academic Press, p. 487-545.

WEBER, K.J., and VAN GEUNS, L.C., 1990, Framework for constructing clastic reservoir simulation models: Journal of Petroleum Technology, v. 42, p. 1248-1253, 1296-1297.

WINOGRAD, I.J., and THORDARSON, W., 1975, Hydrogeologic and hydrogeochemical framework, south-central Great Basin, Nevada-California with special reference to the Nevada Test Site: U.S. Geological Professional Paper 712-C, 126 p.

ZEITO, G.A. 1965, Interbedding of shale breaks and reservoir heterogeneities: Journal of Petroleum Technology, v. 17, p. 1223-1228.

HYDROGEOLOGY AND CHARACTERIZATION OF FLUVIAL AQUIFER SYSTEMS

WILLIAM E. GALLOWAY AND JOHN M. SHARP, JR.

Department of Geological Sciences, The University of Texas at Austin, Austin, Texas 78712-1101

ABSTRACT: Fluvial depositional systems constitute major aquifers in closed and semiclosed terrestrial basins, alluvial valleys and intracontinental sags containing axial river systems, and coastal plains containing fluvial, deltaic shore-zone and sometimes fan and fan-delta systems. Fluvial systems can be grouped into a spectrum defined by trunk-stream bed-load, mixed-load and suspended-load channel types. Each channel type produces a predictable range of external aquifer geometries and internal heterogeneities. Because of the strong contrasts in fluvial systems of permeable, transmissive channel fill facies and confining flood-basin facies, flow patterns are controlled by channel connectivity, which correlates to fluvial system type and overall sand percentage. Bed-load (generally braided) and mixed-load (generally meandering) fluvial systems typically deposit "jigsaw-puzzle" aquifers systems. Heterogeneity increases with decreasing scale of facies units and increasing shale (or mud) in the system. Small meandering rivers and stable mud-rich, low-sinuosity channel systems create "labyrinthine" aquifer systems. In mixed- and suspended-load channel fills, vertical and along-channel textural changes and lithofacies partitioning and the dipping shale baffles in the upper point bar commonly create separate permeability units within a single point-bar sand body. Low-sinuosity suspended-load channels typically consist of a lower active fill and an upper abandonment phase fill—creating two units of differing aquifer properties. Sand-body width-to-thickness ratios are greatest for braided (largely bed-load) channels and are least for low-sinuosity, stable (largely suspended-load) channels and delta distributaries. Facies dimensions control both lateral extent of permeable units in labyrinthine systems and continuity of the more permeable units within jigsaw-puzzle systems.

INTRODUCTION

Fluvial and alluvial fan systems constitute major aquifers that provide large volumes of groundwater for municipal and agricultural use. These aquifers also are among the most likely to become planned or accidental repositories of waste. For these reasons, understanding the natural patterns of ground-water flow and predicting contaminant migration in these systems is quite important. The first part of this paper summarizes a relatively simple framework for fluvial facies recognition that both is grounded in geomorphic concepts and is readily applicable using subsurface (well based) data. The generalized fluvial types are related to external geometry and internal heterogeneities that control groundwater flow. The second part of the paper discusses the hydrostratigraphy and flow systems typical of principal types of alluvial basin fills.

INTERPRETATION AND HYDROSTRATIGRAPHIC CHARACTERIZATON OF FLUVIAL SYSTEMS

A fluvial depositional system, or the major segments of a large fluvial system, can be categorized in terms of the dominant channel type (bed load, mixed load, or suspended load) and erosional or aggradational nature of the trunk or main-stem stream (Schumm, 1977; Galloway and Hobday, 1996). Thus, classification can be based on geo-metric and compositional attributes of the framework sands, which are, in turn, important characteristics of the system as a potential conduit or reservoir for fluids. The dominant channel type of the trunk stream is a first-order basis for interpretation and categorization of the fluvial system.

Channel fills constitute the permeable framework of most fluvial systems. Channel-body geometry and facies organization reflect the nature of sediment load and consequent channel type (Fig. 1) (Schumm, 1977; Galloway, 1981; Galloway and Hobday, 1996). Channel fills are formed in bed-load, mixed-load and suspended-load river systems. Bed-load-dominated systems are commonly braided and produce sheet or broad, tabular belts of permeable channel-fill deposits. Mixed-load systems are meandering and form irregular belts. Suspended-load systems are highly sinuous to anastomosing and tend to form isolated permeable ribbons. Stacking and degree of amalgamation are determined by the combination of channel type and regional aggradation rate. Permeability contrasts can occur at channel-channel contacts because of weathering or clay infiltration into the underlying channel deposit or because muddy sediment was deposited along the scour surface. Thus, individual channel units, even though amalgamated, may remain isolated across the erosive contact. Mud plugs and flood-plain drapes form additional horizontal and vertical flow barriers (Hartkamp-Bakker and Donselaar, 1993).

Bed-Load Fluvial Systems

Bed-load fluvial systems consist dominantly of channel and channel-flank deposits (Fig. 2A); flood-basin facies are typically subordinate and sandy. Associated eolian deposits may be common, even in subhumid climates, because broad expanses of unconsolidated sand are exposed to wind reworking during low water levels. The water table in interchannel areas may be quite deep because such systems commonly occur in the topographically elevated proximal reaches of a sedimentary basin, and the sandy flood-plain and channel sediments are highly permeable. Sheetflood deposits and tabular crevasse splays contain significant amounts of bed-load sediment. Sand percentages for thick stratigraphic intervals along channel axes commonly range between 50 and 90 percent. Braided stream models and many coarse-grained, chute-modified meander belt models document various types of bed-load channel fills.

The channel-fill sequence is dominated by sand (Fig. 1). Coarse sand and gravel are commonly present but are not necessary components; the sequence could consist exclusively of medium to fine sand.

A bed-load channel is characterized by a high width-to-depth ratio and a tendency to erode its banks, resulting in a tabular or belted sand body (Fig. 2A). Amalgamated, multilateral sand belts result from rapid lateral shifting of bed-load channels and may even produce fluvial sheet sands. Low-sinuosity channels are characterized by relatively uniform depths of scour along the base, which may be reflected in the low relief on the base of the sand body and by preservation of laterally continuous sheet or tabular, remnant flood-plain mud units.

Internally, bed-load channel-fill sequences reflect the dominance of bed accretion units, such as longitudinal, transverse, lateral and chute bars, which produce a channel fill consisting of multiple, interlensed depositional units of varying grain size and displaying poorly organized textural sequences. Flood-plain facies are typically sandy.

Mixed-Load Fluvial Systems

Mixed-load channels typically preserve a texturally varied assemblage of flood-basin deposits, including flood-plain muds and silts or backswamp carbonaceous muds and clays. Sand percentages may be high along principal

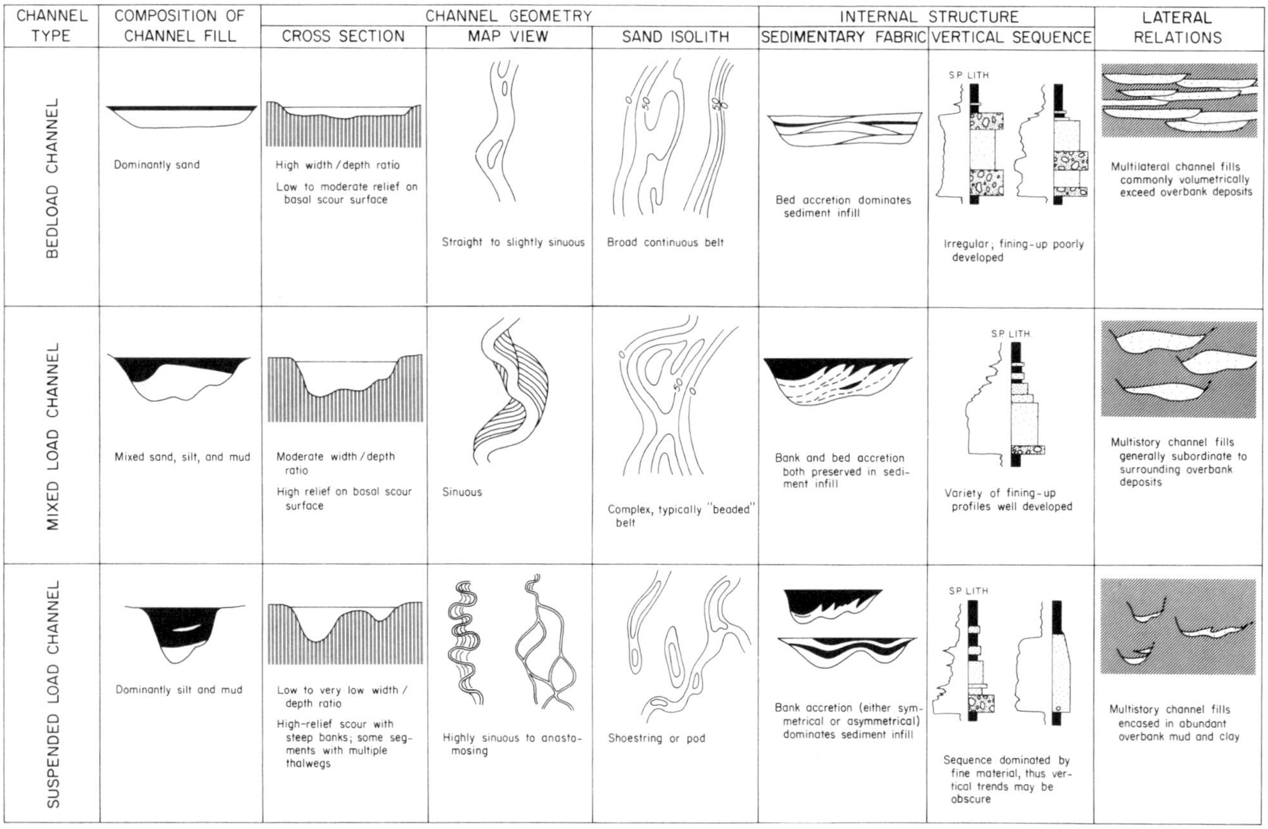

CHANNEL TYPE	COMPOSITION OF CHANNEL FILL	CHANNEL GEOMETRY			INTERNAL STRUCTURE		LATERAL RELATIONS
		CROSS SECTION	MAP VIEW	SAND ISOLITH	SEDIMENTARY FABRIC	VERTICAL SEQUENCE	
BEDLOAD CHANNEL	Dominantly sand	High width / depth ratio. Low to moderate relief on basal scour surface. Straight to slightly sinuous	Straight to slightly sinuous	Broad continuous belt	Bed accretion dominates sediment infill	Irregular; fining-up poorly developed	Multilateral channel fills commonly volumetrically exceed overbank deposits
MIXED LOAD CHANNEL	Mixed sand, silt, and mud	Moderate width / depth ratio. High relief on basal scour surface	Sinuous	Complex, typically "beaded" belt	Bank and bed accretion both preserved in sediment infill	Variety of fining-up profiles well developed	Multistory channel fills generally subordinate to surrounding overbank deposits
SUSPENDED LOAD CHANNEL	Dominantly silt and mud	Low to very low width / depth ratio. High-relief scour with steep banks; some segments with multiple thalwegs	Highly sinuous to anastomosing	Shoestring or pod	Bank accretion (either symmetrical or asymmetrical) dominates sediment infill	Sequence dominated by fine material, thus vertical trends may be obscure	Multistory channel fills encased in abundant overbank mud and clay

FIG. 1.—Geomorphic and depositional attributes of bed-load, mixed-load and suspended-load fluvial systems (from Galloway, 1977).

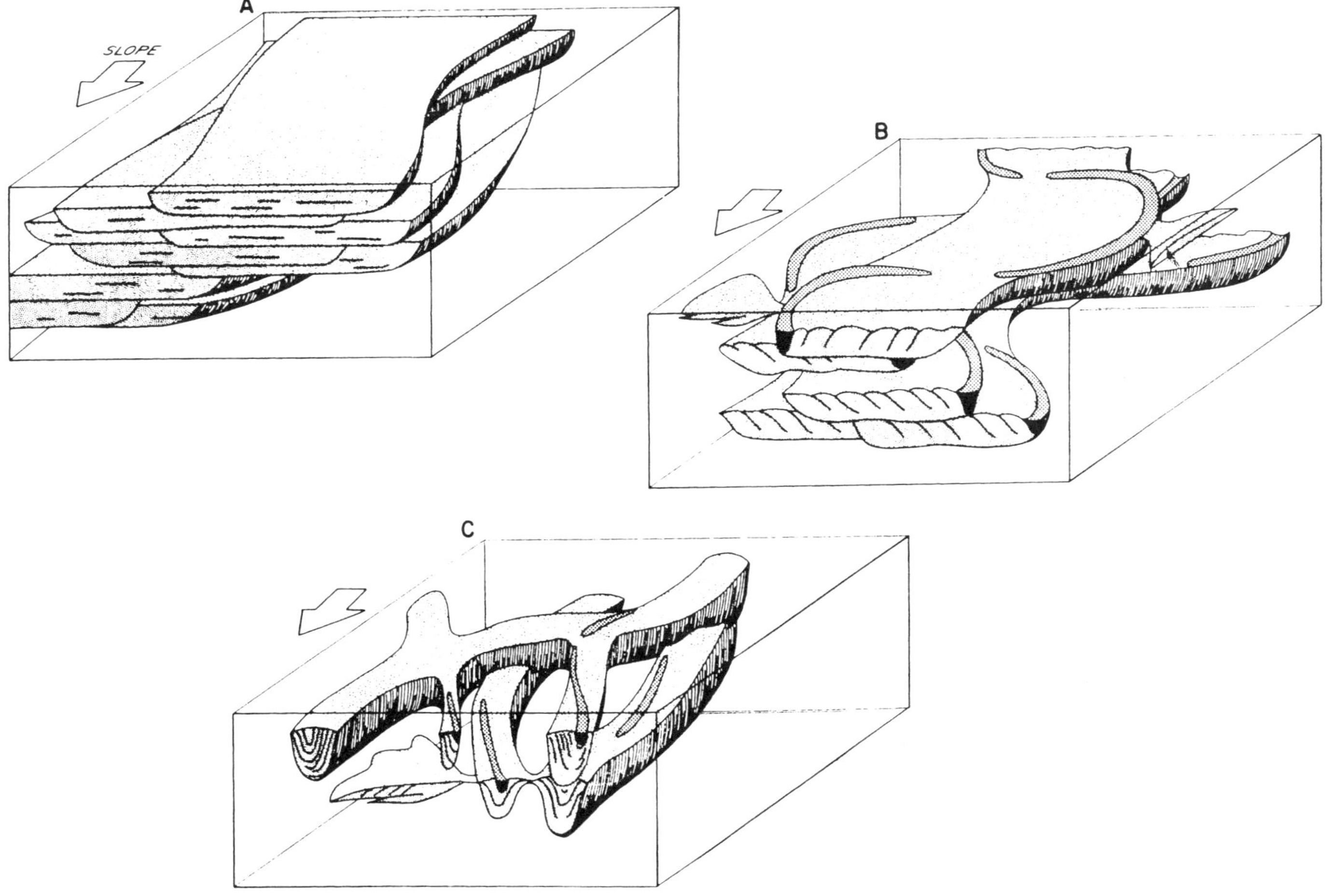

FIG. 2.—Block diagrams showing the contrasting geometry, lateral relationships and internal facies architecture of the framework channel-fill and associated crevasse-splay sand bodies typical of depositing (A) bed-load, (B) mixed-load and (C) anastomosing suspended-load fluvial systems (from Galloway and Hobday, 1996, copyright Springer-Verlag).

fluvial axes where channel-fill units are vertically stacked and locally amalgamated; however, overall sand percentage of thick successions commonly averages between 30 and 60 percent with subordinate silt and clay (Fig. 1). Channel fills are flanked by crevasse splay and levee sands and silts. The water table usually is controlled by the water level in the main channels; thus, high-standing levee and crevasse deposits typically are leached and oxidized. Much of the channel fill remains saturated, and organic debris deposited within lower point bar and channel lag is commonly preserved. Broad, low-lying backswamps also have shallow water tables, and organic debris, peat and syngenetic sulfide minerals are commonly preserved within the mud and clay deposited in this environment. Mixed-load channel fills are typified by the well-known simple point-bar model and its many variations.

Mixed-load channel deposits consist of sand. Channels are sinuous, with variable depths of basal scour along the thalweg. This is reflected in the development of irregular or beaded belts of sand (Fig. 2B) and in the preservation of lenticular, discontinuous flood-basin remnants between channel-fill sequences. A record of mixed bank- and bed-accretion (point bar and channel lag) deposits is preserved in the channel fill, and a repetitive upward-fining sequence typifies most vertical sections through the meander-belt deposit (Fig. 1). Channel meandering and consequent point-bar accretion, combined with stacking and amalgamation of successive channel fills, commonly produces a composite sand body much larger than the original channel. The channel plug provides the best record of true channel dimensions. In modern and ancient mixed-load channels, the abundance of fine, suspended-load sediment (89 to 97 percent of total load, Schumm, 1977) favors deposition and preservation of bounding fine sandy to silty flood-basin muds (Nanson and Croke, 1992).

Suspended-Load Fluvial Systems

Suspended-load fluvial systems have low sand percentages. Flood-basin deposits consist dominantly of backswamp and lacustrine muds and clays. Muddy levees are well developed. Crevasse splays are prominent and may be volumetrically important repositories of sand. Sand-isolith maps commonly show complex anastomosing or distributary patterns (Fig. 1). The low gradients, near base-level position and low-permeability sediments typical of suspended-load fluvial systems result in generally shallow water tables that intersect the land surface along channels and in extensive backswamp and lacustrine basins. Considerable organic debris, bedded peat and syngenetic sulfides commonly are preserved.

Suspended-load paleochannels are narrow and confined (Fig. 1); erosion occurs primarily at the base, and surrounding clay-rich sediments form stable, steep channel banks. Channel-fill deposition occurs dominantly along the banks and may be one-sided in highly sinuous channel segments or symmetrical in straight channel segments.

The dominant channel-fill sediment ranges from very fine sand to silt and clay, but coarse sediment may form the core of the channel fill. Vertical sequences may fine upward or show little vertical variation if the range of grain sizes is limited (Fig. 1).

Sand-body geometry is highly lenticular in cross section and forms sinuous or anastomosing patterns (Fig. 2C). Channel-fill units typically are encased in fine-grained flood-basin deposits and tend to stack vertically creating a labyrinthine aquifer system.

A hierarchy of channel types commonly coexists within a single fluvial system. For example, flood-runoff and crevasse channels flanking the main channel likely will be enriched in suspended load. Tributaries can range from bed- to suspended-load type channels, depending on their provenance and gradient. The trunk channel itself may vary in type where it is locally affected by sediment input, vertical erosion into unusually sandy or muddy substrates, neotectonic deformation or anthropogenic effects. Large, integrated fluvial systems display a systematic down-flow evolution of process, morphology and depositional facies.

In addition to the framework of channel-fill facies, a typical fluvial system consists of variable proportions of crevasse splay, levee, flood-basin and abandoned channel-plug deposits. Some facies, such as levee, plug or splay facies, may be poorly developed or preserved, depending on fluvial channel type and aggradation rate. Following identification of the trunk channel facies, recognized by its sharp base, relatively coarse-grained composition, thickness and geometry, surrounding facies can be delineated. In addition to differing channel-fill facies, ancient aggrading bed-load, mixed-load and suspended-load fluvial systems each display characteristic facies associations and early hydrologic history.

Characterization of Fluvial Aquifers

Outcrop and field studies have characterized facies architectures, permeability distributions, heterogeneities and fluid flow patterns of fluvial sand bodies. Braided (generally bed-load) and meandering (generally mixed-load) fluvial reservoirs typically deposit jigsaw-puzzle (terminology of Weber and van Geuns, 1990; see Galloway and Sharp, this volume) aquifers systems (Jordan and Pryor, 1992; Martin, 1993). Heterogeneity increases with (1) decreasing scale of facies units and (2) increasing mud/shale in the system. Small meandering rivers and stable mud-rich, low-sinuosity channel systems create labyrinthine aquifer systems (Fielding and Crane, 1987; Dreyer, 1990; Dreyer et al., 1990; Bryant and Flint, 1993; Davies et al., 1993).

Some bed-load fluvial systems of braided or coarse-grained meander-belt origin deposit remarkably homogeneous sand aquifers with thin, sporadic silty shale bands and no evidence of subaerial diagenetic alteration. These units show minimal vertical or lateral variability. Other braided channel facies display pronounced mesoscopic heterogeneity because of the presence of cemented zones, discontinuous horizontal shale baffles and lateral and vertical lithofacies variability (Davies et al., 1993; Høimyr et al., 1993; Martin, 1993), but their aggregate permeability is high because baffles are discontinuous. Mixed- and suspended-load channel fills are more heterogeneous and complex. Vertical and along-channel textural changes and lithofacies partitioning and dipping shale baffles in the upper point bar can potentially create separate permeability units within a single point-bar sand body. Low-sinuosity suspended-load channels commonly consist of a lower active fill and an upper abandonment phase fill—creating two units of differing properties (Weber, 1982).

Outcrop studies (Fielding and Crane, 1987; Dreyer, 1990, 1993; Cowan, 1991; Lowry and Raheim, 1991; Bryant and Flint, 1993; Barton, 1994) have defined typical width-to-thickness ratios for common fluvial sand bodies (Fig. 3). Ratios are greatest for braided (largely bed-load) channels and least for low-sinuosity, stable (largely suspended-load) channels and delta distributaries. Facies dimensions control both lateral extent of permeable units in

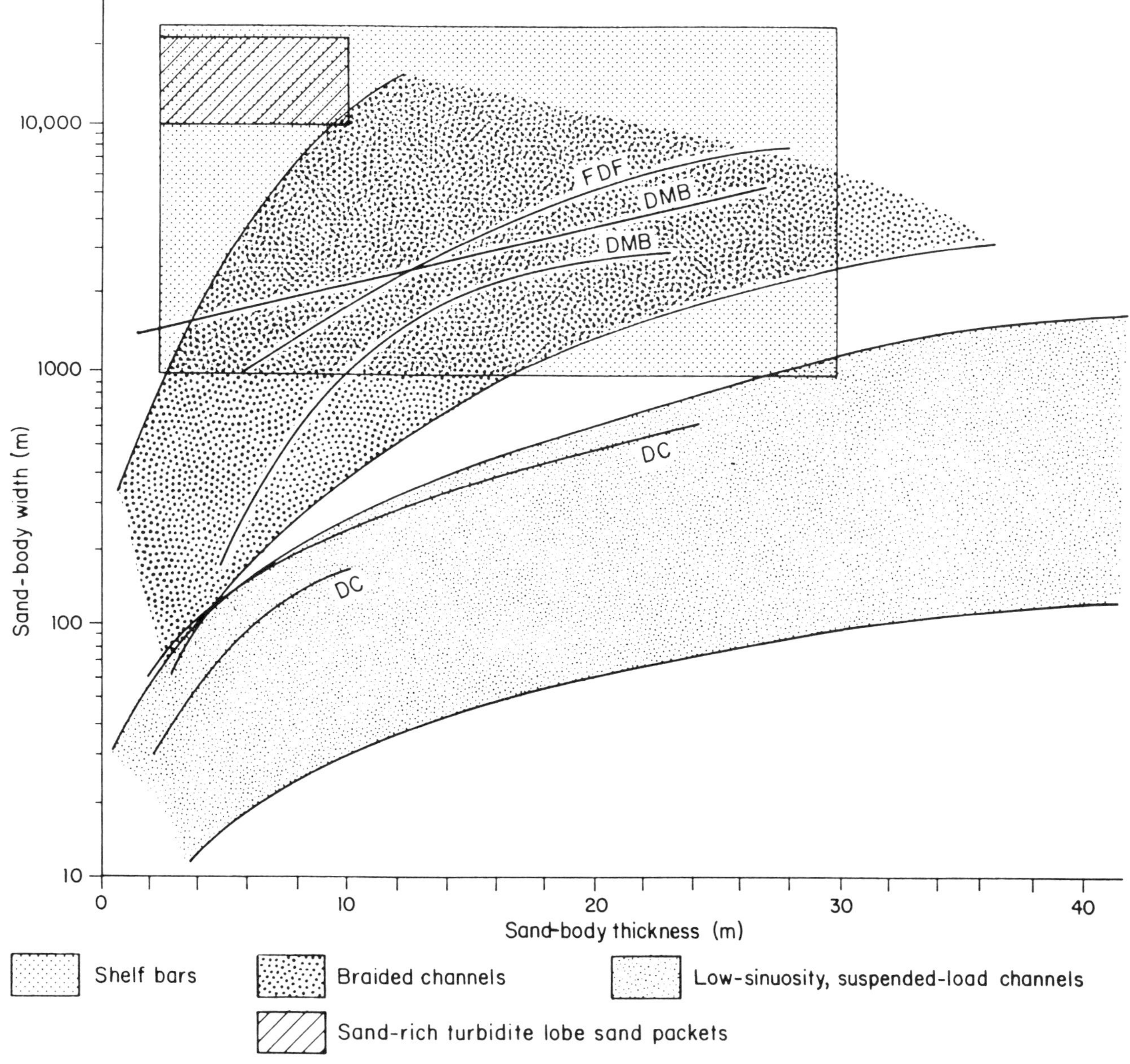

FIG. 3.—Cross plot of typical ranges of sand-body thickness and width based on outcrop studies of various depositional systems. Braided, bed-load channels show high width-to-thickness ratios, similar to those of many marine sand facies. Deposits of suspended load channels have lowest range width-to-thickness ratios for all sand bodies studied. Meandering, mixed-load channels lie between these extremes. DC = distributary channel, DMB = distributary mouth bar, FDF = fan delta front (from Galloway and Hobday, 1996, copyright Springer-Verlag; compiled from various sources).

labyrinthine systems and continuity of the more permeable units within jigsaw-puzzle systems.

Because of the strong contrasts in systems of permeable, transmissive facies (channel fill) and confining facies (flood-basin), flow patterns are controlled by channel-belt connectivity (Fogg, 1990). Connectivity can be correlated with overall sand percentage. Computer simulations (Bridge and Mackey, 1993) suggest that channel units are relatively isolated where the proportion of channel facies (sand) is less than 40 percent; connectivity increases at proportions between 40 and 75 percent; and channels are highly connected where the proportion exceeds 75 percent. Fogg (1986) analyzed a mixed-load fluvial aquifer and found that sand bodies are isolated where sand is less than 20 percent and effectively amalgamated where sand exceeds 60 percent. He demonstrated the significance of interconnectivity by a series of simulations testing lateral interconnectivity of sand bodies (Fig. 4). A recent model

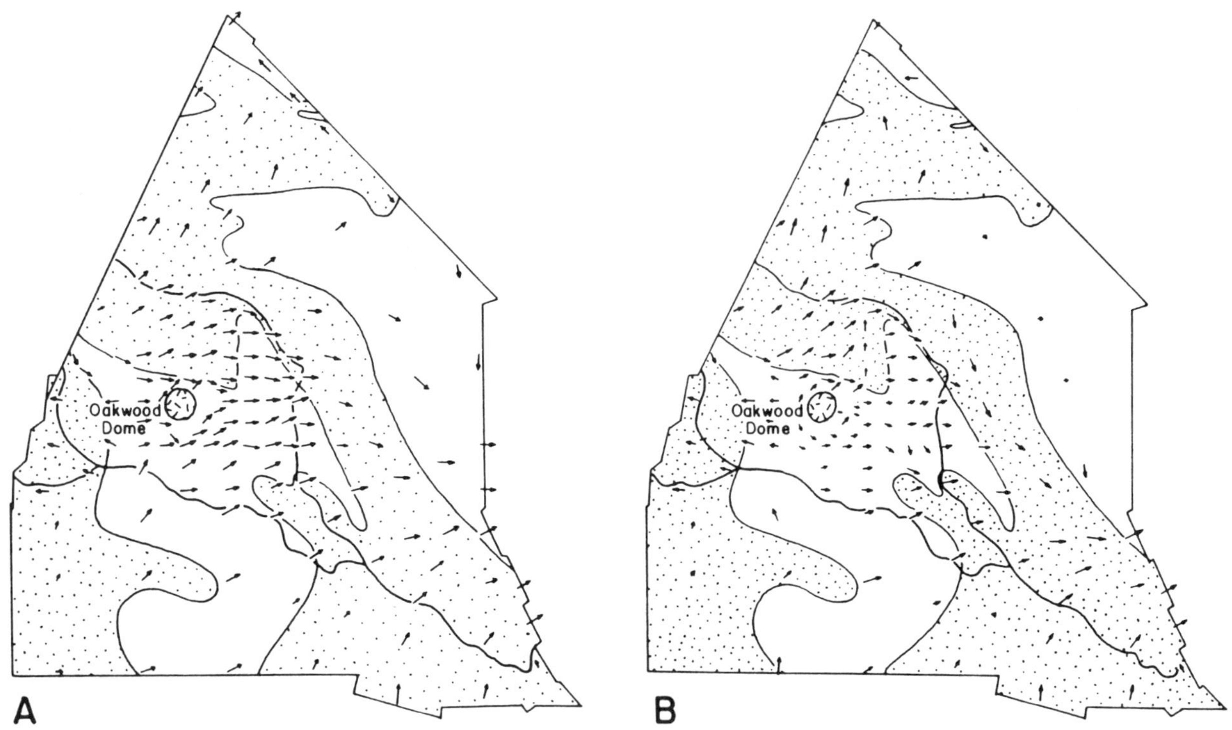

FIG. 4.—Comparison of groundwater flow simulations prepared to predict potential contaminant transport from a salt dome. In A, the aquifer is presumed to consist of interconnected sand bodies. In B, transmissive channel sand bodies are assumed to be unconnected in shaded areas where sand percentage is below 20 percent. Flood-plain and splay deposits control permeability in the low sand areas. Note the very different results of predicted rates and directions of flow (modified from Fogg, 1986).

study (Jones et al., 1995) indicates that the number of discrete channel fills stacked to form the channel belt has a major influence on flow within jigsaw-puzzle fluvial belts.

The hydraulic characteristics of fluvial sediments can be strongly influenced by near-surface geomorphologic processes. For instance, paleosols may form widespread horizontal layers of low permeability. Soils with horizons of chemical or mechanical accumulation of pore-filling material reduce permeability and can limit downward percolation of pollutants (Meehan and Schlemon, 1994).

TYPICAL GROUNDWATER FLOW PATTERNS

We now review the hydrostratigraphy (Seaber, 1988) and typical groundwater flow patterns of three common alluvial basin fills:
1. Closed and semiclosed terrestrial basins containing combinations of alluvial fan, lacustrine, fluvial and eolian systems;
2. Alluvial valleys and basins containing fluvial and alluvial fan systems; and
3. Coastal plains containing alluvial fan, fluvial, deltaic and shore-zone systems.

Closed and Semiclosed Terrestrial Basins

Hydrogeology of closed or partially closed terrestrial basin fills has been studied in numerous settings (Davis et al., 1959; Hardie et al., 1978; Wilkins, 1986; Anderson, 1986; Back et al., 1988; Davis, 1988; Duffy and Al-Hassan, 1988; Belitz and Heines, 1990). Such basin fills exhibit many common elements (Fig. 5, A and B). Commonly, recharge occurs along the basin periphery or in adjacent uplands, lateral flow may typify intermediate environments, and discharge occurs in the basin center. Groundwater flow is largely centripetal, reflecting the enclosing topography. Climate, which can vary from humid to arid and be topographically zoned in intermontane basins, plays a major role in determining rates and locations of recharge and discharge. In closed basins, discharge is either by springs or base-flow into lakes, where it evaporates, or by evapotranspiration through a shallow veneer of lacustrine or alluvial plain sediment. Of course, discharge also can occur through pumping. In partially closed basins, significant discharge may occur by stream flow or by interbasin flow through confined axial alluvial system facies or high-permeability (generally karstic) rocks in the adjacent uplands.

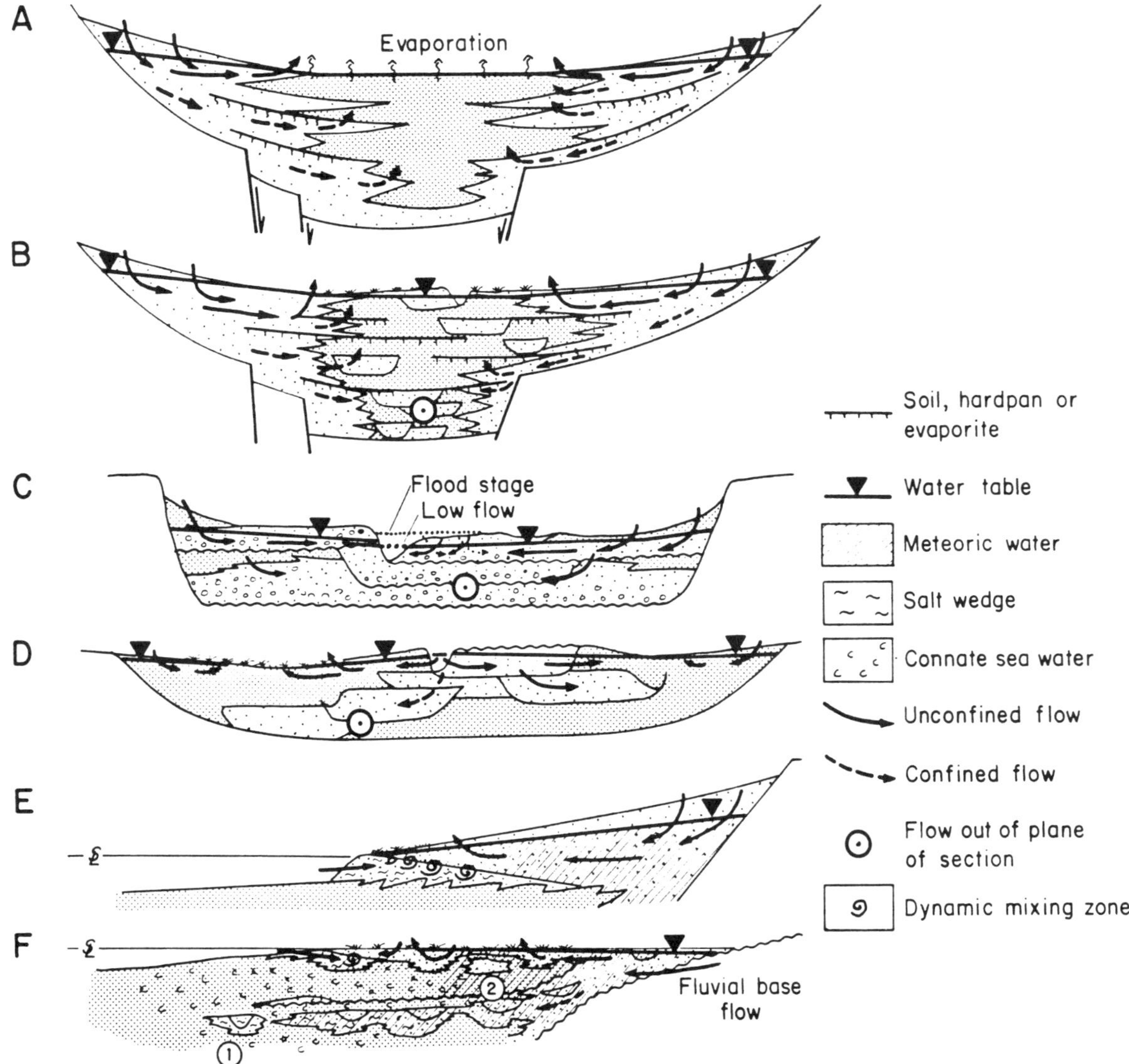

FIG. 5.—Typical syndepositional groundwater flow systems in alluvial basin settings. A. Closed terrestrial basin with alluvial fan and playa lake systems. B. Semienclosed alluvial basin with marginal fan and axial trunk stream. C. Erosional alluvial valley fill. D. Depositional alluvial basin fill. E. Narrow coastal plain with alluvial fan and fan delta systems. F. Broad, low-gradient coastal plain with fluvial and delta systems (modified from Galloway and Hobday, 1996, copyright Springer-Verlag).

Hydraulic conductivity in closed alluvial basins is complex but organized. Marginal facies commonly are coarse, massive and highly conductive; grain size and conductivity decrease basinward (Fig. 6). Basin-center fluvial and lacustrine facies are commonly fine grained and stratified, but they may contain a core of highly conductive, axially oriented sand facies if a perennial fluvial system or large, permanent lake with well-developed shore-zone facies is or was present. Great spatial aquifer variability may exist. Vertical conductivity can vary by up to seven orders of magnitude, and cross-formational flow may be prominent in the hydrogeologic system. Topography produced by depositional, erosional or tectonic processes may create local to intermediate-scale flow systems within the basin. The major sedimentary systems within terrestrial basins are composed of alluvial fan, eolian and axial and/or lacustrine fluvial facies.

Alluvial fan systems.—

Alluvial fan morphology results in a predictable distribution of recharge, lateral flow and discharge zones in the shallow, unconfined to moderately confined fan aquifer sands (Fig. 7). The topographically high proximal fan is commonly the area of deepest water table and active

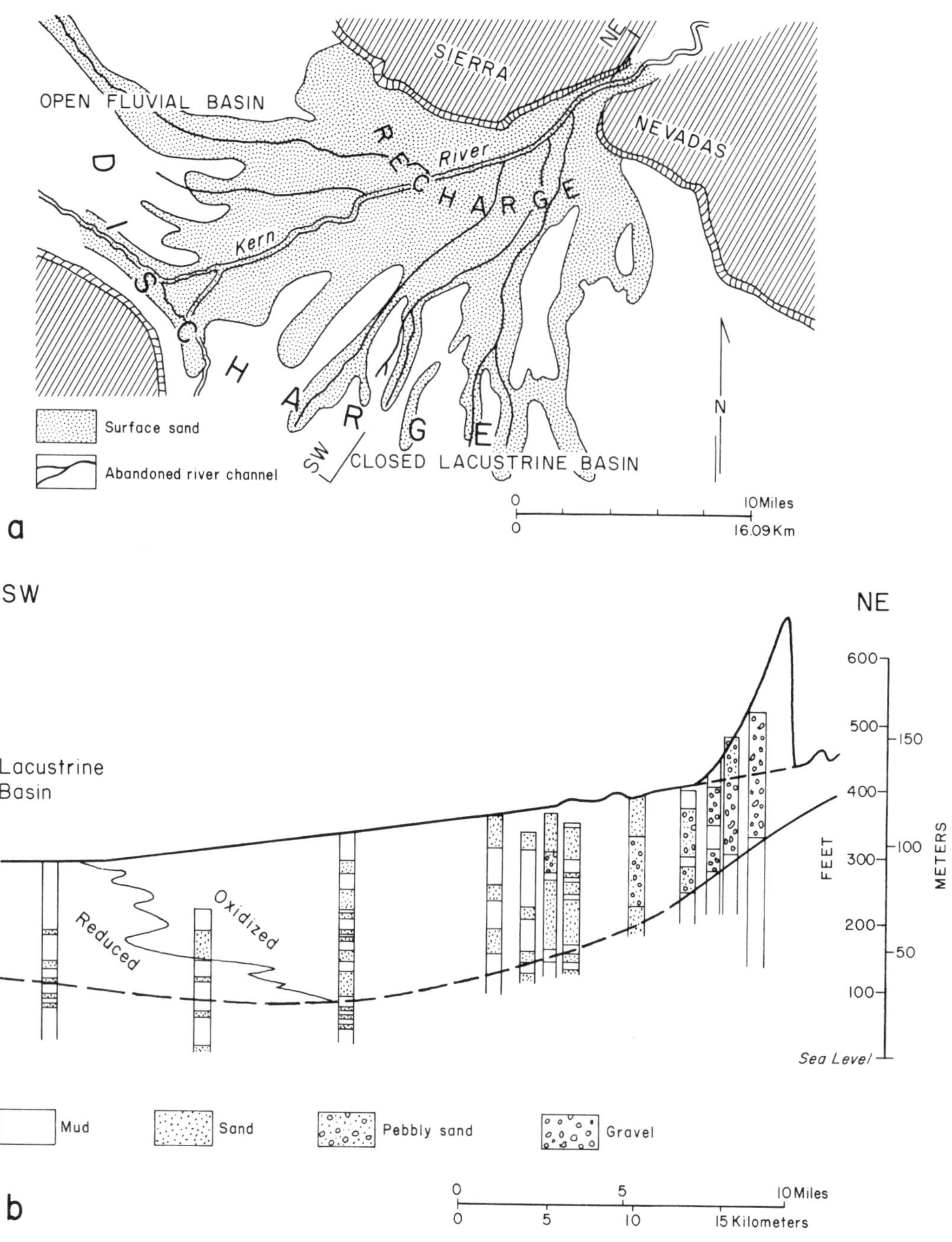

FIG. 6.—Surficial sediments (A) and axial lithofacies profile (B) of the late Quaternary-Holocene Kern Fan, San Joaquin Valley, California. Fan sediments have been oxidized by active groundwater flow (from Galloway and Hobday, 1996, copyright Springer-Verlag).

groundwater recharge. Recharge is greatest from perennial streams, but leakage from intermittent streams, flow from fractured uplands and precipitation on the fan surface may be secondary sources. In arid or semiarid areas pedogenic calcite (caliche) may greatly restrict or eliminate recharge in fans, even in the proximal areas (Darling et al., 1995). Fan-head trenches and fan channels become sites of groundwater discharge, as exemplified in the post-pluvial Quaternary Rio Grande fan, San Luis Valley, Colorado (Fig. 7). The distal fan, which is an area of abruptly decreasing topographic and hydraulic gradient, is a discharge area where the water table is shallow or even emergent. Here distal and interfan sequences interfinger with marsh, swamp, mud-flat or evaporative playa facies. In open basins, such as the San Luis Valley, discharge constitutes a major source of base flow into the through-flowing fluvial

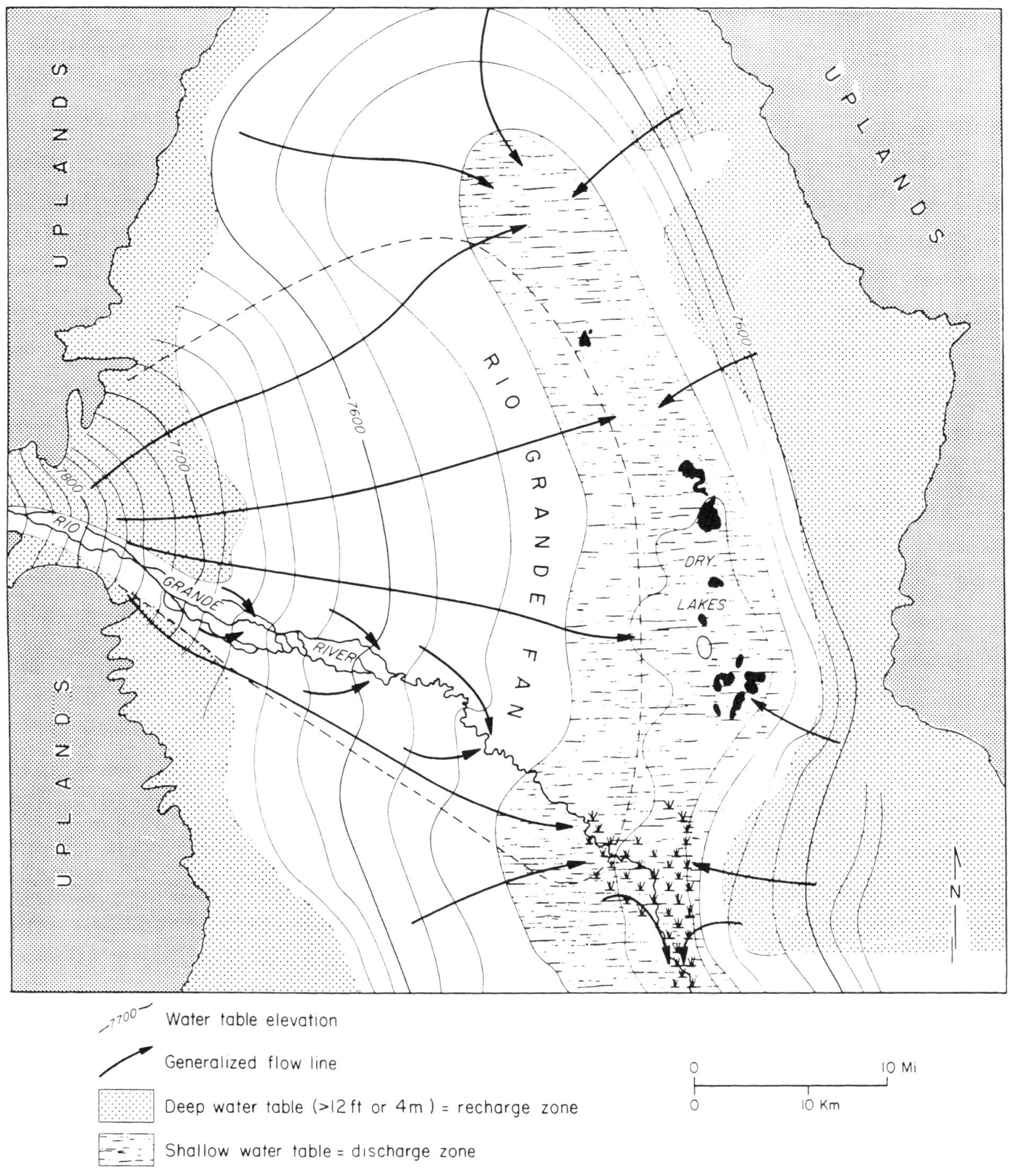

FIG. 7.—Hydrogeology of the Rio Grande fan, San Luis Valley, Colorado. The modern underfit stream is incised in the fan, creating a local discharge axis across the fan. Regional groundwater flow discharges along the basin floor (from Galloway and Hobday, 1996, copyright Springer-Verlag).

system (Fig. 5B). Even in semiarid environments, fan toes may be sites of relatively luxuriant plant growth and preservation of organic debris. Conversely, if evaporation rates are too high, salts become concentrated in less permeable facies around the fan fringe (Deveral and Gallanthine, 1989; Darling et al., 1995).

Alluvial fan systems, particularly perennial stream fans, may form highly transmissive aquifers. Framework sand bodies are coarse, thick, well interconnected and oriented down the fan grade parallel to the general hydraulic gradient. The great agricultural areas of the desert southwestern United States, such as the San Joaquin and San Luis Valleys, rely on the tremendous volumes of groundwater contained in alluvial fan and associated depositional systems. The very high rates of recharge commonly lead to extensive oxidation of upper and midfan deposits, particularly in arid settings were the water table is comparatively deep (Fig. 6) (Davis et al., 1959). Oxidation of detrital ferromagnesian minerals and mechanical dispersal of colloidal products favor a geologically rapid formation of red beds (Walker, 1967). Shallow water tables and regional discharge may insulate the distal fan and associated facies from oxidation. In wet climates highly vegetated fans with well-developed humic soils and shallow ground water may be little affected by early postdepositional oxidation.

Axial Fluvial Systems.—

These commonly develop along the axes of semi-enclosed basins. Where present at the surface the fluvial sediments may serve as either recharge or discharge areas. Along ephemeral or perennial streams, flood-plain facies may form confining units and isolate deeper channel fill (Fig. 5B), fan, and eolian facies. Recharge through the overbank deposits results in rapid soil formation with prominent accumulation of clays and oxides in the B horizon. Poorly drained soils in discharge zones develop hardpans of carbonate, silicate, and iron minerals that restrict and localize discharge (Fig. 5B). Climate change or tectonic rejuvenation may cause the formation and subsequent burial of mature paleosol horizons that enhance permeability stratification in distal fan and fluvial facies.

Associated Depositional Systems.—

Dune deposits form highly permeable units that may be interbedded with fluvial deposits. These units are nearly homogeneous and isotropic, although where present, interbedded interdune flat and playa deposits may severely reduce vertical conductivity. Draa topography may create local flow systems, with discharge by evapotranspiration and precipitation of soluble salts concentrated in interdraa flats. Eolian sands form extensive, highly permeable layer-cake aquifers (Chandler et al., 1989).

Fine-grained, low-permeability sediments deposited in lakes and playas form confining beds that bound terminal fan, eolian and fluvial deposits. Precipitation of salt layers and cemented zones in discharging or ponded waters may create a highly stratified hydrostratigraphy of very low vertical permeability. Consequently, most discharge is concentrated around the periphery of the central lacustrine system within arid and semiarid basins (Fig. 5A). Flow complexities arise where evaporatively concentrated brines

create a shallow density inversion, which may lead to brine reflux downward into underlying aquifers and outward into buried fan aquifers (Duffy and Al-Hassan, 1988). Alternatively, deflation may effectively remove salts and spread them as aerosols over a wide area (Brown and Sharp, 1992). If present, coarse, well-sorted shore-zone facies may form on the margins of larger, permanent lakes in humid basins. These facies are highly permeable and may have significant thickness.

Alluvial Valleys and Basins

Large topographic valleys formed by alternating periods of fluvial incision and intermittent deposition (Fig. 5C), and elongate intracontinental sags filled with aggradational alluvial deposits (Fig. 5D) constitute major hydrologic basin types (Grannemann and Sharp, 1979; Heath, 1982; Sharp, 1988; Larkin and Sharp, 1992; Foster and Chilton, 1993). The general course of their major streams may be consistent over extended geological time (Potter, 1978). These systems are characterized by thick, permeable sandy and gravelly channel-fill facies in contact with a perennial stream and containing important but easily contaminated shallow groundwater resources. Hydraulic conductivities commonly range from 10^1 to 10^3 m/d and decrease upward.

Hydrostratigraphy.—

In an erosional valley fill, the coarsest sediment with maximum conductivity lies above the valley floor, forming a water-table aquifer locally confined by a topstratum of floodplain, soil and marginal colluvial deposits. Terraces may serve as secondary (and occasionally perched) aquifers, although pedogenic processes may restrict their permeability. Valley walls likely serve as hydraulic boundaries and, together with the contained coarse fluvial facies and valley geometry, strongly focus the flow fields within incised valley fills. Figure 8, for example, illustrates the highly predictable effect of valley morphology and lithofacies pattern on migration of a chloride excursion from a disposal pit. Even where permeable facies are juxtaposed, weathering profiles, diagenetic and infiltrated clay, and colluvial drapes may retard flow from the uplands into the valley fills.

In depositional alluvial basins, axial fluvial systems produce a succession of channel-fill belts that are variably isolated by flood-plain fines. Transmissivity varies with the type of sediment transported by the trunk stream. Bed-load system sediments have high primary permeabilities; suspended-load system sediments have low permeabilities.

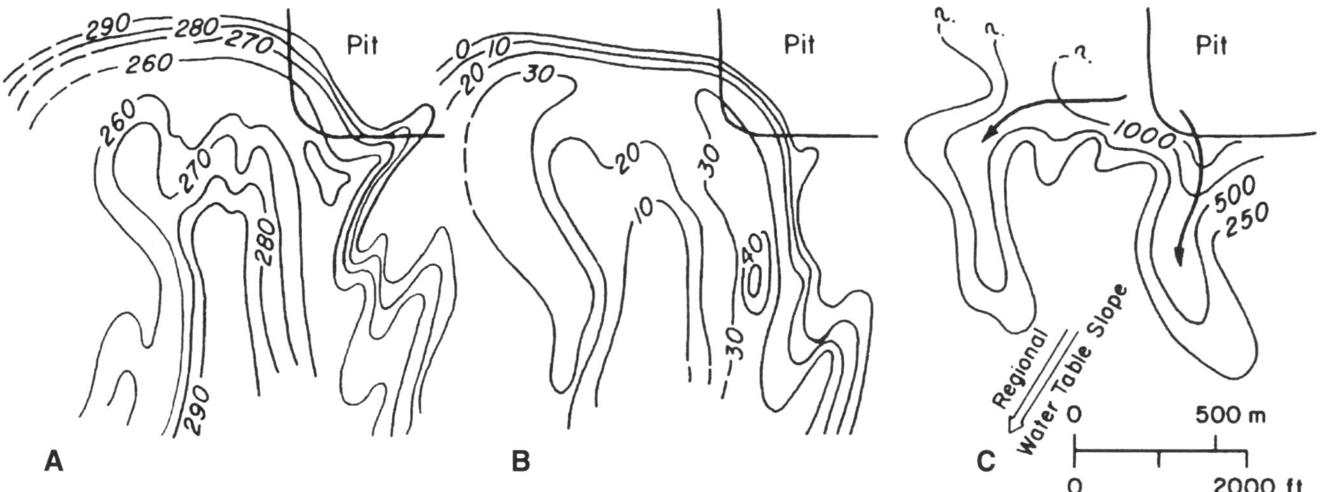

FIG. 8.—Effect of Quaternary valley morphology and facies on the pattern of a chloride excursion from an adjacent waste pit. A. Elevation contours of valley wall. B. Net-sand and gravel isolith map of valley fill. C. Isochloride contour map (from Galloway and Hobday, 1996, copyright Springer-Verlag).

Deeper aquifers may be partially confined, whereas shallow aquifers tend to be hydraulically connected to active streams. These sediments can be extremely heterogeneous and anisotropic because of the internal variability and lenticular geometry of fluvial sand bodies. The general orientation of the permeable facies, however, is subparallel to the course of the alluvial valley.

Groundwater Flow Patterns.—

Groundwater flow within the alluvium may parallel the trunk stream under certain conditions, creating a zone of down-valley underflow (Meinzer, 1923; Sharp, 1988; Larkin and Sharp, 1992). In other cases flow is directly toward the stream as base-flow. End-member examples are shown in Figure 9. The proportion of underflow is great-

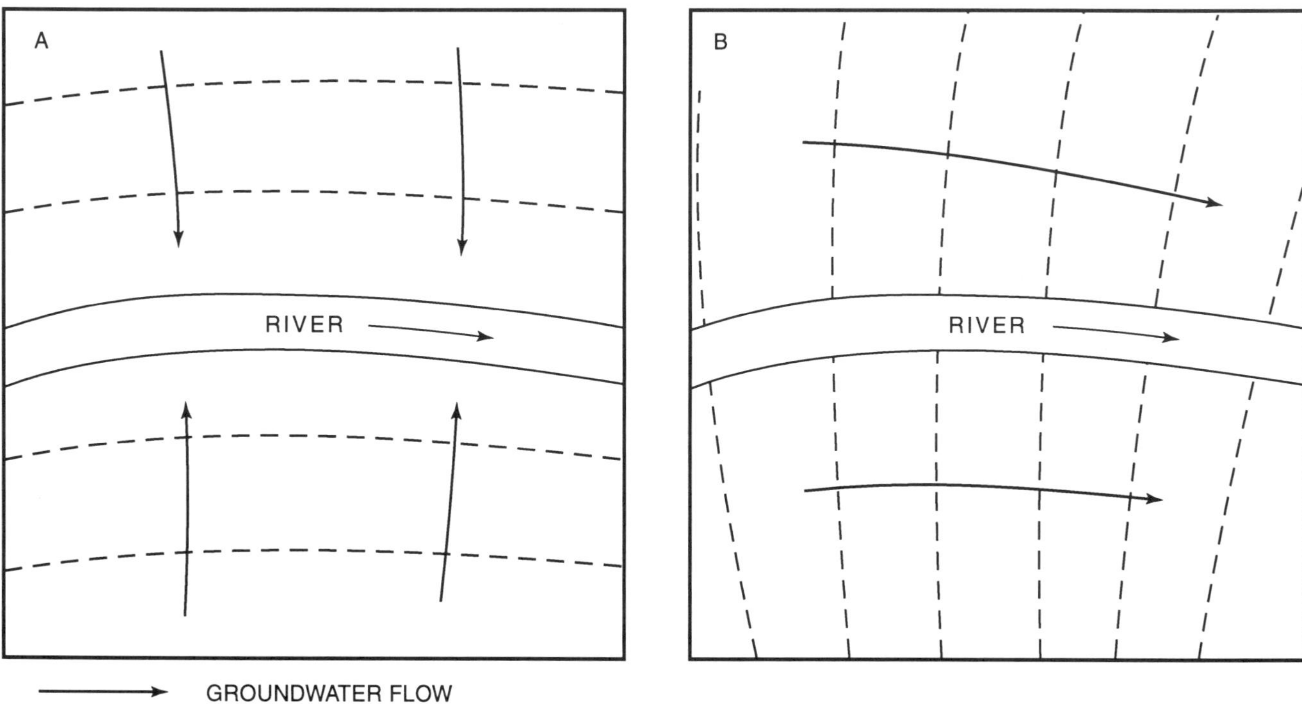

FIG. 9.—Groundwater flow in the alluvial valley fill may be either nearly normal to the stream as base flow (A) or, as underflow (B) (Meinzer, 1923), nearly parallel to the stream (from Larkin and Sharp, 1992).

est in coarse, mixed-load to bed-load systems characterized by shallow, broad channels and steep valley/channel
gradients. Flow systems in alluvial aquifers can be complex and will vary with changes in recharge and discharge.
Basin margins may receive recharge from tributary streams
and direct infiltration of precipitation. Near the channel
belt, flow is highly variable in response to fluctuations in
river level. Discharge commonly dominates around incised
streams; recharge may prevail in elevated aggradational
channels. The relations will alter with stream stage changes,
the permeability of the fluvial sediments and the sediments
on the bed and banks of the streams. Sharp (1988) recognized several patterns (Fig. 10):

1. Zones of rapidly fluctuating groundwater levels. These
   occur close to major streams in response stage changes.
2. Zones of long-term stable groundwater conditions.
   These typically are base-flow situations in areas far
   from the trunk stream or hydraulically connected tributaries. Recharge is predominantly by infiltration of
   precipitation.
3. Zones of predominant down-valley flow (underflow).
4. Zones of groundwater highs. These are created where
   smaller tributary streams lose flow into the alluvium
   or where sheet flow enters the flood plain from the
   surrounding uplands.

5. Zones of groundwater lows. These are created by pumping the aquifers and do not occur naturally.

Coastal Plains and Deltas

The coastal plain is a zone of regional flow with discharge to both major streams and to the ocean. Discharge
to the latter becomes predominant with proximity to the
strandline. Near the strandline a complex zone may be
present in which regional meteoric groundwater discharges
and mixes with seawater. Unconfined, semiconfined and
confined aquifers all may occur in close vertical proximity
(Fig. 5, D, E and F). The hydrogeologic systems reflect
both the physical stratigraphy and facies-controlled transmissivity distribution and the history of coastal deposition
and relative sea-level change (Meisler et al., 1984; Engelen
and Jones, 1986, chapter 7; Custodio, 1987).

Quaternary Coastal Plain Flow Systems.—
Quaternary depositional coastal plains consist of fluvial, deltaic and shore-zone systems. In narrow coastal
plains backed by high relief, alluvial fans and fan deltas
may be prominent geomorphic and depositional elements.
The position of the groundwater discharge zone is determined by regional hydraulic gradients, hydrostratigraphic

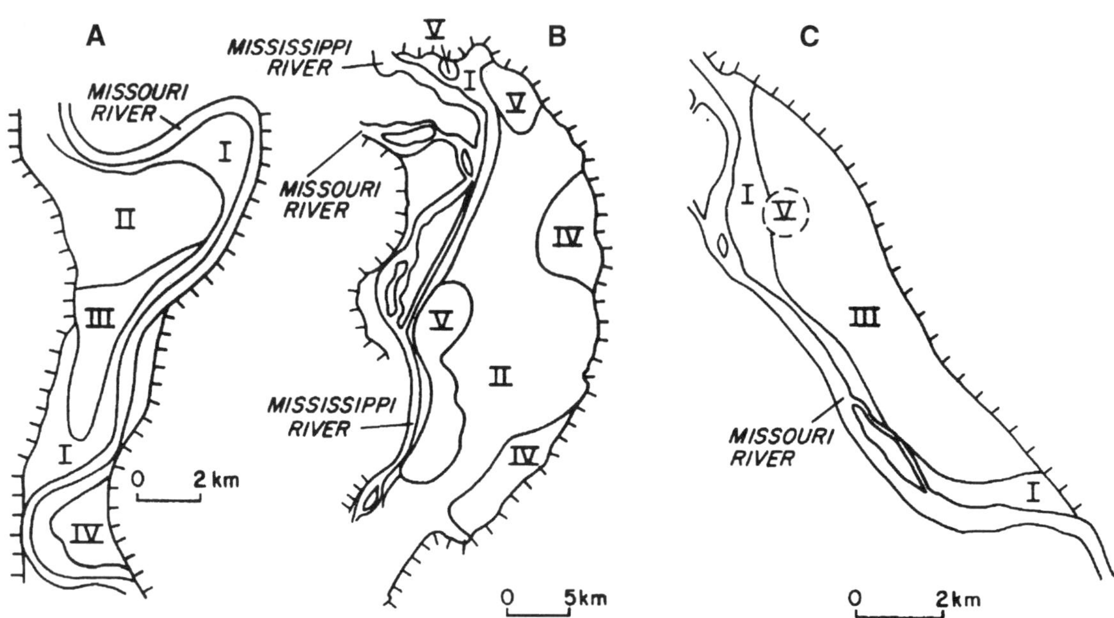

FIG. 10.—Groundwater flow zones in major alluvial (flood-plain) aquifers (from Sharp, 1988; compiled from Schicht and Jones, 1962; U. S. Army Corps of Engineers, 1976; Grannemann and Sharp, 1979; Foreman and Sharp, 1981). I. Zones of rapidly fluctuating groundwater levels controlled by proximity to the river and river stage fluctuations. II. Zones of relative stability, which tend to be far from the river and in which recharge is mostly by precipitation. III. Zones of down-valley flow (or underflow) where the river is subparallel to the valley walls for some distance (other controlling factors are discussed by Larkin and Sharp, 1992). IV. Zones of persistent groundwater mounds, which are a function of influent streams on the flood plain. V. Zones of groundwater lows induced by pumping. Other examples of flow directions in modern fluvial aquifers are presented in Larkin and Sharp (1992).

framework and sea level. Although confined aquifers may discharge water offshore of the strandline, shallow semi-confined or unconfined aquifers discharge near the shore zone. Manifestations of regional discharge include coastal swamps and marshes, shallow lakes, seeps and springs. Within the coastal discharge zone, complex local meteoric flow systems are associated with depositional features such as natural levees, barrier islands and beach-ridge dune fields.

The boundary zone between fresh and salt water is a dynamic interface that reflects topographic head and aquifer stratigraphy. In the classic models of Ghyben (1899) and Herzberg (1901) (i.e., a thick water-table aquifer under steady state and with stagnant sea water), the depth of fresh water is about 40 times the head above mean sea level. In all unconfined flow systems, a salt wedge penetrates inland beneath shallow fresh water (Fig. 5, E and G). Extent of salt-wedge penetration is greatest within thick, highly conductive fluvial or shore-zone facies and is restricted in thin or low-conductivity units. Penetration is enhanced in facies whose greatest permeabilities occur at the top of the aquifer. Thus, observed salt-wedge penetration in unconfined aquifers is most extensive along active or abandoned distributary channel or fluvial axes. Penetration also is common beneath topographically low areas like coastal lakes, marshes, and mud flats. It is least beneath high dune or beach ridges. Depositional topographic highs, such as beach ridges (Fig. 5G), form zones of recharge for local, unconfined flow systems.

On narrow, steep coastal plains, fan and braid-plain delta systems create thick, conductive water-table aquifers that have high potential for recharge in the inner coastal plain (Fig. 5E). Despite the relatively steeply sloping water table, extensive salt-wedge penetration characterizes fan axes. Penetration is limited in the finer, interfan delta facies.

Confined Coastal Plain Aquifer Systems.—
Fresh-water-, uranium- and petroleum-bearing Oligocene and early Miocene sequences of the northwestern Gulf coastal plain provide examples of the complex flow dynamics and hydrochemical evolution typical of fluvial aquifers within a large, depositionally active basin (Galloway, 1977; Smith et al., 1980; Galloway, 1982; Morton and Land, 1987; Kreitler et al., 1990; Harrison and Summa, 1991). As is typical of subsiding coastal plains, older Tertiary aquifers of the Gulf Coast have experienced erosional exhumation along their outcrop belt due to basin-margin uplift compounded by sea-level lowstands during the Quaternary. Greatest topographic relief occurs where valleys

of extrabasinal rivers cut across the strike-oriented outcrop belt. The aquifers dip Gulfward beneath increasing thicknesses of younger sediments, and fluvial systems grade into major deltaic systems in the deep subsurface. Thus, deeper waters show limited effects of exhumation, and aquifers retain waters characteristic, in part, of the burial phase compactional regime.

Shallow aquifers consist of the deposits of major mixed-load and bed-load fluvial systems. Pump-test data show that the average conductivity of bed-load channel-fill sequences is about twice that of mixed-load sand bodies. Both are much more conductive than bounding mudstone and splay facies (Galloway et al., 1982b). At depth, the fluvial systems grade into sand-rich deltaic and shore-zone systems.

Interpretation and synthesis of vertical and lateral head, temperature, resistivity and hydrochemical distributions for groundwaters of the lower Miocene Oakville aquifer reveal flow patterns characteristic of a coastal plain setting (Fig. 11). Factors shown to influence flow directions and volumes include (1) outcrop distribution and topography, (2) distribution and orientation of the permeable channel axes (as delineated by net-sand isolith maps), (3) presence and location of small-displacement, strike-parallel faults and salt diapirs, some of which appear to serve as loci for discharge of deeper, confined aquifers, and (4) coastward burial and confinement of the fluvial sands (Smith et al., 1980; Galloway, 1982).

Local, intermediate and regional flow systems are present (Fig. 9). Recharge occurs by infiltration along the outcrop and through the thin, updip edge of overlying confining units. Regional flow extends from the recharge zone basinward along sand axes toward a broad discharge belt that lies as deep as several thousand feet below sea level. There, waters percolate across a broad area of the overlying confining unit and into shallower aquifers. Waters typically evolve downflow from a Ca^{2+}-HCO_3^- hydrochemical facies to a Na^+-Cl^- facies. Hydrochemical anomalies characterized by unusually abundant Cl^- or SO_4^{2-} suggest active leakage of water from underlying aquifers and typically are encountered along fault zones. Intermediate flow cells, characterized by flow along strike and discharge to the surface or into alluvium of major river valleys, interrupt regional flow for distances of as much 20 miles (30 km) along strike (Fig. 9). Topographic relief of approximately 100 ft (30 m) or less between interfluve highs and valley flood plains affects the slope of the potentiometric surface and, hence, the groundwater flow to depths of as much as 700 ft (210 m).

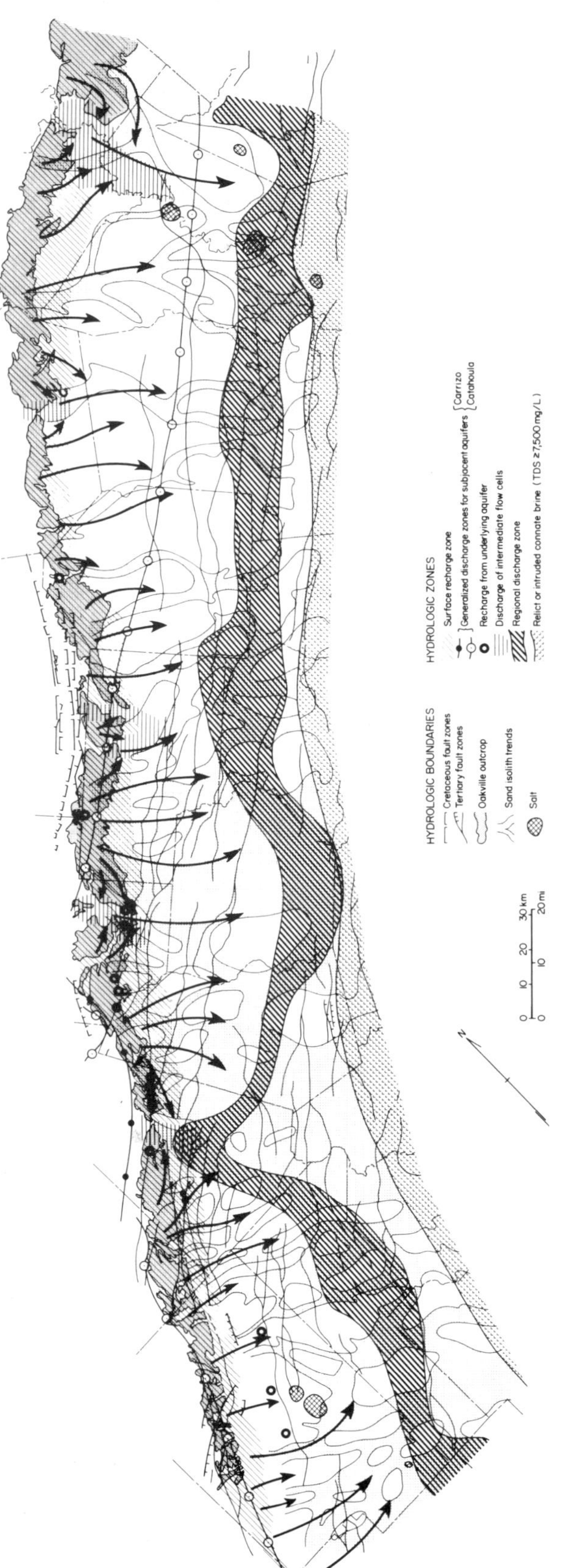

CONCLUSIONS

Interpretation and mapping of fluvial depositional systems and their component framework sand facies provide a description of the three-dimensional distribution of hydraulic parameters; the trend, geometry and spatial relationships of transmissive conduits; and insights into flow boundaries and patterns extant during the syndepositional flow. Syndepositional topography and surface-water flow patterns condition flow geometry in active systems. Post-depositional structural and topographic evolution of the basin determine the changing boundary and head conditions that must be inferred to interpret later flow patterns in increasingly confined alluvial aquifers. The distributions of natural and introduced solutes and of alteration zones reflect this historical interaction between the physical geology and the evolving fluid flow systems.

REFERENCES

ANDERSON, T.W., 1986, Study in southern and central Arizona and parts of adjacent states, *in* Sun, R.J., ed., Regional Aquifer System Analysis Program of the U.S. Geological Survey, Summary of Projects 1978-1984: U.S. Geological Survey Circular 1002, p. 107-115.

BACK, W., ROSENSHEIN, J.S., and SEABER, P.R., eds., 1988, Hydrogeology: Geological Society of America, The Geology of North America, 524 p.

BARTON, M.D., 1994, Outcrop characterization of architecture and permeability structure in fluvial-deltaic sandstones, Cretaceous Ferron Sandstone, Utah: Unpub. Ph.D. dissertation, University of Texas, Austin, 262 p.

BELITZ, K., and HEINES, F.J., 1990, Character and evolution of the groundwater flow system in the central part of the Western San Jaoquin Valley, California: U.S. Geological Survey Water-Supply Paper 2348, p. 7-17.

BRIDGE, J.S., and MACKEY, S.D., 1993, A theoretical study of fluvial sandstone body dimensions, *in* Bryant, I.D., and Flint, S.S., eds., The Geological Modeling of Hydrocarbon Reservoirs and Outcrop Analogues: International Association of Sedimentologists Special Publication 15, p. 213-236.

BROWN, T.J, and SHARP, J.M., JR., 1992, A model for the effects of point-source emissions of aerosols on groundwater systems: Journal of Applied Hydrogeology, v. 3, p. 33-46.

BRYANT, I.D., and FLINT, S.S., 1993, Quantitative clastic reservoir geological modeling—Problems and perspectives, *in* Bryant, I.D., and Flint, S.S., eds., The Geological Modeling of Hydrocarbon Reservoirs and Outcrop Analogues: International Association of Sedimentologists Special Publication 15, p. 3-20.

CHANDLER, M.A., KOCUREK, G., GOGGIN, D.J., and LAKE, L.W., 1989, Effects of stratigraphic heterogeneity on permeability in eolian sand-

FIG. 11 (left).—Complex pattern of modern groundwater flow in the Oakville Sandstone, a confined coastal plain aquifer system. Regional hydraulic gradient is to the southwest. Fluvial channel axes, cross-cutting faults and modern outcrop topography all influence flow patterns (from Galloway et al., 1982a).

stone sequence, Page Sandstone, northern Arizona: American Association of Petroleum Geologists Bulletin, v. 73, p. 658-668.

Cowan, E.J., 1991, The large-scale architecture of the fluvial Westwater Canyon Member, Morrison formation (upper Jurassic), San Juan Basin, New Mexico, *in* Miall, A.D., and Tyler, N., eds., The Three-Dimensional Facies Architecture of Terrigenous Clastic Sediments and its Implications for Hydrocarbon Discovery and Recovery: Society of Sedimentary Geology, Concepts in Sedimentology and Paleontology, v. 3, p. 80-93.

Custodio, E., 1987, Salt-fresh water interrelationships under natural conditions, *in* Custodio, E., and Bruggeman, G.A., eds., Groundwater Problems in Coastal Areas: Geneva, UNESCO, p. 14-96.

Darling, B.K., Hibbs, B.J., Dutton, A.R., and Sharp, J.M., Jr., 1995, Isotope hydrology of the Eagle Mountains area, Hudspeth County, Texas—Implications for development of ground-water resources, *in* Hotchkiss, W.R., Downey, J.S., Gutentag, E.D., and Moore, J.E., eds., Water Resources at Risk: Minneapolis, Minn., American Institute of Hydrology, p. SL12-SL24.

Davies, D.K., Williams, B.P.J., and Vessell, R.K., 1993, Dimensions and quality of reservoirs in low and high sinuosity channel systems, Lower Cretaceous Travis Peak formation, east Texas, USA, *in* North, C.P., and Prosser, D.J., eds., Characterization of Fluvial and Aeolian Reservoirs: Geological Society of America Special Publication 73, p. 95-121.

Davis, G.H., 1988, Western alluvial valleys and the High Plains, *in* Back, W., Rosenshein, J.S., and Seaber, P.W., eds., Hydrogeology: Geological Society of America, The Geology of North America, v. O-2, p. 283-300..

Davis, G.H., Green, J.H., Olmstead, F.W., and Brown, D.W., 1959, Groundwater conditions and storage capacity in the San Joaquin Valley, California: U.S. Geological Survey Water-Supply Paper 1468.

Deveral, S.J., and Gallanthine, S.K., 1989, Relation of salinity and selenium in shallow groundwater to hydrologic and geochemical processes, western San Jaoquin Valley, California: Journal of Hydrology, v. 109, p. 125-149.

Dreyer, T., 1990, Sand body dimensions and infill sequences of stable, humid climate delta plain channels, *in* Boller, A.T., Berg, E., Hjelmeland, O., Kleppe, J., Torsaeter, O., and Aasen, J.O., eds., North Sea Oil and Gas Reservoirs II: Norwegian Institute of Technology, Graham and Trotman, p. 337-351.

Dreyer, T., 1993, Geometry and facies of large-scale flow units in fluvial-dominated fan-delta-front sequences, *in* Ashton, M., ed., Advances in Reservoir Geology: Geological Society of America Special Publication 69, p. 135-174.

Dreyer, T., Scheie, A., and Walderhaug, O., 1990, Minipermeameter based study of permeability trends in channel sand bodies: American Association of Petroleum Geologists Bulletin, v. 74, p. 359-374.

Duffy, C.J., and Al-Hassan, S., 1988, Groundwater circulation in a closed desert basin: topographic scaling and climatic forcing: Water Resources Research, v. 24, p. 1675-1688.

Engelen, G.B., and Jones, G.P., 1986, Developments in the analysis of groundwater flow systems: International Association of Hydrological Sciences Bulletin, v. 163, p. 67-106.

Fielding, C.R., and Crane, R.C., 1987, An application of statistical modeling to the prediction of hydrocarbon recovery factors in fluvial reservoir sequences, *in* Etheridge, F.G., Flores, R.M., and Harvey, M.D., eds., Recent Developments in Fluvial Sedimentology: Society of Economic Paleontologists and Mineralogists Special Publication 39, p. 321-327.

Fogg, G.E., 1986, Groundwater flow and sand body interconnectedness in a thick multiple-aquifer system: Water Resources Research, v. 22, p. 679-694.

Fogg, G.E., 1990, Architecture and interconnectedness of geologic media—Role of low-permeability facies in flow and transport, *in* Neuman, S.P., and Nevetniaks, I., eds., Hydrogeology of Low-Permeability Environments: Special Symposium of 28th International Geological Congress, Washington, D.C., p. 19-40.

Foreman, T.L., and Sharp, J.M., Jr., 1981, Hydraulic properties of a major alluvial aquifer—An isotropic, inhomogeneous system: Journal of Hydrology, v. 53, p. 247-258.

Foster, S.S.D., and Chilton, P.J., 1993, Groundwater systems in the humid tropics, *in* Bonnell, M., Hufschmidt, M.M., and Gladwell, J.S., eds., Hydrology and Water Management in the Humid Tropics: Cambridge, Cambridge University Press, p. 261-269.

Galloway, W.E., 1977, Catahoula Formation of the Texas Coastal Plain—Systems, composition, structural development, ground-water flow history, and uranium distribution: University of Texas, Bureau of Economic Geology Report of Investigations 87.

Galloway, W.E., 1981, Depositional architecture of Cenozoic Gulf Coastal Plain fluvial systems: Society of Economic Paleontologists and Mineralogists Special Publication 31, p. 127-155.

Galloway, W.E., 1982, Epigenetic zonation and fluid flow history of uranium-bearing fluvial aquifer systems, South Texas uranium province: University of Texas, Bureau of Economic Geology Report of Investigations 119.

Galloway, W.E., Henry, C.D., and Smith, G.E., 1982a, Depositional framework, hydrostratigraphy, and uranium mineralization of the Oakville Sandstone (Miocene), Texas Coastal Plain: University of Texas, Bureau of Economic Geology Report of Investigations 113.

Galloway, W.E., Hobday, D.K., and Magara, K., 1982b, Frio Formation of Texas Gulf Coastal Plain—Depositional systems, structural framework, and hydrocarbon origin, migration, distribution, and exploration potential: University of Texas, Bureau of Economic Geology Report of Investigations 122.

Galloway, W.E., and Hobday, D.K., 1996, Terrigenous Clastic Depositional Systems, 2d ed.: Berlin, Springer Verlag, 489 p.

Ghyben, W.B., 1899, Notes in verband met Voorgenomen Put boring Nabji Amsterdam, Tijdschr.: The Hague, Koninhitk. Inst. Ingrs.

Grannemann, N.G., and Sharp, J.M., Jr., 1979, Alluvial hydrogeology of the lower Missouri River Valley: Journal of Hydrology, v. 40, p. 85-99.

Hardie, L.A., Smoot, J.P., and Eugster, H.P., 1978, Saline lakes and their deposits—A sedimentological approach, *in* Matteran, A., and Tucker, M.E., eds., Modern and Ancient Lake Sediments: International Association of Sedimentologists Special Publication 2, p. 7-41.

Harrison, W.J., and Summa, L.L., 1991, Paleohydrology of the Gulf of Mexico Basin: American Journal of Science, v. 291, p. 109-176.

Hartkamp-Bakker, C.A., and Donselaar, M.E., 1993, Permeability patterns in point bar deposits, Tertiary Loranca Basin, central Spain, *in* Bryant, I.D., and Flint, S.S., eds., The Geological Modeling of Hydrocarbon Reservoirs and Outcrop Analogues: International Association of Sedimentologists Special Publication 15, p. 157-268.

Heath, R.C., 1982, Classification of ground-water systems of the United States: Ground Water, v. 20, p. 393-401.

Herzberg, B., 1901, Die Wasserversorgung einiger Nordseebaser: J. Gasbeleucht und Wasserversorgung, v. 44, p. 815-819.

Høimyr, O., Kleppe, A., and Nystuen, J.P., 1993, Effects of heterogeneities in a braided stream channel sandbody on the simulation of oil recovery—A case study from the lower Jurassic Stratfjord Formation, Snorre Field, North Sea, *in* Ashton, M., ed., Advances in Reservoir Geology: Geological Society of America Special Publication 69, p. 105-134.

JONES, A., DOYLE, J., JACOBSEN, T., and KJØNSVIK, D., 1995, Which subseismic heterogeneities influence waterflood performance—A case study of a low net-to-gross fluvial reservoir, *in* de Hann, H. J., ed., New Developments in Improved Oil Recovery: Geological Society of America Special Publication 84, p. 5-18.

JORDAN, D.W., and Pryor, W.A., 1992, Hierarchical levels of heterogeneity in a Mississippi river meander belt and application to reservoir systems: American Association of Petroleum Geologists Bulletin, v. 76, p. 1601-1624.

KrEITLER, C.W., AHKTER, M.S., and DONNELLY, A.C.A., 1990, Hydrochemical characterization of Texas Frio Formation water for deep-well injection of chemical wastes: Environmental Geology and Water Science, v. 16, p. 107-120.

LARKIN, R.G., and SHARP, J.M., JR., 1992, On the relationship between river basin geomorphology, aquifer hydraulics, and groundwater flow direction in alluvial aquifers: Geological Society of America Bulletin, v. 104, p. 1608-1620.

LOWRY, P., and RAHEIM, A., 1991, Characterization of delta front sandstones from a fluvial-dominated delta system, *in* Lake, L.W., Carroll, H.B., Jr., and Wesson, T.C., eds., Reservoir Characterization: London, Academic Press, p. 665-676.

MARTIN, J.H., 1993, A review of braided fluvial hydrocarbon reservoirs, *in* Best, J.L., and Bristow, C.S., eds., Braided Rivers: Geological Society of America Special Publication 75, p. 333-367.

MEEHAN, R.L., and SCHLEMON, R.J., 1994, The sequence stratigraphy of fluvial depositional systems—The role of floodplain sediment storage—A comment: Sedimentary Geology, v. 92, p. 287-288.

MEINZER, O.E., 1923, Outline of ground-water hydrology with definitions: U.S. Geological Survey Water-Supply Paper 494, 71 p.

MEISLER, H., LEAHY, P.P., and KNOBEL, L.L., 1984, Effect of eustatic sea-level changes on saltwater-freshwater in the northern Atlantic coastal plain: U.S. Geological Survey Water-Supply Paper 2255, p. 1-28.

MORTON, R.A., and LAND, L.S., 1987, Regional variations in formation water chemistry, Frio Formation (Oligocene), Texas Gulf Coast: American Association of Petroleum Geologists Bulletin, v. 71, p. 191-206.

NANSON, G.C., and CROKE, J.C., 1992, A genetic classification of floodplains: Geomorphology, v. 4, p. 459-486.

POTTER, P.E.,1978, Significance and origin of big rivers: Journal of Geology, v. 86, p. 3-33.

SCHICHT, R.J., and JONES, E.G., 1962, Ground-water levels and pumpage in East St. Louis area, Illinois: Illinois State Water Survey Report of Investigations 44, 39 p.

SCHUMM, S.A.,1977, The Fluvial System: New York, John Wiley & Sons.

SEABER, P.R.,1988, Hydrostratigraphic units, *in* Back, W., Rosenshein, J.S., and Seaber, P.W., eds., Hydrogeology: Geological Society of America, The Geology of North America, v. O-2, p. 9-14

SHARP, J.M., JR., 1988, Alluvial aquifers along major rivers, *in* Back, W., Rosenshein, J.S., and Seaber, P.R., eds., Hydrogeology: Geological Society of America, The Geology of North America, v. O-2, Ch. 35, p. 273-282.

SMITH, G.E., GALLOWAY, W.E., and HENRY, C.D., 1980, Effects of climatic, structural, and lithologic variables on regional hydrology within the Oakville aquifer of South Texas: Society of Mining Engineers of AIME, 4th Annual Uranium Seminar, p. 3-17.

U.S. ARMY CORPS OF ENGINEERS, 1976, American Bottoms Area—Preliminary Groundwater Analysis: U.S. Army Corps of Engineers, St. Louis, 30 p.

WALKER, T.R., 1967, Formation of red beds in ancient and modern deserts: Geological Society of America Bulletin, v. 78, p. 353-368.

WEBER, K.J., 1982, Influence of common sedimentary structures on fluid flow in reservoir models: Journal of Petroleum Technology, v. 34, p. 665-672.

WEBER, K.J., and VAN GEUNS, L.C., 1990, Framework for constructing clastic reservoir simulation models: Journal of Petroleum Technology, v. 42, p. 1248-1253, 1296-1297.

WILKINS, D.W., 1986, Southwest alluvial basin regional aquifer system, *in* Sun, R.J., ed., Regional Aquifer System Analysis Program of the U.S. Geological Survey, Summary of Projects 1978-1984: U.S. Geological Survey Circular 1002, p. 107-115.

SIMULATION OF GEOLOGIC PATTERNS: A COMPARISON OF STOCHASTIC SIMULATION TECHNIQUES FOR GROUNDWATER TRANSPORT MODELING

TIMOTHY D. SCHEIBE AND CHRISTOPHER J. MURRAY
Pacific Northwest National Laboratory, Richland, WA 99352

ABSTRACT: Stochastic models have been used extensively to represent uncertainty in the spatial distribution of aquifer properties and its impact on prediction of groundwater flow and transport behavior. Because natural porous media are often strongly heterogeneous and exhibit complex spatial structure, it is not possible for any model to completely characterize the spatial distribution of aquifer properties. Several models have been proposed, each of which relies on specific assumptions regarding the character of natural spatial structure. In this study we have performed a direct comparison of the performance of a number of stochastic simulation methods, using a geologically realistic synthetic dataset as the basis for such a comparison. The methods are evaluated in a Monte Carlo sense through comparison of the predicted distributions of a variety of measures of flow and transport behavior.

The results indicate that classical (second-order) stochastic models can be expected to provide biased and nonconservative predictions of many practical measures of flow and transport behavior and to underestimate the uncertainty in those predictions. These empirical results are supported by theoretical arguments based on statistical entropy. Models that preserve the spatial continuity of geologic facies to a greater degree provide better predictions of flow and transport behavior.

INTRODUCTION

Quantitative models of groundwater flow and solute transport attempt to predict the behavior of complex physical systems. At least two sources of uncertainty or error arise in the predictions obtained from such models:

- Parameter uncertainty: Natural heterogeneity, combined with limited measurement data, leads to incomplete knowledge of the spatial distribution of physical properties.
- Model error: Simplifying assumptions generally are required to reduce complex physical systems into tractable model formulations. To the degree that these assumptions do not adequately represent reality, error can be introduced into the model predictions.

Stochastic simulation methods are used to quantify the predictive uncertainty due to parameter uncertainty by generating many equiprobable "realizations" of the parameter space, each of which honors available data, and submitting each realization to a flow/transport model to obtain a possible prediction of system behavior. As a whole, the distribution of predictions so obtained quantitatively represents the predictive uncertainty arising from parameterization error.

However, the stochastic simulation approach does not quantify the uncertainty arising from model error. Model error can cause the predicted distribution of system behaviors to be inaccurate (biased) and/or to have the wrong degree of precision (i.e., estimated uncertainty). Sources of model error include assumptions made during:

- model formulation (e.g., representation of a discrete system of pores and grains as a continuum),

- data interpretation (e.g., assumed volume of influence of a pumping test), and
- model solution (e.g., discrete numerical approximation of differential equations).

In this study we focus on one important source of model error and attempt to evaluate objectively its influence on predictions of groundwater flow and transport behavior. That source of model error is assumptions regarding the character of heterogeneity in natural porous media. Natural porous media are, in most cases, strongly heterogeneous and highly structured. This structure generally is complex, exists at multiple scales and has a variety of characteristics. An essential step in stochastic simulation models is the selection of a quantitative model of spatial structure. Such a model necessarily is a simplification of the true (complex) structure and therefore will introduce some model error into the predictions obtained. It is important to note that the quantification of parameterization error in stochastic models depends on the assumptions underlying the model of parameter uncertainty and therefore is impacted by model error. That is, the uncertainty that is quantified is that "given the model" and does not include uncertainty related to potential errors in the model itself.

METHODOLOGY

The objective of this research is to demonstrate the impacts of model assumptions regarding the character of natural heterogeneity on predicted distributions of flow and transport behaviors.

To address this objective, a set of "numerical experiments" was performed using a fully known, highly resolved synthetic hydraulic conductivity dataset (a "numerical aquifer") as the baseline for comparison. In each numerical experiment, a number of cross sections from the numerical aquifer were analyzed (using complete information) to parameterize a selected stochastic simulation model. The simulation model then was used to generate a number of unconditional realizations. These were used as input to a two-dimensional groundwater flow and solute-transport model, and predictions of various system behaviors were obtained. The distribution of such predictions over all realizations of the simulation model were compared with the distribution of predictions over the different cross sections of the numerical aquifer. The remainder of this section discusses some of these points in further detail.

A Numerical Aquifer

The baseline selected for comparison of the various stochastic simulation models is a synthetic model of hydraulic conductivity in sandy point-bar sediments developed by Scheibe (1993). This model was developed using a geometric simulation method incorporating sedimentological information as described by Scheibe and Freyberg (1995). It represents spatial variations of properties in terms of geometries of discrete elements, rather than in terms of the continuous two-point correlation functions more widely used in hydrologic modeling. (See Haldorsen and Damsleth [1990] for a discussion of discrete and continuous approaches to stochastic simulation.) The discrete elements represent geological features and processes that were reported in detailed sedimentological field studies by Basumallick (1966), Pryor (1973) and Jackson (1976). The principal features of the numerical aquifer are:

- *trough cross-beds* deposited by subaqueous dune migration;
- *scroll bar deposits* primarily composed of fan-shaped packets of sediment oriented nearly perpendicular to the main channel; and
- a *mud drape* deposited during a period of point bar inactivity by receding flood waters.

These features are defined at a high degree of resolution and at a fairly small scale. The overall dimensions of the cross sections considered in this study are approximately 4.6 by 1.8 m, resolved on a 1,200–by–480 grid. Figure 1 shows one such cross section. Discrete features similar to those occurring in the numerical aquifer can be observed in sedimentary deposits at multiple scales ranging from millimeters (e.g., cross-lamination) to hundreds of meters (e.g., channel sand bodies).

While it is our opinion that the numerical aquifer is a realistic representation of the real system it models, it is nevertheless recognized that it is a simplified model of reality and not true reality. Accordingly, the comparisons drawn here should be viewed as being between continuous and discrete statistical models of spatial structure, not between continuous statistical models and reality.

Use of a synthetic numerical aquifer, as compared to actual field data, offers several advantages. First, the synthetic system is fully known at a high level of detail, whereas even the most detailed field characterizations to date cannot be considered exhaustive. Second, the numerical aquifer allows characterization of flow and transport behavior of several "realizations" of the assumed baseline system, in that several different cross sections of the numerical aquifer were selected so as to be essentially uncorrelated. Third,

FIG. 1.—Spatial distribution of permeability in the cross section of the numerical aquifer used as exhaustive data for model estimation. Light regions are high permeability (maximum = 367 Darcy); dark regions are low permeability (minimum = 1.9 Darcy).

because complete information is available from the numerical aquifer for estimation of stochastic model parameters (such as variograms), any differences in model outcomes can be attributed to differences in the models themselves and not to uncertainty in the model parameters.

Stochastic Simulation Methods

Three different stochastic simulation methods were applied to generate unconditional simulations of the structure observed in the numerical aquifer:

- Sequential Gaussian Simulation,
- Sequential Indicator Simulation,
- 2D Markov Chain Simulation.

All three assume a continuous random field model of spatial structure, each in somewhat different fashion.

Sequential Gaussian Simulation (SGS) (see Deutsch and Journel, 1992) assumes a Gaussian or Multinormal spatial distribution of hydraulic conductivity. This distribution is completely characterized by its mean and covariance (or variogram) function. Given a particular mean and covariance function, the Gaussian model is the maximum-entropy representation of the spatial field. Its tendency to diminish the continuity of ordered spatial patterns and, in particular, the continuity of extreme-valued parameters, has been noted by Journel and Alabert (1989). In addition to its potential application in Monte Carlo studies, the Gaussian model is assumed by many analytical stochastic models of groundwater flow and transport (e.g., Gelhar and Axness, 1983; Dagan, 1984; Neuman et al., 1987). The GSLIB software package (Deutsch and Journel, 1992) was used in this study to implement the Sequential Gaussian Simulation method.

Sequential Indicator Simulation (SIS) is similar to SGS in many respects. The principal difference is that SIS employs indicator cutoffs (thresholds) to define classes of parameter values. In doing so, it provides the capability to define separate models of spatial continuity (covariance functions) for each class, as opposed to one lumped model for all classes. It also allows specification of a complete nonparametric conditional distribution function at each simulated location, as opposed to the mean and variance of an assumed Gaussian distribution. Consequently, SIS is better able to represent the continuity of extreme-valued parameters (Journel and Alabert, 1989, 1990) and avoids the maximum-entropy assumption inherent to SGS. Sequential Indicator Simulation has been used by several investigators to simulate subsurface spatial distributions of facies, rock types or permeability classes (e.g., Desbarats, 1987; Journel and Gomez-Hernandez, 1989; Johnson and Dreiss, 1989; Davis et al., 1993; Murray, 1995; Poeter and Townsend, 1994). The method is implemented here using GSLIB software.

2D Markov Chain Simulation represents the spatial field in terms of subwindow configuration frequencies of defined parameter classes. For example, assume that hydraulic conductivity values are grouped into two classes (high and low values). Consider 2 by 2 square subwindows in an exhaustive uniform grid; 16 possible configurations exist, as shown in Figure 2. In general, G^{nm} possible configurations exist, where G is the number of classes, and n and m are the subwindow dimensions. Knowing the relative frequency of each configuration allows specification of 2D Markovian transition probabilities using the relationships of Monne et al. (1981), who also described and demonstrated a method of generating realizations of a 2D random field characterized by these transition probabilities. Although this method is not well known in the hydrologic literature, it has been applied to characterize and generate 2D spatial patterns in the image analysis and pattern recognition field. A loosely related and more sophisticated method, based on indicator transition probabilities, was applied to simulation of alluvial sediment sequences by Carle and Fogg (1996); however, they used a Markov chain to model coregionalization of geologic units in an SIS procedure employing cokriging (see Carle et al., this volume). The key difference between the Markov chain model and the SGS and SIS models is that it incorporates statistics involving n by m points and is therefore multivariate; the other methods involve statistics only of pairs of points and thus are bivariate. A multivariate approach provides potential to characterize more complex geometries (Deutsch, 1992; Guardiano and Srivastava, 1993). The method used in this study was implemented using software developed by Scheibe (1993).

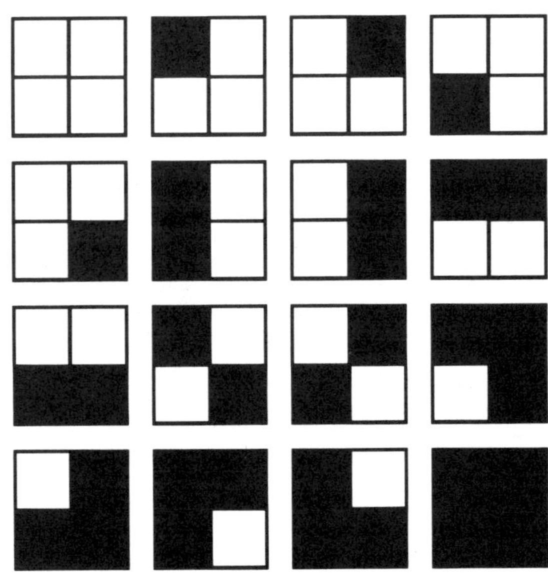

FIG. 2.—Sixteen possible subcell configurations for the case of n = m = 2 and two parameter classes (black and white).

All three of these approaches have been applied in a somewhat naive manner; it would be possible to fine-tune the applications in any number of ways. Each method is really a surrogate for a whole class of methods in which parameterizations and assumptions can be modified to fit specific problems. As such, the comparisons drawn are not so much reflective of the general usefulness of the various methods themselves as of the errors introduced by straightforward ("quick-and-dirty") application of the methods, such as might be undertaken in hydrogeological practice. The goal of this exercise is not to establish which methods are better in a general sense, but rather to demonstrate some of their limitations and define conditions under which they are more or less applicable.

Flow and Transport Modeling

Fifty realizations generated by each of the stochastic simulation models described above were input to a numerical model of groundwater flow and transport. Similarly, twelve cross sections of the numerical aquifer were input to the flow and transport model. These cross sections were selected from locations separated by approximately 0.4 m, greater than the correlation length of 0.26 m, so as to be essentially uncorrelated. The exact locations of the cross sections were selected using a randomized algorithm.

The flow problem was solved using a finite-difference model with a grid size of 1,200 by 480; a conjugate-gradient iterative method was employed to solve the resulting linear system of equations. Once velocities were defined for each grid node, a purely advective, mixed numerical-analytical particle-tracking method (similar to that of Pollock, 1988) was used to compute breakthrough curves. Although local dispersion can significantly affect transport behavior, it was neglected here since the aquifer consists of fairly high-conductivity sediments such that transport is dominated by advection. Previous simulations reported by Scheibe (1993) demonstrate the validity of this assumption in the modeled system. Simple boundary and initial conditions were used: no-flow conditions on top and bottom; specified head conditions imposing flow from right to left under a uniform macroscopic gradient of 0.001. Ten thousand particles were released in an instantaneous vertical line source, distributed vertically in proportion to local water flux rates. These conditions simulate a fully mixed, constant-concentration injection well with a relatively short injection time.

Metrics of Flow and Transport Behavior

Flow and transport model results were summarized, and comparisons drawn, in terms of specific metrics of the flow and transport behavior. The metrics used were:

- total water flux through the cross section;
- early arrival time measured by the 5% particle-arrival quantile;
- median arrival time (50% quantile);
- late arrival time (95% quantile); and
- apparent (effective) longitudinal dispersivity.

These metrics are mostly self-explanatory. Longitudinal dispersivity was estimated from the variance of particle locations at a specified elapsed time using the following equation (Bear, 1979):

$$\alpha_l = \frac{\sigma_x^2}{2\overline{v}t} \tag{1}$$

Where σ_x^2 is the variance of particle locations in the direction of flow, $\overline{v}$ is mean particle velocity and t is the time elapsed since release of an instantaneous pulse of particles. Transverse dispersivity was not estimated because the applied boundary conditions limit transverse spreading.

Related Studies

The current work is related to several previous studies, most closely to the work of Journel and Deutsch (1993), who compared SGS and SIS simulations of a numerical aquifer in terms of predicted flow and transport behaviors. Our work confirms their conclusions and considers a third simulation method (Markov simulation) they did not consider. The numerical aquifer used by Journel and Deutsch (1993) was developed by digitizing an outcrop photograph, whereas the numerical aquifer used here was developed based on detailed sedimentological observations and permeability measurements. Journel and Alabert (1989) compared SGS and SIS simulations for detailed permeability measurements on Berea Sandstone but did not perform flow simulations. Eggleston et al. (1996) performed a comparison of three geostatistical methods (kriging, SGS and simulated annealing) based on field observations at Cape Cod; however, all three methods, as applied by these authors, are based on an underlying assumption of a multi-Gaussian random field. Furthermore, these authors performed comparisons in terms of estimation error and visual appearance but did not perform flow simulations.

MODEL PARAMETERIZATION

Model parameters for all three stochastic simulation methods were determined from a cross section of the numerical aquifer (Fig. 1). The cross section has 1,200 columns and 480 rows, for a total of 576,000 data points, with a "measurement" of permeability available at each point. Figure 3 is a histogram of the permeability data for the cross section. It can be seen that the permeability data are roughly bimodal, with a split between two subpopulations at about 120 D. The mean and median of the lower peak are 42.1 and 32.9 D, respectively. The upper peak has a mean and median of 201.9 and 203.1 D. The full data set has a mean of 98 D and a median of 60.0 D.

Sequential Gaussian Simulation

Sequential Gaussian Simulation assumes that the variable of interest is characterized by a multivariate Gaussian spatial distribution (Deutsch and Journel, 1992). A graphical normal score transform was applied to the permeability data using the GSLIB program NSCORE.F. Although the resulting normal score data have a univariate Gaussian distribution, the transformation does not necessarily transform the data to a multivariate Gaussian distribution. The presence of subpopulations in the permeability frequency distribution (Fig. 3) suggests that the numerical aquifer is not multivariate Gaussian, and examination of the indicator variograms (see below) tends to confirm that suggestion. However, SGS is a well-known and relatively simple algorithm often applied to random fields that do not fit the underlying assumptions, either because of a lack of understanding or because the ease of application and speed of execution of the algorithm outweigh other considerations. The multivariate Gaussian assumption that is the basis of SGS also is the basis of most analytical stochastic theories of groundwater flow and transport.

The variogram of the normal scores was calculated using the GSLIB subroutine GAM2.F along two principal directions of continuity, with the major direction approximately equal to a dip of 10° to the right (see Fig. 1) and the direction of minimum continuity perpendicular to that direction. The selection of a 10° dip to the right as the direction of maximum continuity was based on analysis of exhaustive covariance maps prepared using spectral methods. The variograms were fit using a model consisting of a nugget and the sum of two exponential functions (Isaaks and Srivastava, 1989):

$$\gamma(h) = 0.07 + 0.79 Exp_{28}(h) + 0.14 Exp_{200}(h) \qquad (2)$$

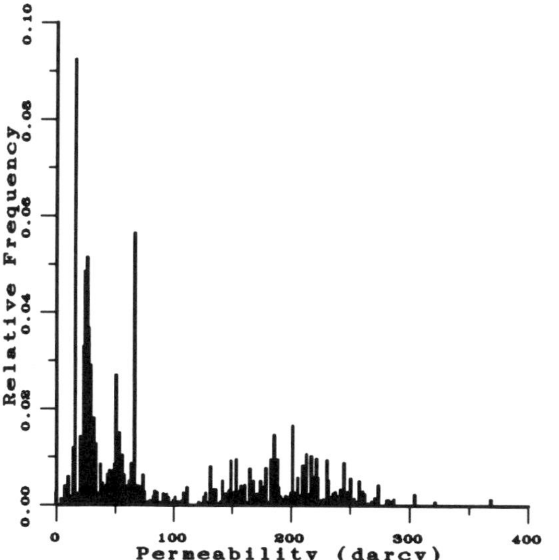

FIG. 3.—Univariate permeability distribution for the selected cross section of the numerical aquifer.

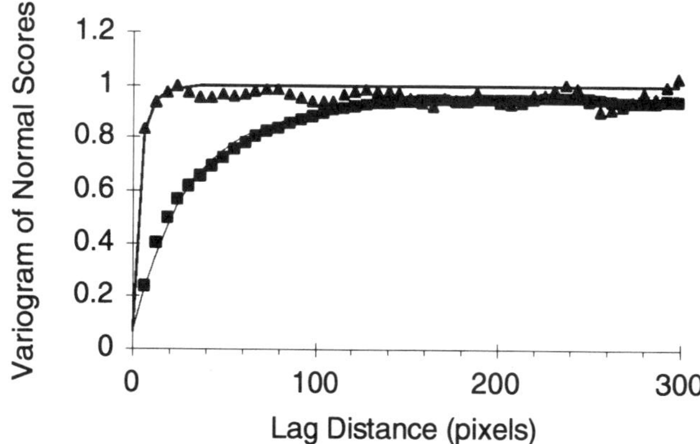

FIG. 4.—Experimental and model variograms of the normal scores of permeability values from a cross section of the numerical aquifer. Squares represent experimental variogram values for a search direction dipping 10° to the right; triangles represent the perpendicular search direction. Solid lines represent the model fit to the experimental variograms; 100 pixels represent 38.1 cm.

where the nugget is 0.07, the sill of the first exponential function is 0.79 with a range of 28 pixels (10.7 cm), and the sill of the second exponential function is 0.14 with a range of 200 pixels (76.2 cm). The anisotropy ratio between the maximum and minimum directions of continuity for the first exponential structure is 9.33 and for the second structure is 20 (Fig. 4).

Sequential Indicator Simulation

Sequential Indicator Simulation can be performed in two modes (Deutsch and Journel, 1992), based on categorical indicators (e.g., facies type) or on indicator transformations of a continuous variable, which was the method employed in this study. The permeability distribution (Fig. 3) was discretized into eight classes by seven indicator cutoffs. The basic scheme was to use one cutoff (120 D) to separate the two populations of the bimodal distribution, then discretize each mode into four classes using the quartiles of the data in that mode. The classes identified by this procedure appeared to provide a reasonable discretization of the permeability data and were closely related to the distribution and geometry of the "sedimentary structures" of the numerical aquifer (see Fig. 5). Exhaustive covariance maps of each indicator class were prepared and then used to guide the interpretation and modeling of the indicator variograms.

One difficulty of which we were aware but was fully exposed by the exhaustive covariance maps is that there are several preferred orientations in the permeability classes. The lowest indicator class has two primary directions of continuity that dip approximately 10° from the horizontal to the left and right (Fig. 6). This is related to the presence of low permeability values on both the stoss and avalanche surfaces of the scroll bars. By the second cutoff, and for most succeeding cutoffs, the continuity of the avalanche face dominates, dipping at about 10° to the right from the horizontal. An additional complexity is the presence of the horizontal clay drape, which imparts a secondary horizontal direction of continuity, especially for the fourth indicator cutoff. In modeling the indicator variograms using GSLIB,

CUTOFF 2: 32.9 D

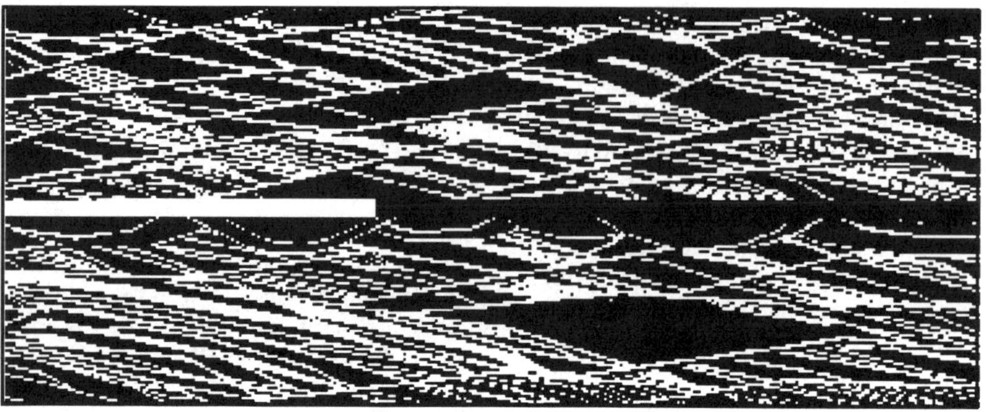

CUTOFF 4: 120.0 D

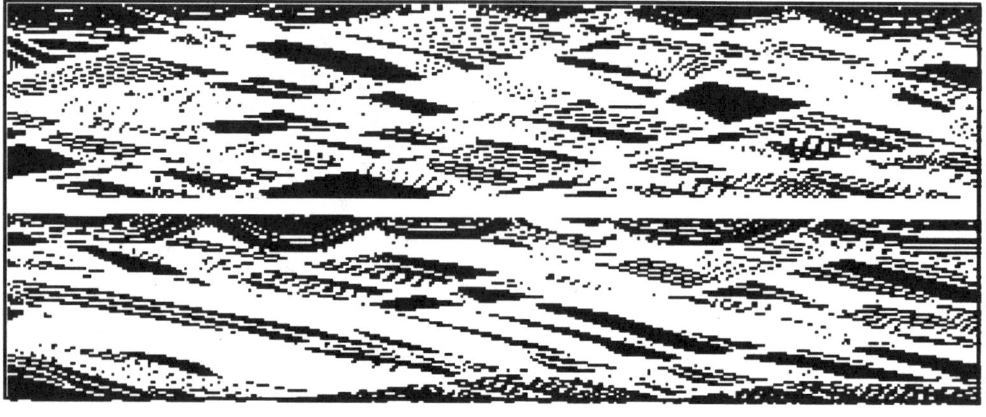

CUTOFF 6: 203.1 D

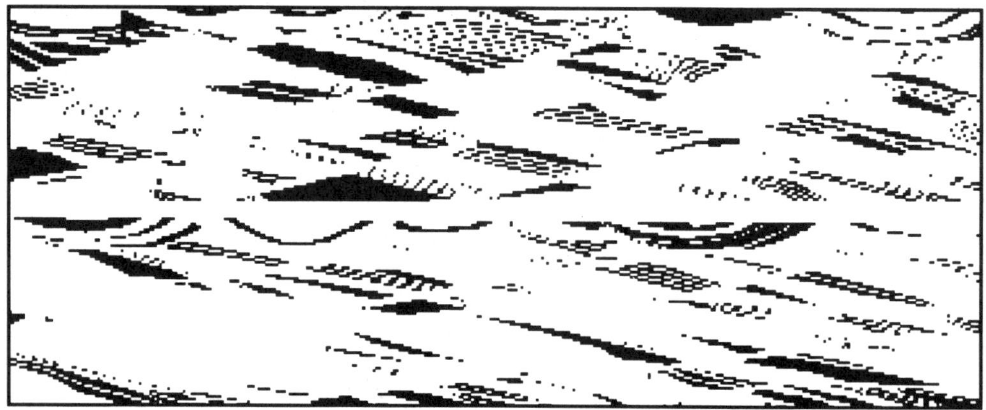

FIG. 5.—Binary images showing the structure at three of the cutoffs used. White regions are below the indicated cutoff permeability value (D = darcy); black are above.

it is not possible to honor nonorthogonal directions of high continuity; so, a decision was made to model the spatial continuity of the avalanche face only for all cutoffs. This does not permit reproduction of the highly continuous clay drape in the sequential indicator simulations (the same problem also exists, of course, for the sequential Gaussian simulations).

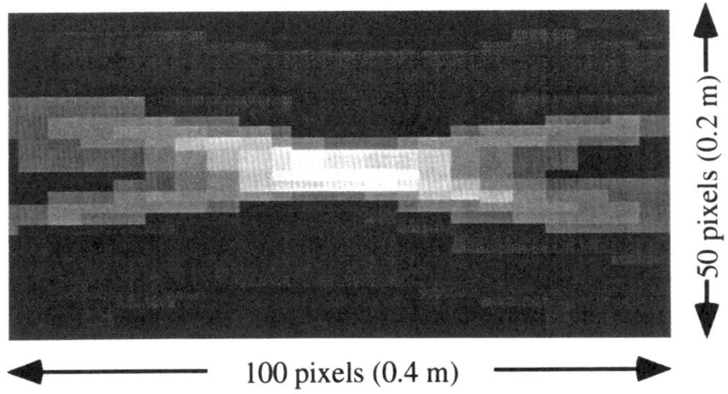

FIG 6.—Center portion of the exhaustive 2D covariance map of the lowest indicator class, showing two primary directions of continuity at approximately 10° to the right and the left. Light tones indicate high correlation; dark tones indicate low correlation.

The indicator variograms were all modeled using a nugget and two exponential structures. During this modeling, an effort was made to avoid drastic jumps in model properties from one indicator cutoff to the next (Deutsch and Journel, 1992), especially with respect to the relative size of the nugget and the anisotropy ratio. The experimental variograms and fitted models for the indicator variograms are portrayed in Figure 7. The indicator classes have a monotonic increase in variogram range so that the higher permeability classes have greater continuity than the lower classes. The trend of increasing spatial continuity for successively higher permeability classes cannot be preserved with a Gaussian model since that model entails greatest continuity for the center of the distribution, with decreasing continuity of the tails.

2D Markov Chain Simulation

Markov chain simulation requires specification of subcell occurrence frequencies of the spatial field. These frequencies were estimated for the training image simply by counting occurrences. To limit storage and computational requirements, hydraulic conductivity was first assigned into four discrete classes (G = 4), and only 2–by–2 (n = 2, m = 2) subcells were considered. Therefore, the total number of possible subcell configurations is $4^4 = 256$. The discretization was performed using an image color-compression algorithm coded in the "v1bgamut" routine of the KHOROS software package (Rasure et al., 1990). This algorithm employs clustering methods to group the continuous data into a specified number of bins (four in this case) in such a manner as to maintain

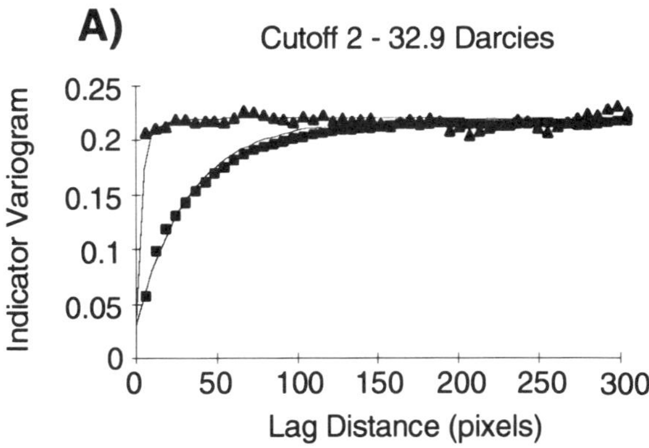

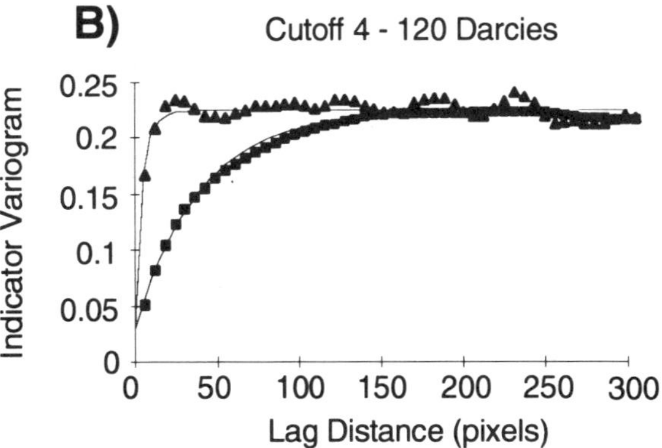

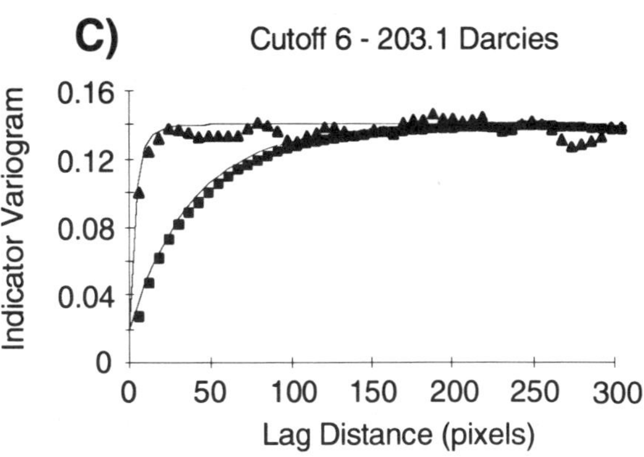

FIG 7.—Experimental and model indicator variograms of the second, fourth and sixth cutoffs of permeability values from a cross section of the numerical aquifer. Squares represent experimental variogram values for a search direction dipping 10° to the right; triangles represent the perpendicular search direction. Solid lines represent the model fit to the experimental variograms; 100 pixels represent 38.1 cm.

the appearance of the image to the greatest degree possible. While the classes so defined are not explicitly related to specific geologic features, the clustering of the histogram tends to lump similar features into the same class. Summary statistics for the classes are shown in Table 1.

TABLE 1.—SUMMARY STATISTICS FOR THE PERMEABILITY (IN DARCY UNITS) OF THE FOUR DISCRETE CLASSES USED IN THE MARKOV CHAIN SIMULATION METHOD

Class	Mean	Standard Deviation
1	40.0838	19.228
2	152.094	24.1141
3	222.312	23.3425
4	304.754	27.9636

The Markov chain method used here is a categorical method in that it generates realizations containing only the specified number of discrete classes (four in this case). Since we need a continuous distribution of permeability for the flow and transport simulations, a second step is necessary to assign a continuous range of permeability values to the spatial distribution. This step can be thought of as the inverse of the clustering that was performed on the initial image to group similar permeability values into discrete classes. This step was performed first by generating several realizations using the SGS method, as described above, but without performing the back-transformation to permeability (e.g., leaving the realizations in standard normal form, with a mean of zero and variance of one). For each Markov chain realization, four SGS realizations were selected at random, one for each class. Then each SGS realization was mapped into those regions of the Markov chain realization containing the corresponding class, transforming the standard normal variates into normal variates with the mean and standard deviations computed from the original image and shown in Table 1. Because four independent SGS realizations were used for each Markov realization, the local permeability variations exhibit spatial correlation within each of the four discrete classes but are uncorrelated across class boundaries. This is effectively a hybrid simulation method (Haldorsen and Damsleth, 1990), wherein the large-scale pattern is described using a discrete method (Markov) and the local variability is described using a continuous method (SGS).

RESULTS AND DISCUSSION

Visual Comparison

The images in Figure 8 visually represent the pattern of hydraulic conductivity in a cross section of the numerical aquifer and one realization of the fields created using each of the three stochastic simulation approaches. In these images the gray scale represents hydraulic conductivity, with light tones representing high conductivity and dark tones representing low conductivity.

Visual comparison of the cross section from the numerical aquifer and the three simulated cross sections clearly demonstrates the significance of the differences among the various model approaches. Two conclusions are evident:

1. The SGS model significantly decreases the continuity of the spatial field, as compared to both the baseline numerical aquifer and the other two models considered (SIS and Markov).
2. All three simulation models are unable to represent the multiple scales and types of structure exhibited by the numerical aquifer. This is not surprising since features such as the mud drape (continuous horizontal feature near the middle of the numerical aquifer cross section) are nonstationary, whereas all three models assume stationarity of model statistics.

The images in Figure 9 show the position of particles after the same amount of elapsed time in an example realization of each simulation method and one cross section of the numerical aquifer. While the center of mass of the particles is approximately the same in all four cases, the degree and character of spreading about that center markedly differ. The effects of reduced continuity in the SGS model is evidenced by the clear reduction in preferential pathway flow, as compared to the numerical aquifer. While the SIS and Markov models do not preserve the full degree of continuity exhibited by the numerical aquifer, they do have significantly greater preferential flow than the SGS model.

Distribution of Flow and Transport Metrics

The distributions of metrics of predicted flow and transport behavior are presented as box plots in Figure 10. All distributions have been normalized to the mean of the distribution of the numerical aquifer cross sections.

The plots show that the clear visual differences among the models translate into significant differences in predicted behavior. However, the magnitudes of the differences depend on the particular behavior of interest. In the case of global flux (which is also representative of effective hydraulic conductivity), all models gave very similar predictions, although the SGS model exhibited a smaller degree of variability in predicted outcomes than the other three. The results from all models also were reasonably similar for median and late arrival time metrics, although the SGS model again generally was most different and tended to diminish the outcome variability. However, early arrival times were overpredicted (i.e., predicted to be later than in the numerical aquifer) by all three models; dispersivities were underpredicted by all three models. The SGS model had the worst performance in comparison to the numerical aquifer,

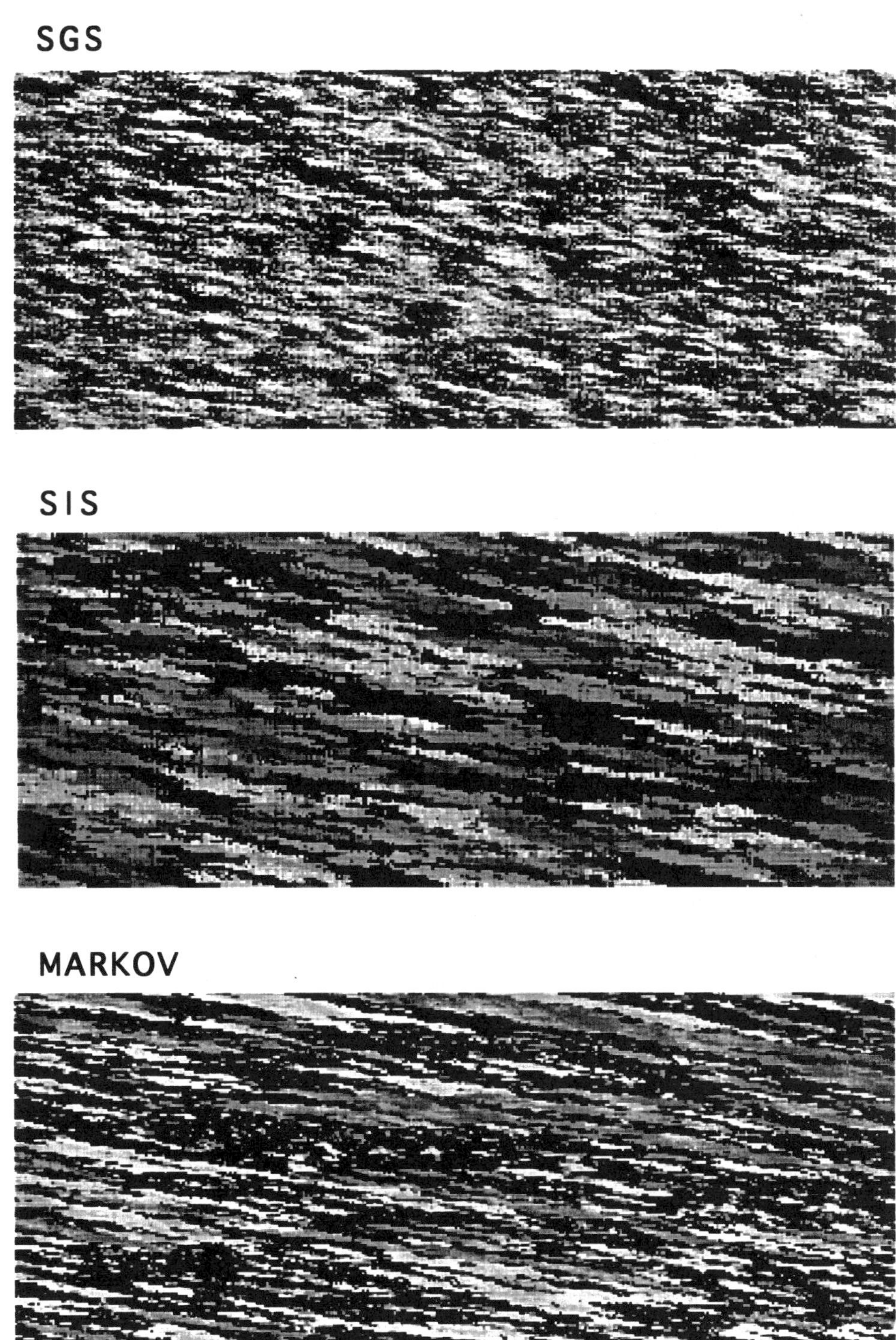

FIG. 8.—Permeability maps of one realization of each of the three stochastic simulation methods, for visual comparison to Figure 1.

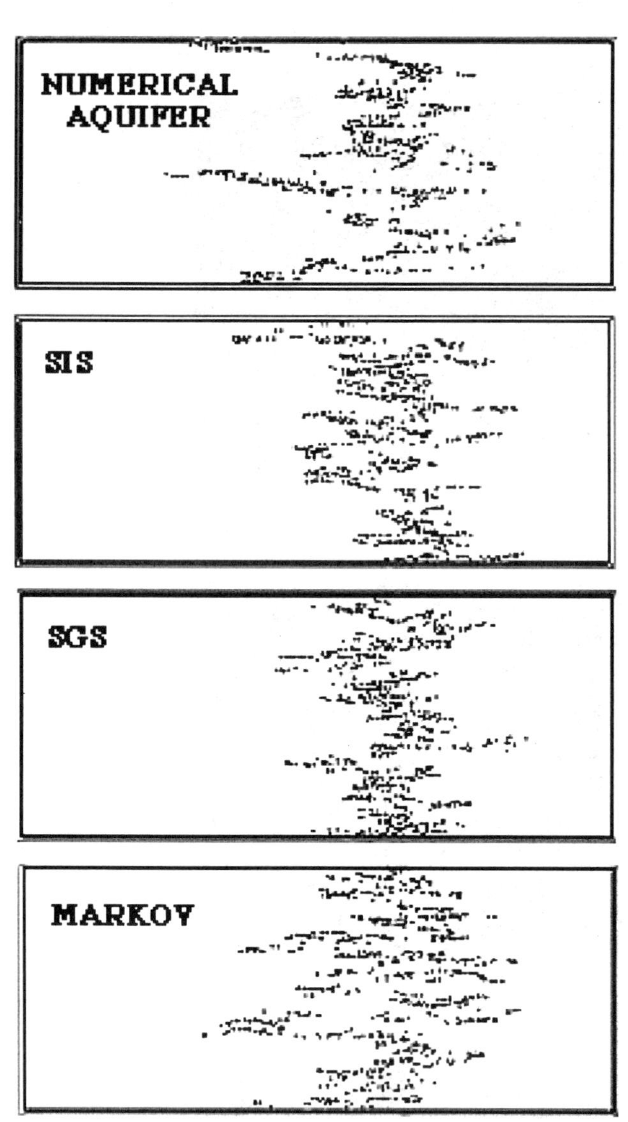

FIG 9.—Snapshots of particle locations at the same time in one realization of each of the three simulation methods and in one cross section of the numerical aquifer. Particles are moving from right to left.

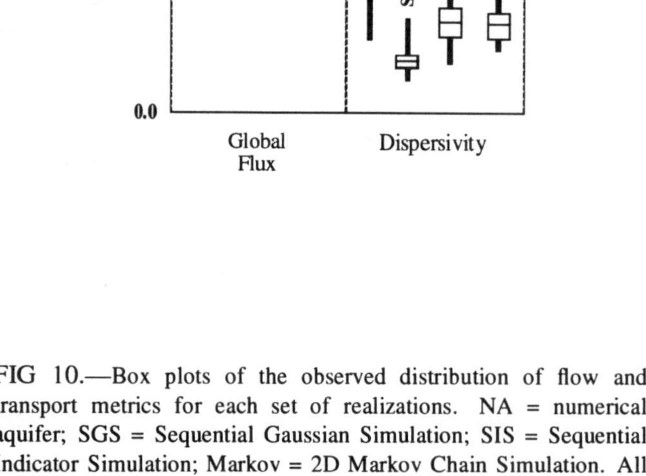

FIG 10.—Box plots of the observed distribution of flow and transport metrics for each set of realizations. NA = numerical aquifer; SGS = Sequential Gaussian Simulation; SIS = Sequential Indicator Simulation; Markov = 2D Markov Chain Simulation. All distributions have been normalized to the mean of the distribution observed in the cross sections of the numerical aquifer.

and the SIS and Markov models were approximately equal to one another.

One conclusion that can be drawn is that certain model predictions (bulk flow rates, mean arrival times) are less sensitive than others (extreme arrival times and dispersivities) to the model of spatial structure selected. A second conclusion is that continuous statistical models (and in particular SGS) tend to provide nonconservative estimates of sensitive model behaviors in cases where highly structured heterogeneity exists (such as in the numerical aquifer). We term these estimates "nonconservative" because of typical applications of groundwater models; overestimating early arrival times (predicting them later than actual) could have significant consequences for subsurface waste disposal facilities, and underestimating late arrivals often plagues "pump-and-treat" remediation designs.

Spatial Entropy Considerations

Recently, investigators have applied the principle of maximum entropy to groundwater parameter-estimation problems (Kitanidis, 1990; Woodbury and Ulrych, 1993). This principle states that, when faced with characterizing a probabilistic distribution under uncertainty, one should choose the distribution that maximizes the entropy given any constraining information. The idea behind the principle is that in this manner the estimates so obtained represent the maximum amount of uncertainty (are conservative), given the available information, and do not inject extraneous information (arbitrary assumptions) into the estimation process. Many stochastic groundwater problems are approached with the assumption that the parameter of interest (e.g., hydraulic conductivity) is normally (or log-normally) distributed, and that the mean, variance and covariance of the distribution are estimable from data. Under such conditions, the distribution that maximizes the entropy *of the spatial random field* is the Gaussian or Multinormal distribution, as used by SGS.

However, the results of our work confirm that, although the SGS method leads to maximum entropy in the parameter space, it tends to give predictions that have low entropy of the distribution of predicted outcomes (see also Journel and Deutsch [1993], who consider this issue in greater detail). In most model applications, the model prediction is of primary concern, and the parameterization is simply a means to achieve a prediction. In such a case, the principle of maximum entropy is not invalidated — it simply ought to be applied at the relevant decision point (the model outcomes) and not at some intermediate point (the model parameterization). In fact, models of spatial variability that have lower entropy than the Gaussian model lead to greater entropy in the predicted outcomes in many cases. Note that spatial entropy is in many ways the antithesis of spatial structure; choosing a model that maximizes spatial entropy (i.e., SGS) is equivalent to arbitrarily minimizing any structure beyond that explicitly characterized by the covariance function. However, we know from geological information and principles that natural porous media often are strongly structured and in a manner not fully captured by a second-order stationary covariance function.

CONCLUSIONS

Summary

The results of this study clearly demonstrate that different model conceptualizations of the spatial structure of natural porous media can lead to quite different predictions. In particular, continuous bivariate stochastic models of structure tend to diminish the impacts of connected pathways and flow barriers and smooth the resulting predictions (diminish variability) as compared to discrete geometric models.

While neither type of model is a perfect representation of reality (perhaps a hybrid approach is best-see Haldorsen and Damsleth, 1990), the geometric and strongly structured nature of many sedimentary deposits should motivate caution in application of traditional bivariate statistical models of structure. This is particularly true when the predictions of interest are strongly impacted by the existence of connected

pathways (such as early arrival times) or low-conductivity barriers.

Therefore, when selecting a stochastic model of spatial structure, one should carefully consider the geological character of the porous medium being represented as well as the type of prediction to be made using the model. The estimate of uncertainty provided by a particular model represents the parametric error *given that model*, but it does not represent the possible limitations in the ability of the model itself to be representative of a complex natural system.

Future Work

Two clear areas for extension of this work are evident. The first is to perform similar numerical examples using a three-dimensional model system. It is well known that limiting the dimensionality of a heterogeneous flow system has a significant effect on the flow pattern by constraining flow to a single plane. Allowing flow to move unconstrained in three dimensions provides greater potential for development of preferential pathways and can be expected to accentuate the effects seen here.

The second area is to consider the effects of conditioning data on the results. Kitanidis (1995) pointed out that using conditioning data may diminish the distinction between predictions made using Gaussian and non-Gaussian models. The application of conditioning data is in itself complex because of the diversity of scales at which hydraulic conductivity data typically are available (e.g., air minipermeameter measurements, core samples, slug tests, pumping tests). Definition of preferential flow paths (or barriers) that exist at large scales (in that they are connected over long distances) and simultaneously at small scales (in that they may be narrow in cross-section) may be quite difficult. While large-scale observations may encompass the region containing a preferential pathway, they also average in the effects of the surrounding materials in a way that inhibits resolution of the preferential pathway. Small-scale observations allow resolution of a preferential pathway, but doing so requires many more measurements than are typically available. Geophysical tomography methods, combined with detailed outcrop studies, may offer a means of integrating these multiple scales while yet resolving important features. The authors and other collaborators are currently exploring these issues at a field site in which several types and scales of conditioning data are available.

ACKNOWLEDGMENTS

This research was funded by the Pacific Northwest National Laboratory's Laboratory Directed Research and Development program. Pacific Northwest National Laboratory is operated for the U.S. Department of Energy by Battelle under Contract DE-AC06–76RLO 1830.

REFERENCES

BASUMALLICK, S., 1966, Size differentiation in a cross-stratified unit: Sedimentology, v. 6, p. 35-68.

BEAR, J., 1979, Hydraulics of Groundwater: New York, McGraw-Hill, 567 p.

CARLE, S.F., AND FOGG, G.E., 1996, Transition probablility-based indicator geostatistics: Mathematical Geology, v. 28, p. 453-476.

DAGAN, G., 1984, Solute transport in heterogeneous porous formations: Journal of Fluid Mechanics, v. 145, p. 151-177.

DAVIS, J.M., LOHMANN, R.C., PHILLIPS, F.M., WILSON, J.L., AND LOVE, D.W., 1993, Architecture of the Sierra Ladrones formation, central New Mexico--Depositional controls of the permeability correlation structure: Geological Society of America Bulletin, v, 105, p. 998-1007.

DESBARATS, A.J., 1987, Numerical estimation of effective permeability in sandshale formations: Water Resources Research, v. 23, p. 273-286.

DEUTSCH, C.V., 1992, Annealing techniques applied to reservoir modeling and the integration of geological and engineering (well test) data: Unpub. Ph.D. thesis, Stanford University, Stanford, Calif.

DEUTSCH, C.V., AND JOURNEL, A.G., 1992, GSLIB--Geostatistical Software Library and User's Guide: New York, Oxford University Press, 340 p.

EGGLESTON, J.R., ROJSTACZER, S.A., AND PIERCE, J.J., 1996, Identification of hydraulic conductivity structure in sand and gravel aquifers--Cape Cod data set: Water Resources Research, v. 32, p. 1209-1222.

GELHAR, L.W., AND AXNESS, C.L., 1983, Three-dimensional stochastic analysis of macrodispersion in aquifers: Water Resources Research, v. 19, p. 161-180.

GUARDIANO, F., AND SRIVASTAVA, M., 1993, Multivariate geostatistics--Beyond bivariate moments, *in* Soares, A., ed., Geostatistics Troia '92: Kluwer Academic Publishers, p. 133-144.

HALDORSEN, H.H., AND DAMSLETH, E., 1990, Stochastic modeling: Journal of Petroleum Technology, v. 42, p. 404-412.

ISAAKS, E., AND SRIVASTAVA, M., 1989, An Introduction to Applied Geostatistics: New York, Oxford University Press, 561 p.

JACKSON, R. G., II, 1976, Largescale ripples of the lower Wabash River: Sedimentology, v. 23, p. 593-623.

JOHNSON, N.M., AND DREISS, S. J., 1989, Hydrostratigraphic interpretation using indicator geostatistics: Water Resources Research, v. 25, p. 2501-2510.

JOURNEL, A.G., AND ALABERT, F.G., 1989, Nongaussian data expansion in the earth sciences: Terra Nova, v. 1, p. 123-134.

JOURNEL, A.G., AND ALABERT, F. G., 1990, New method for reservoir mapping: Journal of Petroleum Technology, v. 42, p. 212-218.

JOURNEL, A.G., AND DEUTSCH, C.V., 1993, Entropy and spatial disorder: Mathematical Geology, v. 25, p. 329-355.

JOURNEL, A.G., AND GOMEZ-HERNANDEZ, J., 1989, Stochastic imaging of the Wilmington clastic sequence, *in* Proceedings, SPE Annual Technical Conference, p. 591-606.

KITANIDIS, P.K., 1990, A conceptual framework for the geostatistical approach to the inverse problem, Abstract, *in* Bachu, S., ed., Parameter Identification and Estimation for Aquifer and Reservoir Charaterization: Proceedings of the 5th Canadian/American Conference on Hydrogeology, Calgary, Alberta, Canada, National Water Well Association, p. 275.

KITANIDIS, P.K., 1995, Recent advances in geostatistical inference on hydrogeological variables: Reviews of Geophysics, v. 33 Supplement, p. 1103-1109.

MONNE, J., SCHMITT, F., AND MASSALOUX, D., 1981, Bidimensional texture synthesis by markov chains: Computer Graphics and Image Processing, v. 17, p. 1-23.

MURRAY, C.J., 1995, Identification and 3d modeling of petrophysical rock types, *in* Yarus, J.M. and Chambers, R.L., eds., Stochastic Modeling and Geostatistics--Principles, Methods, and Case Studies: American Association of Petroleum Geologists, Computer Applications 3, p. 323-336.

NEUMAN, S.P., WINTER, C.L., AND NEWMAN, C.M., 1987, Stochastic theory of fieldscale Fickian dispersion in anisotropic porous media: Water Resources Research, v. 23, p. 453-466.

POETER, E., AND TOWNSEND, P., 1994, Assessment of critical flow paths for improved remediation management: Ground Water, v. 32, p. 439-447.

POLLOCK, D.W., 1988, Semianalytical computation of pathlines for finite difference models: Ground Water, v. 26, p 743-750.

PRYOR, W., 1973, Permeability-porosity patterns and variations in some holocene sand bodies: American Association of Petroleum Geologists Bulletin, v. 57, p. 162-189.

RASURE, J.D., ARGIRO, D., SAUER, T., AND WILLIAMS, C., 1990, A visual language and software development environment for image processing: International Journal of Imaging Systems and Technology, v. 2, p. 183-199.

SCHEIBE, T.D., 1993, Characterization of the spatial structuring of natural porous media and its impacts on subsurface flow and transport: Unpub. Ph.D. thesis, Stanford University, Stanford, Calif.

SCHEIBE, T.D., AND FREYBERG, D.L., 1995, The use of sedimentological information for geometric simulation of natural porous media structure: Water Resources Research, v. 31, p. 3259-3270.

WOODBURY, A.D., AND ULRYCH, T.J., 1993, Minimum relative entropy--Forward probabilistic modeling: Water Resources Research, v. 29, p. 2847-2860.

A METHOD FOR CHARACTERIZING HYDROGEOLOGIC HETEROGENEITY USING LITHOLOGIC DATA

GREGORY P. FLACH[1], L. LARRY HAMM[1], MARY K. HARRIS[1]
PAUL A. THAYER[2], JOHN S. HASELOW[3] AND ANDREW D. SMITS[4]

[1] *Westinghouse Savannah River Company, Aiken, South Carolina 29808*
[2] *The University of North Carolina at Wilmington, Wilmington, North Carolina 28403*
[3] *Haselow Engineering, Aiken, South Carolina 29803*
[4] *Science Applications International Corporation, Augusta, Georgia 30901*

ABSTRACT: Large-scale (> 1 m) variability in hydraulic conductivity usually is the main influence on field-scale groundwater flow patterns and dispersive transport. Incorporating realistic hydraulic conductivity heterogeneity into flow and transport models is paramount to accurate simulations, particularly for contaminant migration. Sediment lithologic descriptions and geophysical logs typically offer finer spatial resolution, and therefore more potential information about site-scale heterogeneity, than other site characterization data.

In this study, a technique for generating a heterogeneous, three-dimensional hydraulic conductivity field from sediment lithologic descriptions is presented. The approach involves creating a three-dimensional, fine-scale representation of mud (silt + clay) percentage using a "stratified" interpolation algorithm. Mud percentage then is translated into horizontal and vertical conductivity using direct correlations derived from measured data and inverse groundwater flow modeling. Lastly, the fine-scale conductivity fields are averaged to create a coarser grid for use in groundwater flow and transport modeling.

The approach is demonstrated using a finite-element groundwater flow model of a Savannah River Site solid radioactive and hazardous waste burial ground. Hydrostratigraphic units in the area consist of fluvial, deltaic and shallow marine sand, mud and calcareous sediments that exhibit abrupt facies changes over short distances. For this application, the technique improves estimates of large-scale flow patterns and dispersive transport. The conductivity fields mimic actual lithologic data, providing a more realistic picture of subsurface heterogeneity. Field-observed preferential pathways for contaminant migration are replicated in the simulations without the need to artificially create zones of high conductivity.

INTRODUCTION

Groundwater flow and especially contaminant transport can be affected by many physical, chemical and microbiological factors. Among these, large-scale (>1 m) variability in hydraulic conductivity is the dominant influence on field-scale groundwater flow patterns and dispersive transport, with other factors introducing second-order effects (Brusseau, 1994). Incorporating hydraulic conductivity heterogeneity into flow and transport models is paramount to accurate simulations in most situations, particularly for contaminant migration. Sediment lithoiogic descriptions and downhole geophysical logs typically offer finer spatial resolution, and therefore more potential information about heterogeneity, than most other site characterization data. However, these data do not provide direct information about hydrogeologic properties and can be viewed as "soft" hydrogeologic information. The emergence of powerful computing resources has enabled the routine use of spatially dense, three-dimensional (3D) "soft" hydrogeologic data in 3D flow and transport models. Methods for supplementing the "hard" hydrogeologic data (e.g., pump tests, slug tests, laboratory conductivity data), conventionally used to develop flow and transport models with spatially dense, 3D "soft" hydrogeologic data are the focus of this study. By utilizing widely available and relatively inexpensive core data, these methods should reduce model uncertainty and the amount of field-scale hydrogeologic data required.

Two key elements of these methods are (1) an algorithm for generating a complete 3D representation from discrete, spatially scattered field data, and (2) a method for translating "soft" data into hydraulic conductivity, and other necessary hydrogeologic parameters. Most articles in the literature focus on only one of these components. An exception is Lahm et al. (1995), who presented a general procedure for estimating hydraulic conductivity from porosity (derived from geophysical logs) using a direct correlation and for generating a 3D conductivity field using a Bayesian updating/Cholesky decomposition stochastic approach. Below we review literature falling in the two categories above, emphasizing studies most relevant to the geology of the U.S. Department of Energy's (DOE) Savannah River Site (SRS) (Fig. 1).

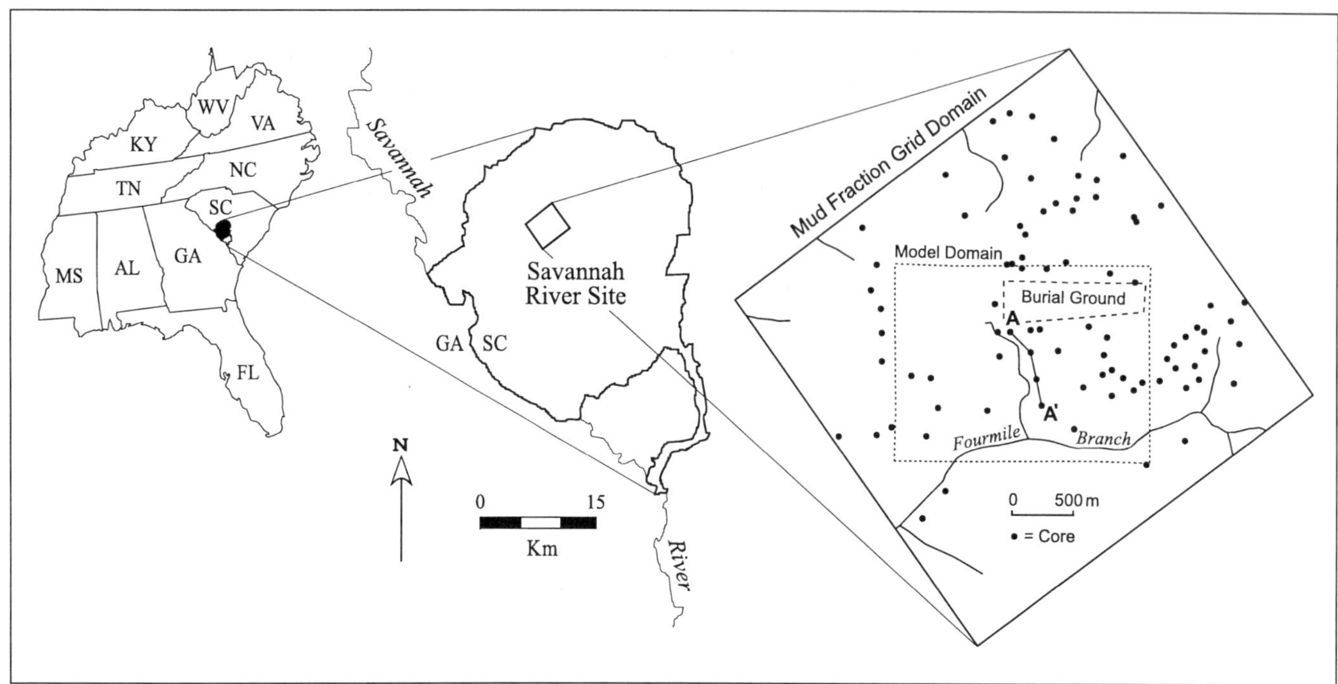

FIG. 1.—Location map of the original Savannah River Site burial ground showing the flow-model domain and available sediment lithologic descriptions from 84 wells.

Interpolation Algorithms

Several methods have been proposed for generating a complete 3D representation from discrete, spatially scattered field data, including deterministic interpolation, kriging, indicator geostatistics, turning bands methods, inverse and inverse conditional analysis, fractal-based approaches and sedimentological models. All these approaches have been used successfully to interpolate data. Each has inherent strengths and weaknesses, which should be considered when selecting an algorithm for a particular application. Deterministic methods, such as inverse distance weighting and spline interpolation, are conceptually simple, straightforward to implement and generally yield good 3D representations of data. However, these methods typically ignore known or inferred information about spatial correlation, which can be used to improve the fit. Also, the uncertainty in interpolated values is difficult to estimate. For these reasons, investigators have pursued geostatistical techniques over the last two decades.

Kriging (e.g., Journel and Huijbregts, 1978; Isaaks and Srivastava, 1989) originated in the mining industry as a geostatistical interpolation method that utilizes information on the covariance of variables in space, in addition to data values. Kriging, which has been widely employed over the last decade or so, offers an improvement over simple interpolation schemes in many instances. The method also provides the uncertainty associated with each interpolated value, which may be important information depending on the application. Although an improved fit can be achieved, the choices of variogram and underlying error distribution are subjective and can lead to a wide range of possible solutions. Also, traditional kriging assumes that the covariance field is stationary or weakly stationary (only dependent upon distance between points), which may not be an appropriate assumption. In this case, additional effort is required to transform the original variable into a stationary one. Like other methods, kriging often performs poorly when hard data are scarce. This is especially true for kriging methods that incorporate anisotropy (covariance dependent upon direction) to account for layering of sediments, for example.

Journel (1986) and Rubin and Journel (1991) attempted to overcome data deficiencies by utilizing "soft" information; they developed the binary indicator method with kriging to generate 3D fields. Phillips and Wilson (1989) and Brannan and Haselow (1993) also developed methods based on geometrical descriptions of geological media and statistical functions to generate fields. These turning bands methods provide additional refinement to the kriging approach but can be labor intensive.

The observation that some natural processes are known to be fractal has motivated investigations of fractal-based algorithms to generate 3D representations of geological systems. Hewett (1986) and Hewett and Behrens (1988), for example, used fractal concepts to describe 3D geological systems with measurements of porosity from geophysical logs. Molz and Boman (1993) later applied the procedure to a hydraulic conductivity data base. Boman et al. (1995) concluded that while fractal-based methods are promising, their application requires further understanding and refinement.

Some alternative methods generate 3D fields using knowledge of the physical processes of sediment deposition. For example, Webb (1994) developed a simulator to mimic a braided stream depositional environment and generate 3D fields. These methods show promise but will require significantly more research to encompass the complex and varied depositional processes that occur during formation of a hydrogeological unit.

Dagan (1985) proposed a method in which the 3D field is determined with the aid of an inverse solution of the groundwater flow equation. The method is termed a "conditional" or "unconditional" inverse simulation, depending upon whether measured data are honored in the interpolated field. Another inverse approach that has been proposed to generate 3D realizations is to solve the inverse problem for both the flow and transport equation to estimate the hydrogeologic properties of an aquifer (e.g., Sun and Yeh; 1990; Wagner and Gorelick, 1987). Hyndman et al. (1994) proposed refining this method using seismic data. However, solving the flow and transport equations simultaneously is nontrivial and requires more development.

Additional research is needed to evaluate and understand 3D interpolation algorithms in a practical sense, as Boman et al. (1995) have done in evaluating inverse distance weighting, kriging and fractal-based methods for generating a 3D conductivity field. Their evaluation indicates that a simple deterministic approach involving successive inverse distance-weighting interpolation by horizontal layer is the preferred approach, based on its relative simplicity and the absence of improved tracer test modeling results using the other methods. Based on the conclusion of Boman et al. (1995), a relatively simple deterministic interpolation method was chosen for this study.

Translation of "Soft" Data into Hydrogeologic Parameters

Methods employed in the petroleum industry for relating intrinsic permeability to borehole geophysical data are described by Lake and Carroll (1986) and Flint and Bryant (1993). Lahm et al. (1995) recently reviewed several efforts to relate permeability to borehole geophysical data, including Timur (1968), Croft (1971), Keys and MacCary (1971), Kelly (1977), Biella et al. (1983), Doveton (1986), Wendt et al. (1986), Urish (1981) and Jorgensen (1989). They also evaluated three methods of estimating conductivity primarily from porosity, as applied to the Milk River aquifer in Alberta, Canada. Lahm et al. (1995) found that a direct correlation between conductivity and porosity produced a lower residual between the measured and predicted datasets than did the methods of Urish (1981) and Jorgensen (1989).

Many predictive methods for estimating hydraulic conductivity from grain-size data have been proposed over the last several decades (Freeze and Cherry, 1979; Domenico and Schwartz, 1990). These methods are most reliable when applied to aquifer zones consisting of clean sand and gravel. Riha (1993) evaluated ten predictive equations as applied to Coastal Plain sediment beneath the SRS and found that none performed adequately. The methods evaluated were those of Hazen (1892), Kozeny-Carman (in Carman, 1937), Krumbein and Monk (1943), Harleman et al. (1963), Masch and Denny (1966), Beard and Weyl (1973), Puckett et al. (1985), Ahuja et al. (1989), Franzmeier (1991) and Jabro (1992). Riha (1993) developed a predictive equation from a small SRS data set, but that correlation has not been tested beyond the data used to develop the correlation.

Kegley et al. (1994) performed over 4,000 minipermeameter tests of Tertiary-age Coastal Plain sediments at the SRS and correlated the geometric mean of measured permeability for each stratigraphic unit within each well to mud fraction estimated by binocular microscope examination. The correlation coefficient of the expression is -0.92.

Present Study

For this study, borehole geophysical data are combined with information from microscopic examination of core to yield qualitative grain-size distribution information in the form of estimated percent gravel, sand and mud. A general approach is presented for generating a heterogeneous, 3D hydraulic conductivity field from estimated mud fraction (silt + clay sized material). The mud-fraction data first are interpolated onto a 3D Cartesian grid using the minimum-tension spline algorithm implemented in EarthVision® version 2.0 (Dynamics Graphics, Inc., Alameda, CA) with minimal "vertical influence" selected. This easily applied interpolation approach is analogous to the successive inverse distance-weighting interpolation by horizontal layer scheme recommended by Boman et al. (1995). The mud-fraction grid is translated into horizontal and vertical conductivity fields using direct correlations of conductivity to mud fraction. The correlations are developed initially from laboratory data and adjusted through inverse groundwater flow modeling. Selection of a direct correlation approach is supported by Kegley et al. (1994) and Lahm et al. (1995). The fine-scale conductivity fields then are transferred to a coarser grid for use in groundwater flow modeling through arithmetic and harmonic averaging. This method is demonstrated in a finite-element groundwater flow model for a radioactive and hazardous waste burial ground at the Savannah River Site.

SITE DESCRIPTION AND HYDROGEOLOGY

The study area is located in the Upper Atlantic Coastal Plain physiographic province in southwestern South Carolina at the U. S. Department of Energy's Savannah River Site (SRS) (Fig. 1). The SRS occupies about 800 km^2 and was set aside in 1950 as a controlled area for the production of nuclear material for national defense and specialized nuclear materials for medical and industrial purposes. The study area is located near the center of the SRS within the 40-km^2 General Separations Area (GSA) (Fig. 1). This area is undergoing intense hydrogeological characterization for RCRA and CERCLA investigations. The focus of this study is the original solid radioactive and hazardous waste burial ground in the GSA. The burial ground received wastes from approximately 1952 to 1972. Primary groundwater contaminants include tritium, trichloroethylene (TCE) and tetrachloroethylene (PCE). Groundwater plumes emanating from the burial ground are migrating to the south toward Fourmile Branch (Fig. 1).

The SRS is underlain by a southeast-dipping wedge of unconsolidated and consolidated sediment consisting of sand, mud, limestone and gravel, which range in age from early Late Cretaceous (Cenomanian) to Holocene (Fallaw and Price, 1995). This study involves the Tertiary-age sediment, principally the Eocene to Miocene sequence (Fig. 2). The Gordon aquifer, the Gordon confining unit and the Upper Three Runs aquifer are the hydrostratigraphic units of interest (Fig. 2).

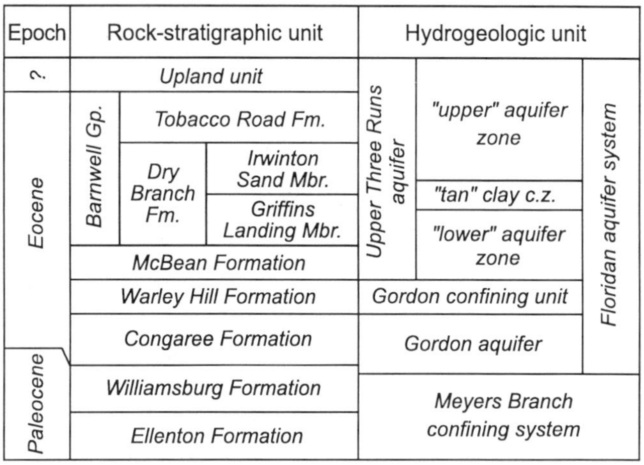

Epoch	Rock-stratigraphic unit			Hydrogeologic unit		
?	*Upland unit*			*Upper Three Runs aquifer*	*"upper" aquifer zone*	*Floridan aquifer system*
Eocene	*Barnwell Gp.*	*Tobacco Road Fm.*				
		Dry Branch Fm.	*Irwinton Sand Mbr.*		*"tan" clay c.z.*	
			Griffins Landing Mbr.		*"lower" aquifer zone*	
	McBean Formation					
	Warley Hill Formation			*Gordon confining unit*		
	Congaree Formation			*Gordon aquifer*		
Paleocene	*Williamsburg Formation*			*Meyers Branch confining system*		
	Ellenton Formation					

FIG. 2.—Lithostratigraphic and hydrostratigraphic nomenclature at the SRS (modified from Aadland et al., 1995).

The Gordon aquifer consists of unconsolidated sand and clayey sand of the Congaree Formation and, where present, the sandy parts of the underlying Williamsburg Formation. The sand is yellowish- to grayish-orange, subrounded to well-rounded, moderately to poorly sorted and medium- to coarse-grained. The Gordon confining unit, which separates the Gordon and Upper Three Runs aquifers, is lithostratigraphically equivalent to the Warley Hill Formation. This unit is composed of fine-grained, silty and clayey sand, sandy clay and clay. The clay is stiff to hard and sometimes fissile. Zones of silica-cemented sand and clay, and calcareous sediment are common.

The Upper Three Runs aquifer includes all sediment from the ground surface to the top of the Gordon aquifer. Lithostratigraphically, the Upper Three Runs aquifer includes the Upland unit, Tobacco Road Sand, Dry Branch Formation and Santee Formation (Fig. 2). The Upper Three Runs aquifer is divided by the "tan clay" confining zone into a "lower" aquifer zone and "upper" aquifer zone. The "lower" aquifer zone consists of various lithologies, including terrigenous sand and clay, calcareous skeletal sand and sandy limestone. The "tan

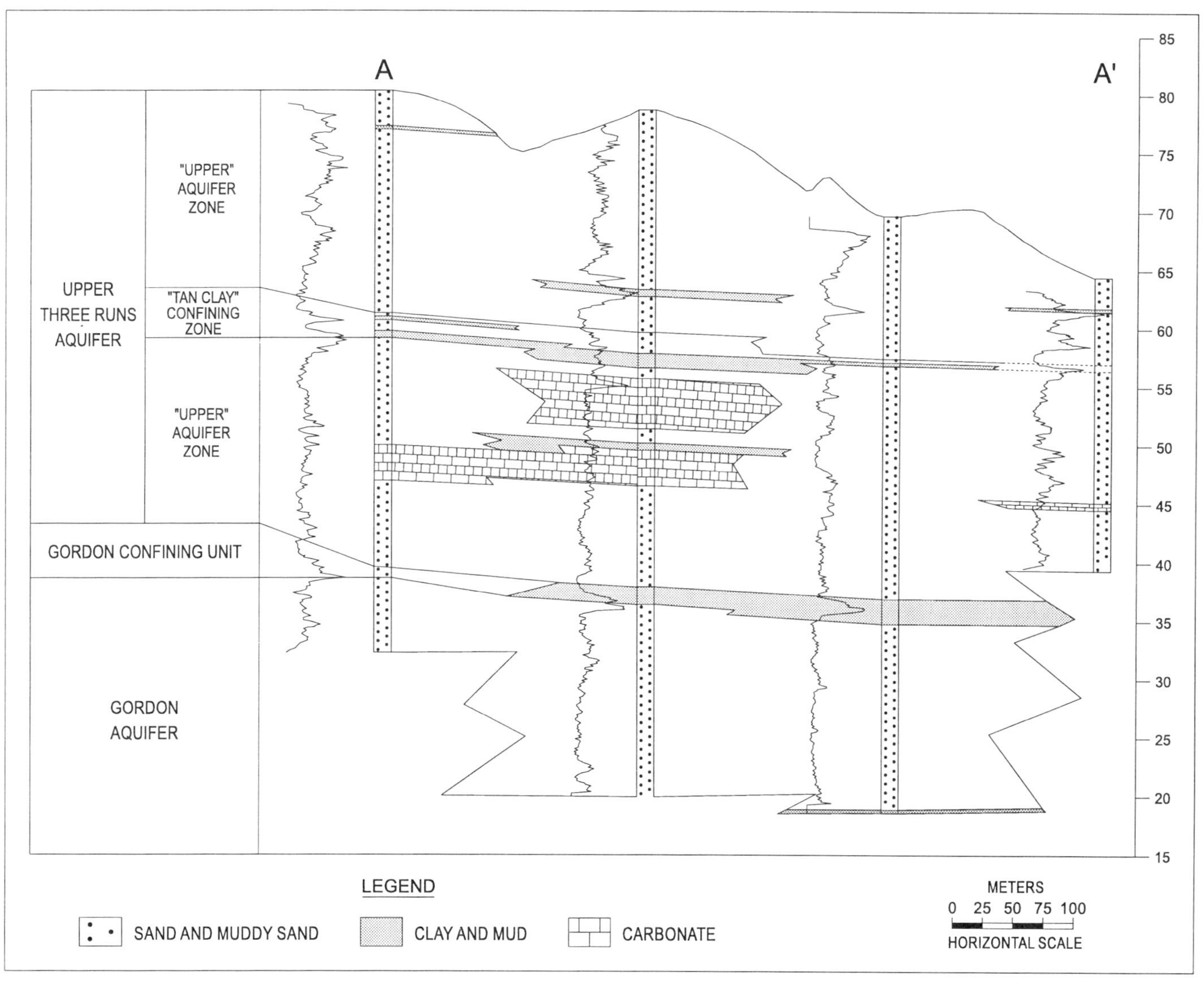

FIG. 3.—Lithostratigraphic and hydrostratigraphic cross section A-A' (Fig. 1).

clay" confining zone contains light-yellowish tan to orange clay and sandy clay interbedded with clayey sand and sand. The "upper" aquifer zone is characterized by sand and clayey sand with minor intercalated clay layers. Gravelly layers are common.

Hydrogeological characterization of the study area indicates that groundwater contamination is mainly contained within the Upper Three Runs aquifer over a lateral distance of about 1 km. Although contaminant transport occurs mainly within this hydrostratigraphic unit, the aquifer exhibits considerable field-scale heterogeneity, as indicated by wide-ranging lithologic and physical characteristics shown in Figure 3.

METHODS

Data from detailed, foot-by-foot descriptions of continuous drill core from 84 boreholes were utilized in this study. Figure 1 shows the areal distribution of the data in relation to the burial ground, and the selected numerical flow model domain. The core descriptions include data on core recovery; degree of induration; color; sedimentary structures; volume percent terrigenous gravel, sand and mud; maximum and modal size of the terrigenous fraction; volume percent carbonate gravel, sand and mud; volume percent cement; volume percent total carbonate sediment; sediment or rock name; grain

123

sorting; volume percent porosity and dominant type; fossil types; and volume percent accessory constituents, including muscovite, glauconite, lignite, sulfides and heavy minerals. The volume percent terrigenous gravel, sand and mud data were used in generating 3D conductivity fields.

Figure 4 outlines the overall methodology used to generate heterogeneous hydraulic conductivity fields from lithologic descriptions. Mud-fraction data initially are interpolated onto a 3D grid using an algorithm that preserves sharp vertical contrasts. The mud-fraction grid is translated into horizontal (K_h) and vertical conductivity (K_v) fields using direct "correlations" of conductivity to mud fraction. Laboratory conductivity measurements on whole core and quantitative grain-size distribution data from sieve analyses are utilized as a guide to development of the correlations. On the first pass, the correlations are based strictly on laboratory conductivity data. On subsequent passes, inverse flow modeling also influences the correlations through a feedback loop; that is, the correlations are adjusted to achieve better flow-model results when compared to hydraulic head data and other targets. The fine-scale conductivity fields are transferred to the coarser groundwater flow-model grid through arithmetic (K_h) and harmonic (K_v) averaging. The process is discussed below in greater detail.

Three-Dimensional Interpolation of Mud Fraction Data

Boman et al. (1995) demonstrated that relatively simple interpolation concepts can be used successfully to generate heterogeneous 3D conductivity fields for flow and transport models. They employed what can be termed successive inverse distance-weighting (IDW) interpolation by horizontal layer, or "stratified IDW" in their terminology. The idea is conceptually to divide the subsurface into horizontal layers of uniform thickness. These layers are arbitrary and do not necessarily conform to formation boundaries or other geologic features. Within each layer, two-dimensional inverse distance weighting is independently applied using only the subset of borehole data lying within that layer. With this stratified IDW approach, smoothing occurs in the horizontal plane while sharp vertical contrasts (if present) are preserved.

A similar effect can be achieved using the minimum-tension spline interpolation algorithm implemented in the EarthVision® software package (ver. 2.0, Dynamic Graphics, Inc., Alameda, CA), by selecting a very small

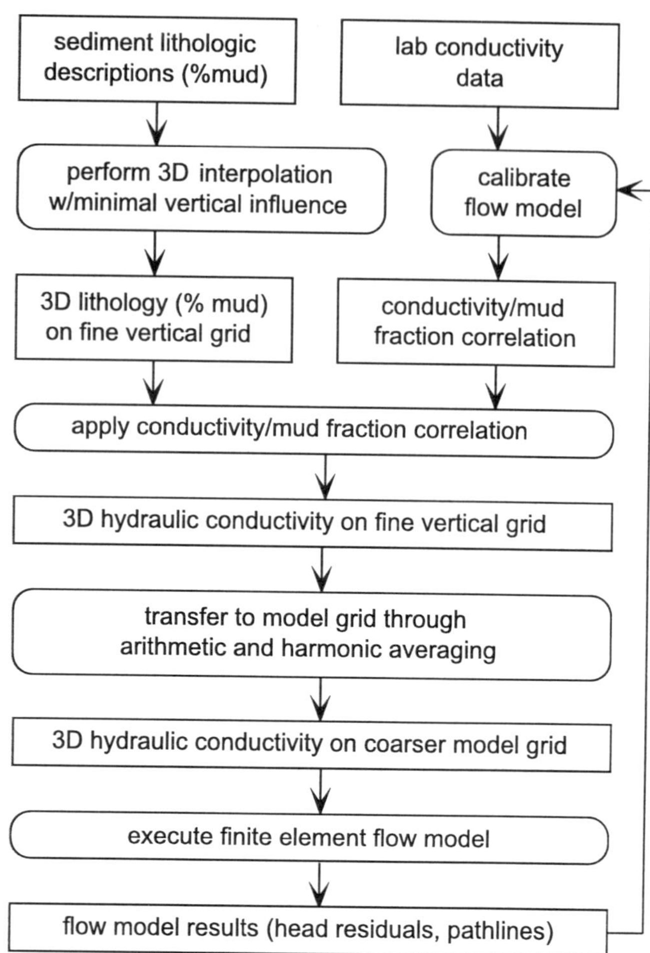

FIG. 4.—Flowchart summarizing the methodology for creating a heterogeneous conductivity field using data from sediment lithologic descriptions.

"vertical influence factor." The vertical influence factor is applied to data lying above and below the horizontal plane passing through the interpolation node of interest. By choosing a very small value, almost no weight is given to data above or below each interpolation node relative to data in the horizontal plane of that node.

In this study the lithologic descriptions from the 84 boreholes depicted in Figure 1 were interpolated onto a 23 by 23 by 251 grid of dimensions 3,353 m (11,000 ft) by 3,353 m (11,000 ft) by 76 m (250 ft). The areal dimensions correspond to the solid line box in Figure 1. The areal resolution was uniformly set to 152 m (500 ft) while vertical resolution was uniformly set to 0.30 m (1 ft), the same as the raw data. In all, 12,626 mud-fraction data points were employed (Fig. 5). A vertical influence factor of 0.01 was selected to preserve vertical contrasts. Interpolated values of mud fraction were constrained to

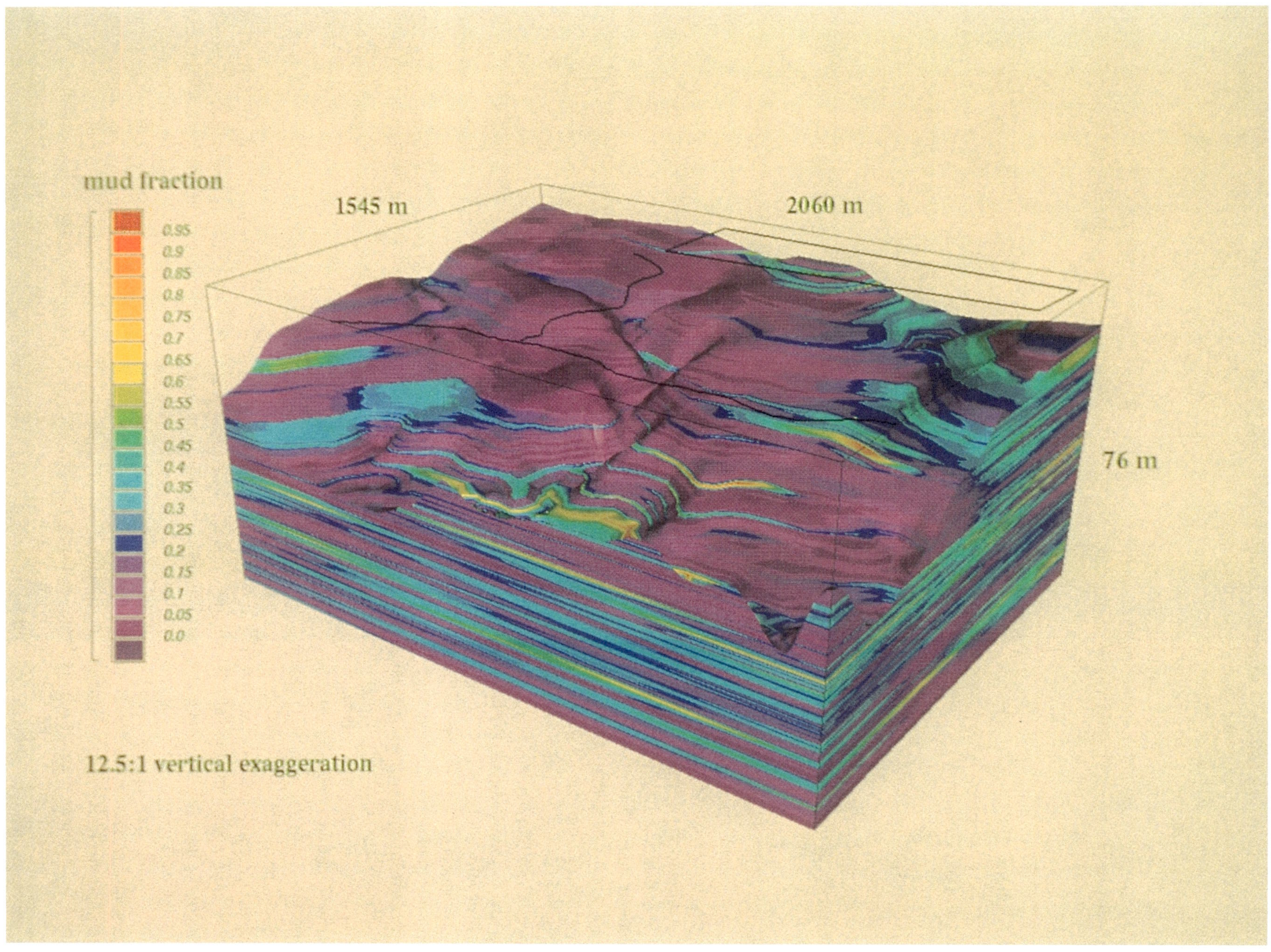

Fig. 5.—Three-dimensional mud-fraction variation within the flow-model domain (Fig. 1) generated with EarthVision[®]'s minimum tension 3D gridding algorithm and a vertical influence factor of 0.01. A total of 12,626 mud-fraction values from the 84 cores depicted in Figure 1 were interpolated.

fall within the physical range of 0 to 1 to remedy under- and over-shoots between or beyond data. At several locations lacking measured data, pseudo-data were added as control points to minimize extrapolation errors. Figure 5 illustrates the portion of the resulting 3D mud-fraction representation contained within the flow-model areal domain (dashed box in Figure 1) and cropped by the ground surface. The lithologic heterogeneity of the subsurface hydrostratigraphic units is easily recognized and corresponds with hand-contoured lithofacies maps and cross sections (Fig. 3) (Thayer et al., 1994). Note that the interpolation process has preserved horizontal stratification of the sediments.

Conductivity Versus Mud-Fraction Correlations

Laboratory measurements of horizontal and vertical conductivity for Tertiary-age sediments at the SRS are available from a number of sources (Bledsoe et al., 1990; Riha, 1993; Kegley et al., 1994; Aadland et al., 1995) (Fig. 6). The data show an overall trend of decreasing conductivity with increasing mud fraction, but they exhibit a great deal of scatter. Estimated mud fractions are available for the entire length of all cores, whereas sieve data are available primarily for transmissive well screen zones (3 to 10 m). In order to utilize the most data, a direct correlation between conductivity and estimated mud fraction was chosen instead of a grain-size distribution data correlation. Kegley et al. (1994) and Lahm et al. (1995) demonstrated success using this approach. The initial conductivity-mud fraction correlations were developed based solely on the laboratory conductivity data. However, the resulting flow simulation did not adequately match hydraulic head, average recharge and stream gain targets. Consequently, the correlations were adjusted manually to improve flow results while maintaining consistency with the laboratory data. The stair-step functions shown in Figure 6 are the final outcome of this iterative flow-model calibration process.

A stair-step functional form was chosen for two reasons. First, the interpolated mud- fraction field shown in Figure 5 contains numerous regions where under- and over-shoots in the initial fit are clipped by EarthVision® to lie within the specified physical range of 0 to 1. These initial under- and over-shoots are located between sparsely distributed data and at the fringes of the data (i.e., extrapolation errors). As a result the interpolated mud-fraction field contains an artificially large number of values at or near 0 and 1. The flat portions of the stair-step functions tend to alleviate this problem by

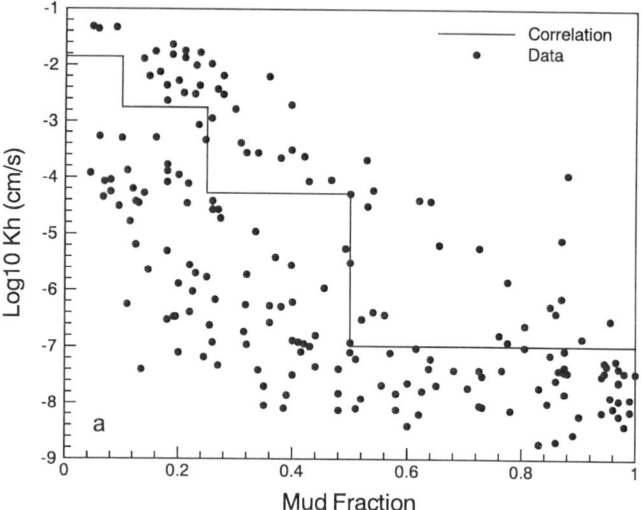

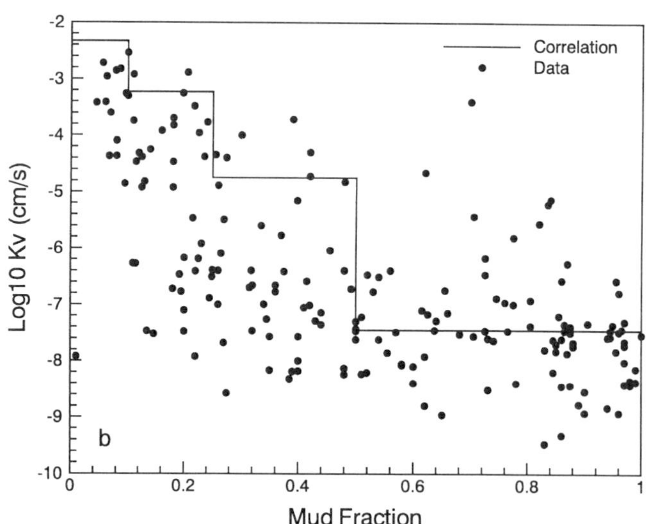

FIG. 6.—Hydraulic conductivity as a function of mud fraction. The data represent laboratory measurements of conductivity from undisturbed core samples and mud fraction measured by sieve analysis. The stair-step correlation line is the result of comparison to the data shown and inverse flow modeling. (a) horizontal conductivity, (b) vertical conductivity.

assigning the same conductivity to all mud fraction values in the vicinity of 0 and 1, respectively. Second, stair-step functions provide the analyst very simple and direct control over conductivity values during model calibration when compared to other potential functional forms. A disadvantage of the correlations is the presence of unrealistically abrupt changes in conductivity. This deficiency is not serious, however, because the abruptness is effectively smoothed out when the conductivity fields are translated to the coarser flow-model grid through the averaging process discussed below.

In order to reduce the degrees of freedom during calibration to a reasonable number, restrictions were invoked early in the process. The correlations were constrained such that the ratio of horizontal to vertical conductivity is 3. This conductivity anisotropy ratio is suggested by the laboratory data of Bledsoe et al. (1990). Also, the junctions between steps of 0.1, 0.25 and 0.50 were held constant during final calibration. These choices yield four independent horizontal conductivity parameters spanning mud-fraction ranges of 0.0 to 0.1, 0.1 to 0.25, 0.25 to 0.50 and 0.50 to 1.0. The flow-model calibration process in this study is analogous to that performed for a conventional "layer cake" model, with horizontal conductivities within mud- fraction ranges taking the place of hydrostratigraphic unit conductivities (or zonal conductivities within a unit).

Transfer to Flow-Model Grid

FACT, a variably-saturated 3D finite-element flow and transport code developed in-house, was chosen for flow simulation. FACT is a derivative of the SAFT3D and VAM3DCG codes developed by HydroGeoLogic, Inc. (Huyakorn et al., 1991; Huyakorn and Panday, 1992). The code solves Richard's equations. FACT assumes that the hydraulic conductivity tensor is aligned with the principal axes of the porous medium and that its diagonal values are specified at the element centroids. In this study the porous medium is assumed to be isotropic in the two horizontal directions, but anisotropic with respect to its vertical direction.

To demonstrate the methodology, a finite-element mesh was selected that differs in extent, orientation and spatial resolution from that of the mud-fraction grid generated above. The extent of the flow model and its orientation relative to the mud-fraction grid are shown in Figure 1 (dashed versus solid boxes). This domain captures the entire groundwater flow field from the burial ground to surface discharge at Fourmile Branch (Fig. 1) and places model boundaries near monitoring wells so that boundary conditions may be more accurately specified. Areal dimensions are 2,060 m (6,760 ft) by 1,545 m (5,070 ft). Eight-noded rectangular "brick" elements, which are restricted to deformations only in the vertical direction, are used. Elemental dimensions in the areal extent were uniformly set to 39.6 m (130 ft) per side. The top of the mesh conforms to the ground surface, while the bottom of the mesh conforms to the top of the highly competent Meyers Branch confining system (Fig. 2). Element heights range from about 0.3 m (1 ft) to 6 m (20 ft) with an average value around 3 m (10 ft). The overall nodal dimensions of the flow grid are 53 by 40 by 30 (52 by 39 by 29 elements). The selected flow grid is much coarser than the mud-fraction grid in the vertical direction but finer in the areal directions.

The process for translating the 3D mud- fraction grid shown in Figure 5 into "composite" elemental conductivity defined over the 3D flow grid involves three steps and the creation of a fine-scale intermediate grid. First, while maintaining the 0.3-m (1-ft) vertical resolution of the mud-fraction grid, mud-fraction values were interpolated onto an intermediate grid whose areal grid points coincide with the element areal centroids of the flow grid. A natural bi-cubic spline interpolation was performed for each horizontal layer, independently. In the next step, mud-fraction values defined on the resulting fine-scale intermediate grid were translated into "local" horizontal and vertical conductivities using the correlations shown in Figure 6. In the final step, these local conductivities on the intermediate grid were vertically averaged over each finite element (typically ten fine-scale layers per element). Figure 7 depicts this multistep process starting from the fine-scale grid. During this last stage, composite conductivities (horizontal and vertical) were computed based on appropriate averaging (arithmetic and harmonic, respectively), where local conductivities are assumed to reflect horizontal layering of aquifer materials that extend over the entire areal extent of each element. For a given finite element these composite (horizontal and vertical) conductivities are expressed by the equations:

$$K_h^{comp} = \frac{\sum\limits_{i=ib}^{i=it} \left(K_h \xi \Delta z\right)_i}{\sum\limits_{i=ib}^{i=it} \left(\xi \Delta z\right)_i} \tag{1}$$

$$K_v^{comp} = \frac{\sum\limits_{i=ib}^{i=it} (\xi \Delta z)_i}{\sum\limits_{i=ib}^{i=it} (\xi \Delta z / K_v)_i} \qquad (2)$$

where

ib $\equiv$ bottom fine-scale layer contained within element,

it $\equiv$ top fine-scale layer contained within element,

Δz_i $\equiv$ vertical height of ith fine-scale layer,

ξ_i $\equiv$ fraction of ith fine-scale layer contained within element,

K_{hi} $\equiv$ local horizontal conductivity of ith fine-scale layer,

K_{vi} $\equiv$ local vertical conductivity of ith fine-scale layer,

K_h^{comp} $\equiv$ composite horizontal conductivity of element,

K_v^{comp} $\equiv$ composite vertical conductivity of element

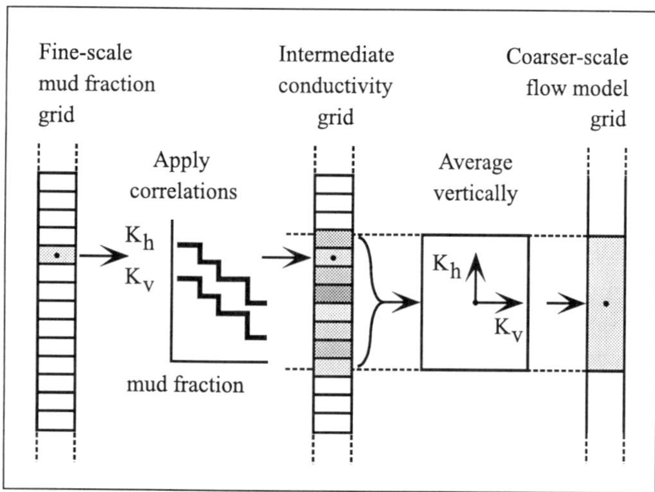

FIG. 7.—Process for translating the fine-scale mud-fraction grid into the coarser scale flow- model conductivity grid.

The resulting horizontal and vertical conductivity fields are shown in Figures 8 and 9. In these figures a wire-frame mesh outlines the finite-element flow grid, and the finite-element centroidal conductivity values have been linearly interpolated. Both conductivity fields are seen to be highly heterogeneous in the upper part of the aquifer system. At the ground surface low-conductivity zones in the top layer of elements in the upper far corner of each grid are defined in order to model low-infiltration zones corresponding to industrial and capped areas.

Recharge and Drain Boundary Conditions

The present methodology permits low-permeability zones to crop out at several locations, as shown in Figure 5, and reflected in Figures 8 and 9. These detailed features, together with a complex, a priori unknown seepline, make conventional manual specification of model recharge and drainage conditions a complicated and tedious process. Where low-permeability zones crop out, a lower infiltration rate should be specified compared to the average recharge so that computed hydraulic heads do not exceed ground elevation. Physically, most rainfall runs off low-conductivity areas, resulting in low infiltration and heads not exceeding ground elevation. Drain boundary conditions are needed wherever computed hydraulic head exceeds the ground elevation so that groundwater may properly discharge at the surface under these head conditions. Conversely, recharge boundary conditions should be specified wherever computed hydraulic head is below the ground surface. Because computed heads are not known beforehand, a manual trial-and-error process is conventionally required to determine whether a drain or recharge boundary condition is appropriate. If a recharge condition is needed, the appropriate local infiltration rate also must be specified.

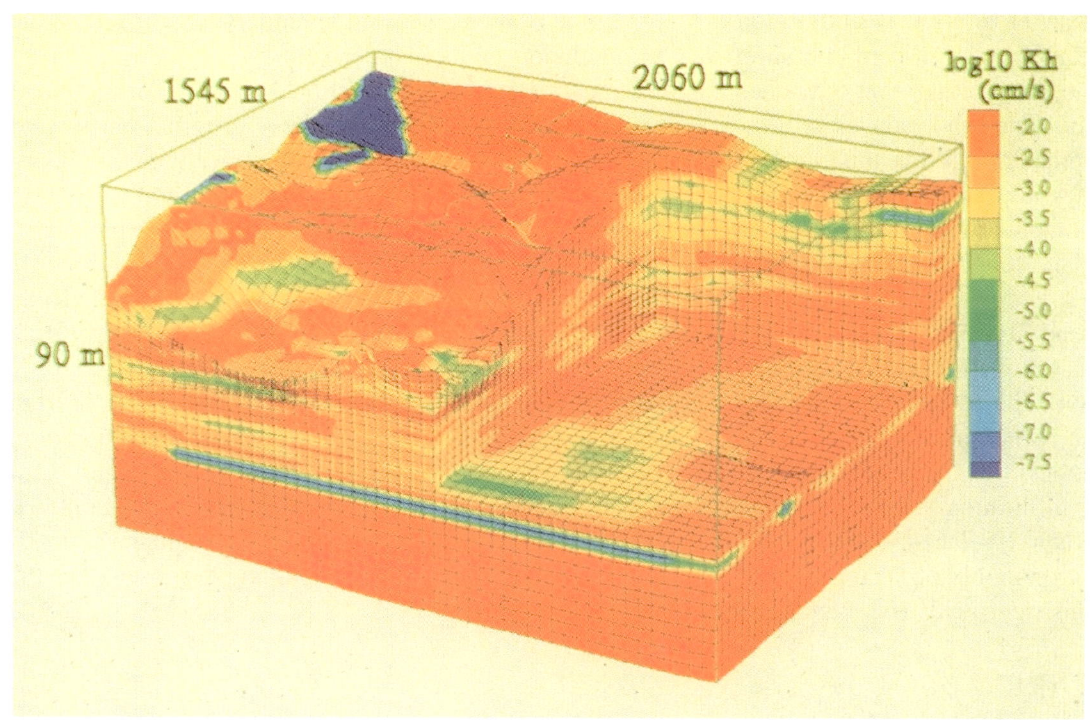

FIG. 8.— Three-dimensional horizontal conductivity distribution on the flow-model finite-element mesh.

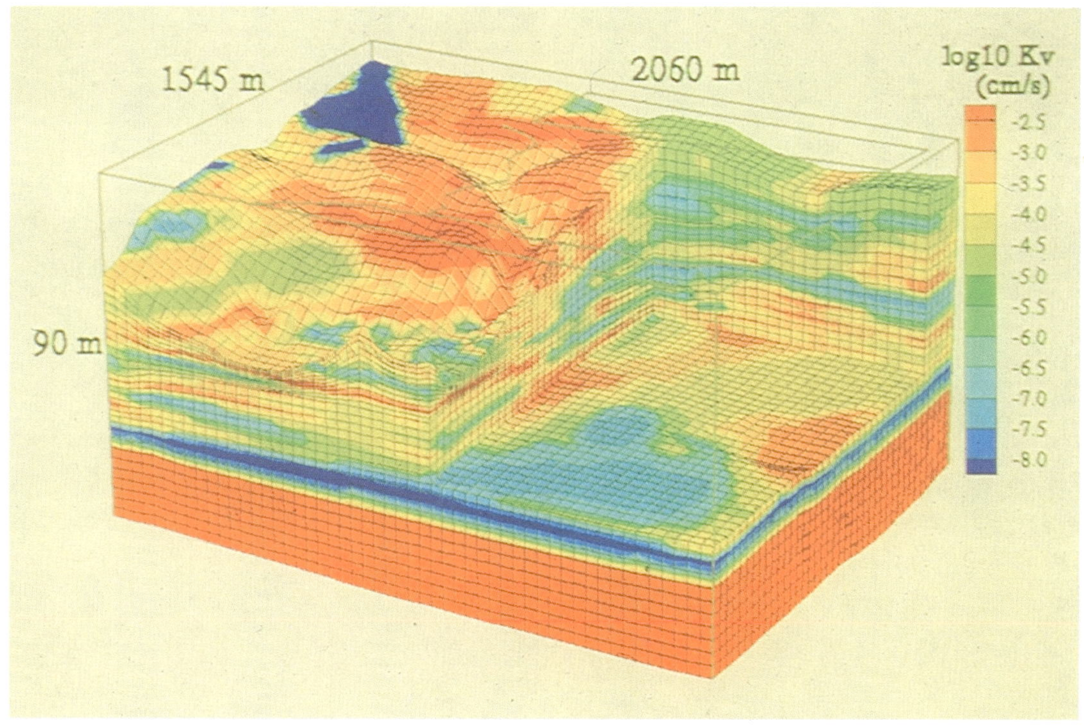

FIG. 9.— Three-dimensional vertical conductivity distribution on the flow-model finite-element mesh.

Huyakorn et al. (1986) implemented an automated process for selecting a drain versus recharge boundary condition following Neuman et al. (1974) and Rulon (1984). The procedure involves a Picard iteration strategy embedded in the flow code that switches between boundary condition types and reduces infiltration as a seepline is approached. With this method convergence difficulties may arise for complex terrain; consequently, additional computational overhead is required to update boundary conditions during the iteration process.

The deficiencies expressed above can be eliminated by combining the concepts of recharge and drainage into a single boundary condition. The basic idea is that locally the surface is either recharging or draining the subsurface, and a continuous transition should exist between these conditions. Infiltration should occur for negative pressure head (water level below the ground surface), and aquifer discharge should occur for positive pressure head. To be consistent with the continuity needs of the Newton-Raphson iterative solver employed in FACT, the overall function representing this "combined" recharge/drain boundary condition should be continuous in its first derivative.

Figure 10 presents a combined recharge/drain boundary condition that meets the above criteria. When water level is well below the ground surface, recharge occurs at the maximum rate permitted locally. As the pressure head approaches zero, recharge is smoothly reduced to zero. For positive pressure head, the surface drains the aquifer at a rate proportional to the pressure head. To the left of the transition zone, the combined recharge/ drain boundary condition is exactly the same as the conventional recharge boundary condition. To the right of the transition zone, the combined recharge/drain boundary condition is identical to a typical drain boundary condition. The transition zone reflects a nonlinear region connecting two limiting linear boundary conditions. The mathematical formulation chosen for this function is as follows:

$$Q_c = \begin{cases} Q_R & \text{for} & \psi \leq \dfrac{3}{2}\hat{\psi} \\[2ex] \dfrac{Q_R}{8}\left[7 - 2x - x^2\right] & \text{for} & \dfrac{3}{2}\hat{\psi} < \psi < \dfrac{1}{2}\hat{\psi} \\[2ex] -M_D\psi & \text{for} & \dfrac{1}{2}\hat{\psi} \leq \psi \end{cases} \tag{3}$$

where

$$\psi = h - z_c, \tag{4}$$

$$Q_R = A_D R_{max}, \tag{5}$$

$$M_D \equiv A_D\left(\frac{K}{b}\right)_D, \tag{6}$$

$$x \equiv 2\left(\frac{\hat{\psi} - \psi}{\hat{\psi}}\right), \tag{7}$$

$$\hat{\psi} \equiv -\frac{Q_R}{M_D}, \tag{8}$$

and

$$h \equiv \text{hydraulic head},$$

$$z_c \equiv \text{elevation of combination boundary condition},$$

Q_c ≡ volumetric source or sink from combined effects of recharge and drainage,

R_{max} ≡ maximum local recharge (43.2 cm/year in this study),

ψ ≡ pressure head,

A_D ≡ area available for recharge and drainage (geometric area),

$(K/b)_D$ ≡ leakance coefficient (18.2 year^{-1} in this study).

Equation (3) represents a two-parameter model requiring the specification of maximum local recharge rate (R_{max}) and surface leakance coefficient $(K/b)_D$. The level of ponding along a seepage face can be adjusted by varying the magnitude of the surface leakance coefficient. Equation (3) is applied at every node over the entire top surface of the flow model in this study. Seepage faces are automatically established during the iterative solution of the nonlinear flow and boundary condition equations.

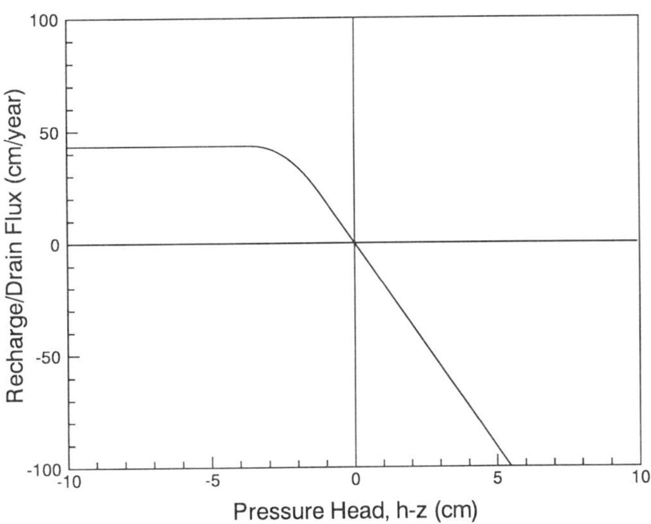

FIG. 10.—Combined recharge/drain boundary condition applied over the entire top surface of the flow-model finite-element mesh.

FLOW MODEL RESULTS AND DISCUSSION

FACT flow-model results for the SRS burial ground are shown in Figures 11a and 12a. The results were generated from the optimal synthetic conductivity fields depicted in Figures 8 and 9. The root-mean-square residual between steady-state model and time-average measured hydraulic heads is 1.0 m (3 ft). Figure 11a shows 3D pathlines originating from the burial ground that have been projected onto a horizontal plane overlaying water saturation at the top surface. The seeplines surrounding Fourmile Branch conform closely to field-observed locations. As illustrated, groundwater pathlines originating from the burial ground converge into a preferred pathway that terminates at a small tributary to Fourmile Branch, consistent with contaminant monitoring data. In addition, the local convergence of pathlines observed in Figure 11a indicates the presence of finer scale preferred pathways. Figure 12a shows 3D pathlines projected onto cross section B-B' in Figure 11a with mud fraction as the background. This cross section coincides with the preferred pathway for groundwater flow from the burial ground and highlights vertical flow patterns. The example pathlines reveal a complex flow field resulting from heterogeneities incorporated into the conductivity field. According to Figure 12a, contaminant migration from the burial ground can be expected to follow multiple distinct pathways within the same aquifer, corresponding to regions of low mud content.

For comparison, a conventional "layer cake" flow model was created by assigning a uniform conductivity to each hydrostratigraphic unit. The resulting models (Figs. 11b, 12b) exhibit significantly different large-scale flow patterns, despite having similar average conductivities and similar overall hydraulic head residuals (1.2 m r.m.s. for the homogeneous layer model). For the heterogeneous model (Fig. 11a), groundwater flows westward in the eastern portion of the burial ground and then southward at the west end. Contaminant monitoring at perimeter wells confirms this overall flow pattern. The westward flow reflects high mud content beneath the entire east end of the burial ground, parts of which are visible in Figure 5. In contrast, the "layer cake" model could not be made to behave in a similar manner for laterally homogeneous properties. Instead, groundwater flowed mostly to the south throughout the entire burial ground (Fig. 11b).

As shown in Figure 12a, groundwater beneath the burial ground in the heterogeneous model flows southward to Fourmile Branch, consistent with monitoring data, which indicate a groundwater flow divide just

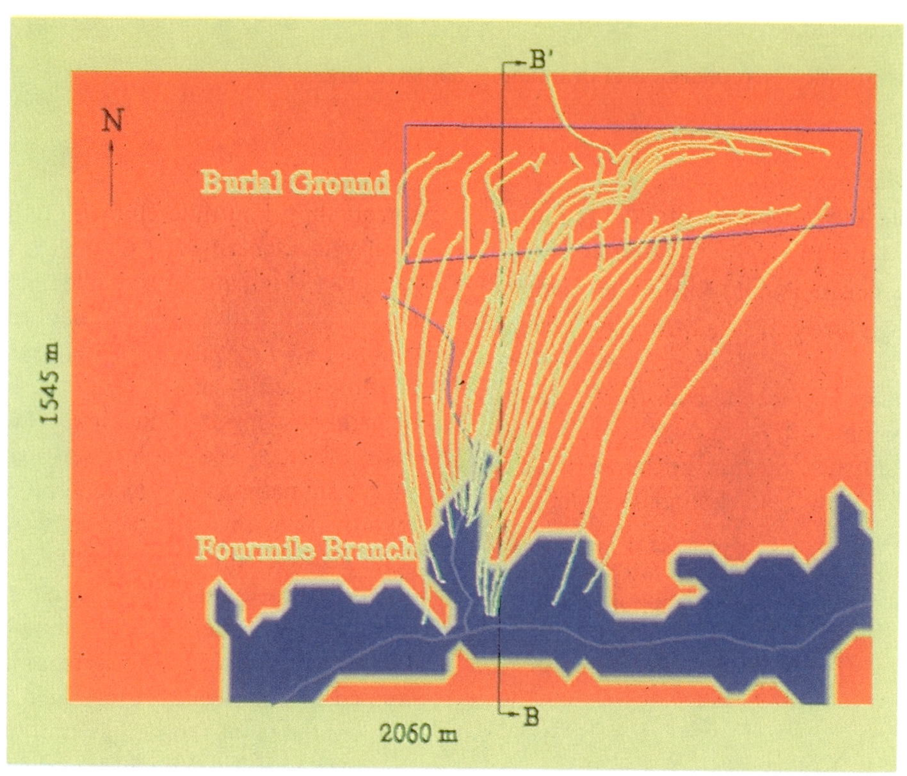

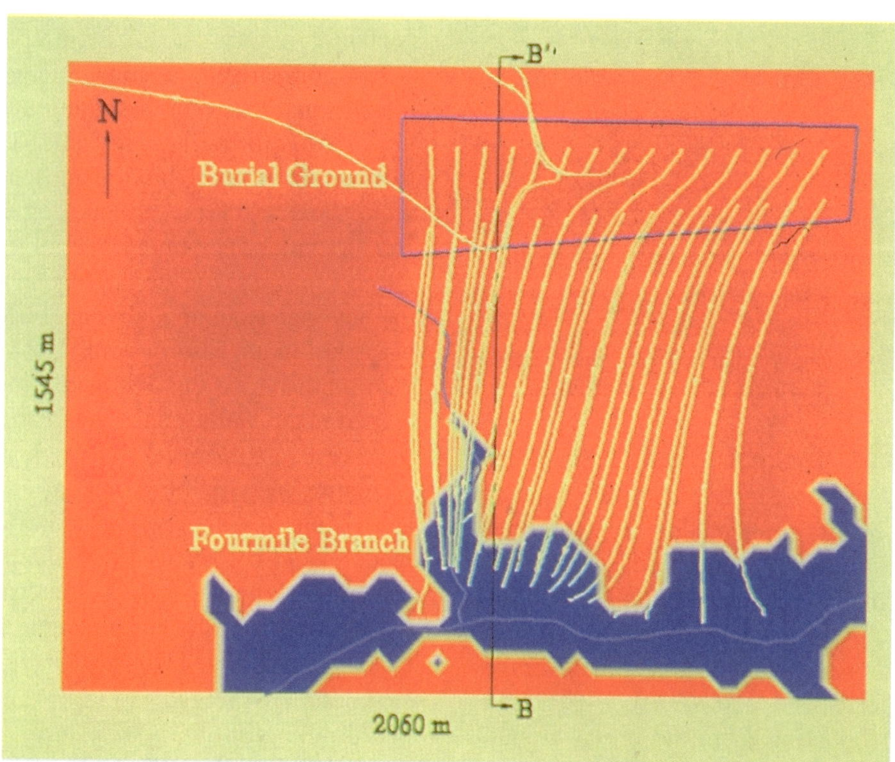

Fig. 11.—Three-dimensional pathlines originating from the SRS burial ground projected onto a horizontal plane overlaying water saturation. Conductivity field (a) (top) depicted by Figs. 8 and 9, (b) (bottom) based on assumption of homogeneous aquifer/aquitard layers.

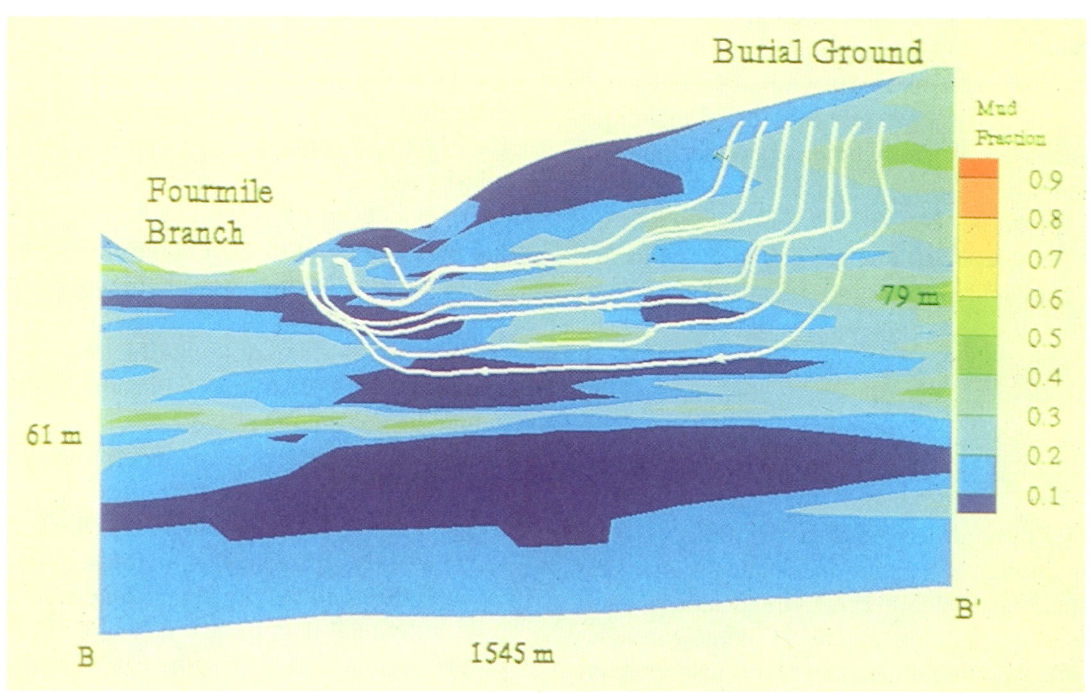

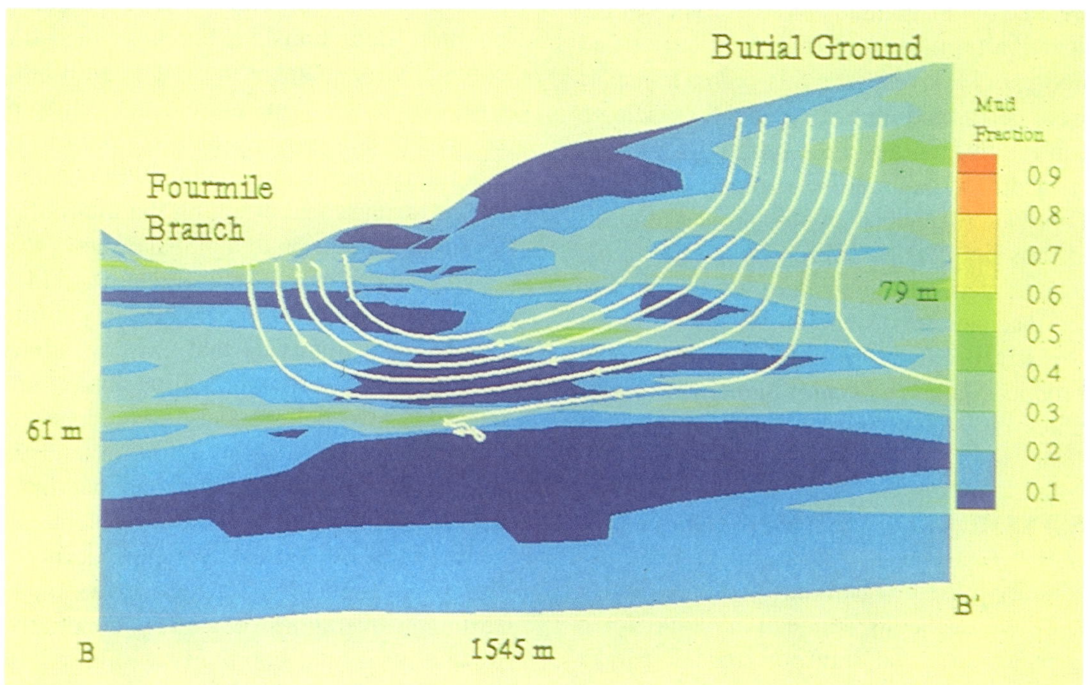

FIG. 12.—Three-dimensional pathlines projected onto cross section B-B' in Figure 11 with mud fraction as the background. Conductivity field (a) (top) depicted by Figs. 8 and 9, (b) (bottom) based on assumption of homogeneous aquifer/aquitard layers.

north of the burial ground. The homogeneous model does not account for the high mud content along the northern edge of the burial ground (Figs. 12a, b) and allows some groundwater to flow northward. The homogeneous model certainly can be improved by manually creating zones of variable K_h/K_v and/or additional layers to reflect the lithologic data, as is typically done. Results similar to those produced by the present methodology also can be obtained with sufficient detail and effort. In this case the present automated technique offers a substantial savings in labor, more so than improved results.

A second probable benefit of the heterogeneous model is the presence of multiple parallel pathways available for contaminant transport, as noted above and shown in Figure 12a. As discussed by Brusseau (1994), for example, contaminant dispersion is enhanced by aquifer heterogeneities at all scales. Heterogeneity produces a varying velocity field that transports portions of the contaminant plume at different rates. Heterogeneity not explicitly incorporated into the flow and transport model through the conductivity field must be accounted for through a separate mechanical dispersion term. The dispersivity value in this term is empirical and scale dependent and requires extensive monitoring data to define accurately. Fine-scale flow and transport models that capture as much aquifer heterogeneity as possible can reduce the magnitude and importance of the empirical dispersion term, leading to more accurate and reliable predictions. To investigate this point, transport simulations were performed for each model by placing a tracer concentration within the burial ground. Unfortunately, a direct, quantitative comparison of the conventional and heterogeneous models proved unfeasible because of the significant differences in large-scale flow patterns exhibited by the models, which masked differences in mechanical dispersion. A contributing factor is the relatively coarse resolution of the model grid, which produces large (masking) numerical dispersion and enables only a modest potential increase in mechanical dispersion

ASSESSMENT OF METHODOLOGY

The methodology presented here has proven effective in generating a heterogeneous conductivity field for a SRS burial ground flow and transport model. Model conductivity values determined from actual lithologic data provide a more realistic picture of subsurface heterogeneity, compared to traditional modeling approaches that produce a small number of relatively homogeneous layers. The approach also improves large-scale flow patterns and dispersive transport compared to conventional methods. Preferential pathways for contaminant migration, inferred from field observations, are replicated in the model without the need to artificially create zones of high conductivity. The concepts are general and can be applied to other sites. The resulting conductivity fields may be used with other finite-element or finite-difference groundwater codes. Nevertheless, improvements and extensions can and should be considered.

As evidenced by large data scatter in Figure 6, translating lithologic information into hydraulic conductivity could be greatly improved. Utilizing information about grain size distribution would help, but Riha's (1993) study suggests that additional factors, such as cementation, pore size distribution, bedding type, and others should be considered as well.

It would be beneficial to incorporate laboratory conductivity measurements more directly into the process for generating a conductivity field. Presently, these data only weakly influence the final conductivity field through the conductivity versus mud-fraction correlation (i.e., individual data points in Figure 6 have little effect on the correlation). The same can be said of conductivity information derived from slug and pump tests, for example. One solution is to omit the mud-fraction interpolation step, translate borehole mud-fraction data to conductivity, augment these data with laboratory and in situ conductivity measurements, and interpolate the composite conductivity data set onto a 3D grid. Then, laboratory and in situ conductivity measurements would have a strong local effect on the interpolated conductivity field, in addition to a global effect.

The interpolated mud-fraction field displayed in Figure 5 appears to underestimate confining zone continuity. Clay intervals that exist in adjacent cores are slightly offset vertically and unconnected in the model; in reality, we believe these should be connected. The "stratified" interpolation algorithm tends not to connect these intervals because they lie within different horizontal gridding layers. This problem is alleviated to some extent by vertical averaging during the transfer of fine-scale conductivity to the flow- model grid. Also, harmonic averaging of vertical conductivities assumes perfect horizontal continuity within a grid element and slightly counteracts underestimation of continuity between elements. The conformal gridding option in the EarthVision® software could help achieve higher confining-zone continuity by incorporating known variations in stratal dip into the gridding process. Selecting a coarser vertical resolution for the interpolated grid would increase interconnectedness, as would a larger vertical influence factor. These options tend to blur vertical contrasts, however, leaving conformal gridding as the preferred approach.

CONCLUSIONS

Based on the flow-modeling results of this study we can conclude:

1. Fine-scale, heterogeneous hydraulic conductivity fields can be successfully generated directly from lithologic data using the general methods described herein.

2. These hydraulic conductivity fields appear to provide a more realistic picture of subsurface hydrologic heterogeneity than conventional "layer cake" modeling approaches.

3. The approach improves large-scale flow patterns and dispersive transport in groundwater modeling. For the SRS burial ground application, field-observed preferential pathways for contaminant migration are replicated without the need to artificially create zones of high conductivity.

ACKNOWLEDGMENTS

This work was supported by Westinghouse Savannah River Company under U.S. Department of Energy Contract No. DE-AC09-89SR18035. We thank Dr. Robert Ehrlich for reviewing a draft of the manuscript.

REFERENCES

AADLAND, R. K., GELLICI, J. A., AND THAYER, P. A., 1995, Hydrogeologic framework of west-central South Carolina: South Carolina Department of Natural Resources, Water Resources Division Report 5, 200 p.

AHUJA, L. R., CASSEL, D. K., BRUCE, R. R., AND BARNES, B. B., 1989, Evaluation of spatial distribution of hydraulic conductivity using effective porosity data: Soil Science, v. 148, p. 404-411.

BEARD, D. C. AND WEYL, P. K., 1973, Influence of texture on porosity and permeability of unconsolidated sand: American Association of Petroleum Geologists Bulletin, v. 57, p. 349-369.

BIELLA, G., LOZEJ, A., AND TABACCO, I., 1983, Experimental study of some hydrogeological properties of unconsolidated porous media: Ground Water, v. 21, p. 741-751.

BLEDSOE, H. W., AADLAND, R. K., AND SARGENT, K. A., 1990, Baseline hydrogeologic investigation — Summary report (U): Aiken, S.C., Westinghouse Savannah River Company Report WSRC-RP-90-1010, 40 p.

BOMAN, G. K., MOLZ, F. J., GUVEN, O., 1995, An evaluation of interpolation methodologies for generating three-dimensional hydraulic property distributions from measured data: Ground Water, v. 22, p. 247-258.

BRANNAN, J. R., AND HASELOW, J. S., 1993, Compound random field models of multiple scale hydraulic conductivity: Water Resources Research, v. 29, p. 365-372.

BRUSSEAU, M. L., 1994, Transport of reactive contaminants in heterogeneous porous media: Reviews of Geophysics, v. 32, p. 285-313.

CARMAN, P. C., 1937, Fluid flow through granular beds: Transactions, Institution of Chemical Engineers, London, v. 15, p. 150-156.

CROFT, M. G., 1971, A method of calculating permeability from electric logs: U. S. Geological Survey Professional Paper 750-B, p. B265-B269.

DAGAN, G., 1985, Stochastic modeling of groundwater flow by unconditional and conditional probabilities — The inverse problem: Water Resources Research, v. 21, p. 65-72.

DOMENICO, P. A., AND SCHWARTZ, F. W., 1990, Physical and Chemical Hydrogeology: New York, John Wiley & Sons, 824 p.

DOVETON, J. H., 1986, Log Analysis of Subsurface Geology, Concepts and Computer Methods: New York, John Wiley & Sons, 273 p.

FALLAW, W. C., AND PRICE, V., 1995, Stratigraphy of the Savannah River Site and vicinity: Southeastern Geology, v. 35, p. 21-58.

FLINT, S. S., AND BRYANT, I. D., EDS., 1993, The Geological Modeling of Hydrocarbon Reservoirs and Outcrop Analogues: International Association of Sedimentologists, Special Publication 15, 269 p.

FRANZMEIER, D. P., 1991, Estimation of hydraulic conductivity from effective porosity data for some Indiana soils: Soil Science Society of America Journal, v. 55, p. 1801-1803.

FREEZE, R. A., AND CHERRY, J. A., 1979, Groundwater: Englewood Cliffs, N. J., Prentice-Hall, 604 p.

HARLEMAN, D. R. F., MELHORN, P. F., AND RUMER, R. R., JR., 1963, Dispersion-permeability correlation in porous media: Journal of the Hydraulics Division, Proceedings of the American Society of Civil Engineers, v. 89, p. 67-85.

HAZEN, A., 1892, Some physical properties of sands and gravels with special reference to their use in filtration: Lawrence, Mass., 24th Annual Report of the State Board of Health of Massachusetts for 1892, 24 p.

HEWETT, T. A., 1986, Fractal distributions of reservoir heterogeneity and their influence on fluid transport: SPE Paper 15386, presented at the 61st SPE Annual Technical Conference and Exhibition, New Orleans, La., 16 p.

HEWETT, T. A., AND BEHRENS, R. A., 1988, Conditional simulation of reservoir heterogeneity with fractals, in Proceedings, 63d SPE Annual Technical Conference and Exhibition, Houston, Texas: SPE Paper 18326, p. 645-660.

HUYAKORN, P. S., AND PANDAY, S., 1992, VAM3DCG, variably saturated analysis model in three-dimensions with preconditioned conjugate gradient matrix solvers — Documentation and user's guide, ver. 2.4: Herndon, Virginia, HydroGeoLogic, Inc., 274 p.

HUYAKORN, P. S., PANDAY, S., AND BIRDIE, T., 1991, SAFT3D, subsurface analysis finite element model for flow and transport in 3 dimensions, ver. 1.3 — Documentation and user's guide: Herndon, Virginia, HydroGeoLogic, Inc., 288 p.

HUYAKORN, P. S., SPRINGER, E. P., GUVANASEN, V., AND WADSWORTH, T. D., 1986, A three-dimensional finite-element model for simulating water flow in variably saturated porous media: Water Resources Research, v. 22, p. 1790-1808.

HYNDMAN, D. W., HARRIS, J. M., AND GORELICK, S. M., 1994, Coupled seismic and tracer test inversion for aquifer property characterization: Water Resources Research, v. 30, p. 1965-1977.

ISAAKS, E. H., AND SRIVASTAVA, R. M., 1989, Applied Geostatistics: New York, Oxford University Press, 561 p.

JABRO, J. D., 1992, Estimation of saturated hydraulic conductivity of soils from particle size distribution and bulk density: Transactions of the American Society of Agricultural Engineers, v. 35, p. 557-560.

JORGENSEN, D. G., 1989, Using geophysical logs to estimate porosity, water resistivity, and intrinsic permeability: U.S. Geological Survey Water-Supply Paper 2321, 24 p.

JOURNEL, A. G., 1986, Nonparametric estimation and qualitative information —The soft kriging approach: Mathematical Geology, v. 15, p. 268-286.

JOURNEL, A. G., AND HUIJBREGTS, C. J., 1978, Mining Geostatistics: New York, Academic Press, 600 p.

KEGLEY, W. P., FALLAW, W. C., SNIPES, D. S., BENSON, S. M., AND PRICE, V., JR., 1994, Textural factors affecting permeability at the MWD well field, Savannah River Site, Aiken, South Carolina: Southeastern Geology, v. 34, p. 139-161.

KELLY, W. E., 1977, Geoelectric sounding for estimating aquifer hydraulic conductivity: Ground Water, v. 15, p. 420-425.

KEYS, S. W., AND MACCARY, L. M., 1971, Application of borehole geophysics to water-resources investigations: U.S. Geological Survey Techniques of Water-Resources Investigations, Book 2, 123 p.

KRUMBEIN, W. C., AND MONK, G. D., 1943, Permeability as a function of the size parameters of sedimentary particles: American Institute of Mining and Metallurgical Engineers, Technical Publication 1492, p. 153-163.

LAHM, T. D., BAIR, E. S., AND SCHWARTZ, F. W., 1995, The use of stochastic simulations and geophysical logs to characterize spatial heterogeneity in hydrogeologic parameters: Mathematical Geology, v. 27, p. 259-278.

LAKE, L. W., AND CARROLL, H. B., JR., EDS., 1986, Reservoir Characterization: Orlando, Academic Press, 659 p.

MASCH, F. D., AND DENNY, K. J., 1966, Grain size distribution and its effect on the permeability of unconsolidated sands: Water Resources Research, v. 2, p. 665-677.

MOLZ, F. J., AND BOMAN, G. K., 1993, A fractal-based stochastic interpolation scheme in subsurface hydrology: Water Resources Research, v. 29, p. 3769-3774.

NEUMAN, S. P., FEDDES, R. A., AND BRESLER, E., 1974, Finite element simulation of flow in saturated-unsaturated soils considering water uptake by plants: Report for Project ALO-5WC-77, Hydrodynamics and Hydraulic Engineering Laboratory, Technion, Hafia, Israel, 104 p.

PHILLIPS, F. M., AND WILSON, J. L., 1989, An approach to estimating hydraulic conductivity spatial correlation scales using geological characteristics: Water Resources Research, v. 25, p. 141-143.

PUCKETT, W. E., DANE, J. H., AND HAJEK, B. F., 1985, Physical and mineralogical data to determine soil hydraulic properties: Soil Science Society of America Journal, v. 49, p. 831-836.

RIHA, B. D., 1993, Predicting saturated hydraulic conductivity for unconsolidated soils from commonly measured textural properties: Unpub. M.S. Thesis, Clemson University, Clemson, S.C., 87 p.

RUBIN, Y., AND JOURNEL, A. G., 1991, Simulation of non-Gaussian space random functions for modeling transport in groundwater: Water Resources Research, v. 27, p. 1711-1721.

RULON, J., 1984, The development of multiple seepage faces along heterogeneous hillsides: Unpub. Ph.D. Thesis, University of British Columbia, Vancouver, Canada, 161 p.

SUN, N.-Z., AND YEH, W. W.-G., 1990, Coupled inverse problems in groundwater modeling, part 1 — Sensitivity analysis and parameter identification: Water Resources Research, v. 26, p. 2507-2525.

THAYER, P. A., SMITS, A. D., HARRIS, M. K., AMIDON, M. B., AND LEWIS, C. M., 1993, Hydrostratigraphic maps of the General Separations Area (GSA), Savannah River Site (SRS), Aiken, South Carolina, Phase II (U): Aiken, S. C., Westinghouse Savannah River Company Report WSRC-RP-94-40, 86 p.

TIMUR, A., 1968, An investigation of permeability, porosity, and residual water saturation relationships for sandstone reservoirs: Log Analyst, v. 9, p. 8-17.

URISH, D. W., 1981, Electrical resistivity-hydraulic conductivity relationships in glacial outwash aquifers: Water Resources Research, v. 17, p. 1401-1408.

WAGNER, B. J., AND GORELICK, S. M., 1987, Optimal groundwater quality management under parameter uncertainty: Water Resources Research, v. 23, p. 1162-1174.

WEBB, E. K., 1994, Simulating the three-dimensional distribution of sediment units in braided-stream deposits: Journal of Sedimentary Research, Section B, v. 64, p. 219-231.

WENDT, W. A., SAKURAI, S., AND NELSON, P. H., 1986, Permeability prediction from well logs using multiple regression, in Lake, L. W., and Carroll, H. B., Jr. eds., Reservoir Characterization: Orlando, Academic Press, p. 181-221.

GEOSTATISTICAL ANALYSIS OF FACIES DISTRIBUTIONS: ELEMENTS OF A QUANTITATIVE FACIES MODEL

DAVID F. DOMINIC, ROBERT W. RITZI, JR., EDWARD C. REBOULET, AND AMBER C. ZIMMER

Department of Geological Sciences and Center for Ground Water Management
Wright State University, Dayton, Ohio 45435

ABSTRACT: Predicting flow and contaminant transport in sedimentary aquifers requires three-dimensional, quantitative characterizations of facies distributions, including the proportions, mean lengths and spatial correlation of the facies. Sedimentary facies analysis focuses on these same characteristics and, when coupled with geostatistical methodologies, can provide the necessary quantification of facies distributions as well as ideas about the origin and typical geometry of facies, what we call *quantitative facies models*. Because they incorporate an understanding of depositional processes, quantitative facies models must be based upon facies distributions that are genetically related. In aquifer systems composed of facies having sharp permeability contrasts, and in which indicator geostatistics can be utilized, genetically related facies assemblages are the appropriate statistical populations.

We illustrate these points by considering a portion of the Miami Valley aquifer system in southwestern Ohio, where three facies assemblages can be defined. These assemblages, which differ in mean facies proportions, mean facies thicknesses and lateral facies dimensions are attributed to different combinations of depositional processes associated with Pleistocene glaciation.

INTRODUCTION

Hydrogeologists increasingly look to sedimentologists for information about sedimentary aquifers, especially those in which contaminant transport is of concern. Hydrogeologists generally need three-dimensional, quantitative characterizations of sediments and sedimentary rocks that can be directly incorporated into flow and transport models. As Carle and Fogg (1996) have pointed out, these quantitative characterizations include proportions, mean length and spatial correlation or juxtapositioning of differing sediment bodies or facies. Facies analysis, a common approach to studying sedimentary deposits, focuses on these same characteristics but generally in a qualitative way. Although the aim of facies analysis is a clear description of the deposits, that description often is used to interpret the processes and environments of deposition. In contrast, hydrogeologists need to predict the occurrence of specific facies at specific locations (within a modeled domain, for example). Schwartz (1977), Anderson (1990) and Debarats (1990) have shown that transport through assemblages of facies with distinctly different hydraulic properties cannot be appropriately modeled unless the individual facies and the preferential flow pathways created by the boundaries between them are represented. Furthermore, hydrogeologists often need to quantify the probability at which specific facies occur (e.g., see Freeze et al., 1990). Anderson (1990) clearly illuminated the gap between what sedimentologists have produced and what hydrogeologists now need.

In recent years, a body of work has been growing seeking to fill that gap. A promising approach is to quantify aspects of facies distributions at sites where data are abundant in the hope that the generalized results will be useful at analogous sites where data are sparse. We refer to these generalized results as *quantitative facies models*, a term we define and illustrate more completely below. One of the main issues we arrive at and focus on in this paper is that developing quantitative facies models requires identifying statistically meaningful populations. We have performed new analyses in the glacigenic Miami Valley aquifer in order to illustrate this issue.

Heterogeneity and Facies

In most sedimentary reservoirs, heterogeneity is created by the complex arrangement of discrete regions, within which rock or sediment properties, and the hydraulic properties they control, are relatively uniform (Anderson, 1991). These regions vary in scale depending on how narrowly their uniformity is defined. In some cases slight but potentially important differences in porosity and permeability occur at the scale of individual beds, whereas elsewhere the differences within entire stratigraphic formations may be less important than the contrasts between them. In that the scope and objectives of an investigation reasonably dictate the degree of uniformity and therefore the scale of these regions, they are best categorized as facies, a term not restricted to any specific scale.

Koltermann and Gorelick (1996) reviewed many approaches to representing heterogeneity in sedimentary aquifers. They divided these methods into three categories based on whether they seek to (1) imitate the processes of groundwater flow or sedimentary deposition (process-imitating methods), (2) imitate the spatial arrangement of aquifer materials (structure-imitating methods), or (3) describe the distributions of hydrologic properties based on local and regional data combined with conceptual facies models (descriptive methods). Any of these approaches

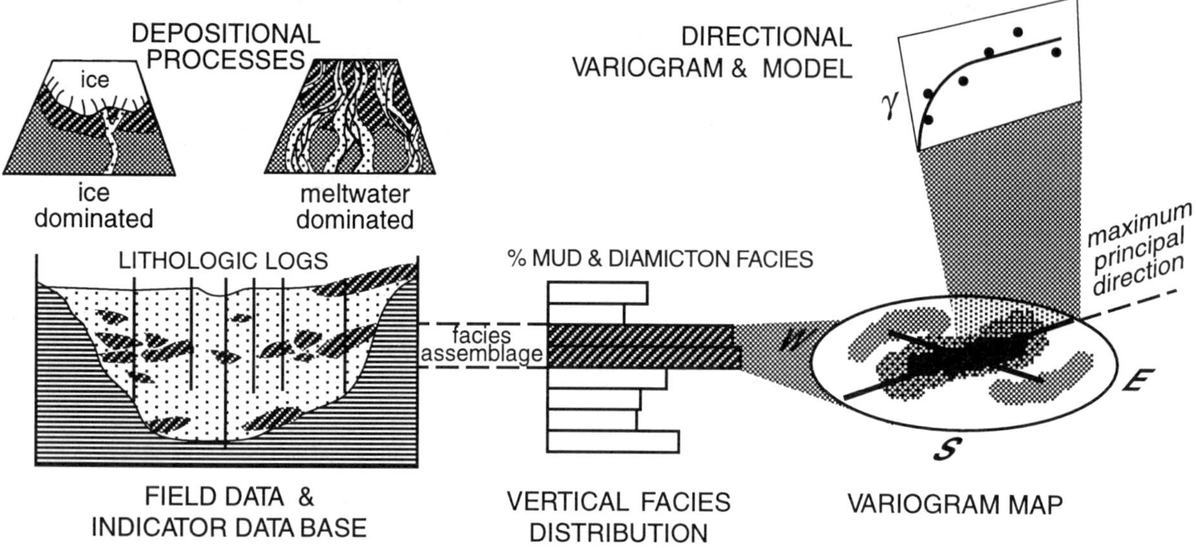

FIG. 1—Conceptual approach to defining elements of a quantitative facies model. As shown schematically in the lower left, this buried-valley aquifer system contains complexly intermixed sand and gravel (stippled) and mud and diamicton (diagonal ruled) facies. Lithologic logs provide the basis for representing facies using an indicator approach. Relevant data populations are identified as areas in which characteristics (facies proportions, mean facies lengths) are similar. The spatial correlation of facies within populations is estimated by computing variograms (as shown on the right) or transition probabilities. Because these data populations are interpreted to be facies assemblages, the geostatistical results support interpretations about the original depositional processes (upper left).

may be used independently, but in order to generate plausible representations of heterogeneity, both process-imitating and structure-imitating methods should be guided by models of aquifer stratigraphy, which is the focus of descriptive methods. Therefore, descriptive methods can contribute to many efforts seeking to model heterogeneity.

When data are sparse and/or facies are complexly distributed, it is generally not possible to identify precise locations of individual facies. In these cases conceptual models of facies may be used either to generalize the description of facies occurrences for a particular site (local facies models) or to help predict the probable occurrences of facies at a new site (generic facies models). Whatever the scale of individual facies, it is often useful to view them as part of larger scale groupings or facies assemblages, which in turn may occur in predictable sequences at a regional scale. Conceptual models of facies are possible at each of these scales. They may be useful for generalizing or predicting (1) the geometry of individual facies, (2) the proportions and spatial distribution of facies within facies assemblages, or (3) trends in the ways facies assemblages replace one another vertically and/or laterally.

When data are numerous, geostatistical methodologies can directly provide useful descriptions of heterogeneity. In aquifer systems composed of two generalized facies having a sharp permeability contrast, the indicator geostatistical approach can be used (e.g., Johnson and Dreiss, 1989). Sharply contrasting facies can be represented by an indicator random variable, $I(x)$,

which assumes a value of either zero (0) or one (1), according to which of the two facies is present at location x. When corrected for the effect of data clustering (e.g., Deutsch and Journel, 1992), the mean of $I(x)$ estimates the proportion of the facies coded as one (1). The spatial correlation of facies can be derived from indicator variograms or transition probabilities (Carle and Fogg, 1996). With the spatial correlation thus defined, indicator kriging can be used to compute the probability that facies will occur at any location (e.g., Johnson and Dreiss, 1989; Ritzi et al., 1994). Alternatively, various simulation algorithms (Journel and Alabert, 1989; Gomez-Hernandez and Srivastava, 1990; Deutsch and Journel, 1992) can be used to generate representative facies distributions.

At specific sites, available lithologic data may be insufficient for geostatistical analyses but may support generalized interpretations of depositional processes. In these cases it is helpful to augment the limited data with additional information. Indeed, Wingle and Poeter (1993) showed that stochastic simulations based on even a small number of data accurately reflected the underlying facies distribution, provided that information on spatial correlation was available in the form of a variogram model. Importantly, they showed that this was true even when a variogram model was characterized by ranges of its parameters (nugget, range, anisotropy in the principal directions) and not by exact values, so long as "expert judgment" could identify the existence and probable direction of anisotropy. Thus, these results confirm the value of what Fogg (1990, p. 2) called the "imaginative geologic interpretation."

Because the data available are limited at most sites, some form of geologic interpretation often is needed. Sedimentary facies models provide interpretive help but often are too qualitative to be of much value. However, when developed from data-rich sites, facies models can include estimates of the proportions of facies and measures of spatial correlation as well as ideas about the origin and typical geometry of facies. Such *quantitative facies models* represent not only an aid to geologic interpretation at analogous sites but the "expert judgment" to quantify that interpretation.

Figure 1 outlines our view of a quantitative facies model. Two key elements of this approach are the geostatistical descriptions (facies proportions, models of spatial correlation expressed as variograms or transition probabilities, anisotropy) and an understanding of the depositional processes. Because the geostatistical descriptions quantify the heterogeneity resulting from specific processes of deposition, they can be used to estimate heterogeneity at sites where similar depositional processes are inferred to have operated. Thus, geostatistical descriptions provide practical links between generalized analogs and site-specific investigations. Likewise, understanding relevant depositional processes helps predict similarities to or deviations from the general model due to local conditions.

Studies following similar approaches at a regional scale include those by Johnson and Dreiss (1989) and Johnson (1995), who evaluated the spatial correlation of hydraulic properties in sediments associated with an alluvial fan; Fogg (1986, 1989), who evaluated two different fluvial environments; and Ritzi et al. (1994, 1995, 1996) and Dominic et al. (1996), who studied a glacigenic aquifer. In contrast, Davis et al. (1993) studied the spatial correlation of permeability in fluvial sediments exposed in a three-dimensional outcrop. Despite the differences in scale, all these studies are similar in linking interpretations of depositional processes with geostatistical results to strengthen the overall description of heterogeneity.

While this approach is not new, it does systematize to some degree the integration of geologic and stochastic methodologies promoted by Fogg (1990). As a synthesis of geostatistical data and sedimentological interpretation, quantitative facies models should be useful at sites where the available information (both hard and soft data) support interpretations about depositional setting but are insufficient to define spatial correlation directly. Another aspect of this integration is that it encourages interpreting the sedimentary data in light of the geostatistical results, in what may be called *statistical stratigraphy*. Johnson (1995, p. 3218) noted that sedimentary deposits are created by a "stochastic assemblage of influences" and that hydrostratigraphic boundaries have both random and structured qualities. Geostatistical methodologies help reveal the structure within sedimentary deposits that might otherwise be masked by their random aspects. Whether any consistency in structure exists among sedimentary aquifers can be judged only after comparing results from numerous examples created by similar depositional processes.

DISCUSSION

Facies and Hydrostratigraphic Facies

Facies are often distinguished by the type of characteristics used to define them: lithofacies are based upon lithologic properties; petrofacies are based upon properties measured in geophysical logs. Units defined in terms of hydraulic properties have been called hydrostratigraphic facies (Fogg, 1986) or hydrogeologic facies (Anderson, 1989). It is important to note, however, that very little of the subsurface is ever sampled in enough detail to define hydraulic properties directly. Therefore, the idea of hydrostratigraphic facies (strictly defined) has more value in theory than in practice. More abundant than measured hydraulic properties are observations of lithofacies, whose characteristics, such as grain size and sorting, exert strong controls on permeability or hydraulic conductivity. The results of Fogg (1986) and Davis et al. (1993), among others, support the idea that sedimentary facies can define the spatial arrangement of hydraulic properties. Thus, hydrostratigraphic facies often can be inferred from lithofacies. In addition, Johnson and Dreiss (1989) showed how several restrictively defined lithofacies can be combined into fewer hydrostratigraphic facies.

Facies Assemblages and Data Populations

An important aspect in developing facies models is identifying vertical intervals and lateral regions that are genetically related. Where lithofacies can be narrowly defined, facies assemblages composed of several lithofacies may be identified and attributed to the operation of differing depositional processes within single environments. In these cases, facies assemblages are the genetically related units of interest. In other cases, the lithofacies are so extremely diverse and/or complexly arranged that it is unhelpful to maintain the full range of diversity, and broader scale distinctions may be more useful. For example, Johnson and Dreiss (1989) generalized various lithofacies of alluvial fan deposits into just two facies, based on grain size (and inferred permeability). Yet even where facies are broadly defined, their distributions within genetically related facies assemblages are the aspects of heterogeneity critical both for interpreting their origin and for modeling groundwater flow. In these cases, different facies assemblages may contain the same two (or more) facies but differ in their proportion, scale, and spatial arrangement.

Seen in the light of geostatistics, facies assemblages correspond to populations of facies occurrences (data populations). This suggests that a fundamental step in geostatistical analysis should be identifying facies assemblages. Furthermore, geostatistical results have value for interpretation, as well as description, only when they come from single data populations. We illustrate this important point below with results from the Miami Valley aquifer system.

ILLUSTRATION

The Great Miami River and its tributaries in southwestern Ohio overlie an extensive aquifer system composed of unconsolidated Pleistocene and Holocene sediments, which partly fill a previously eroded bedrock valley (Figure 2). The Miami Valley aquifer system is similar to other buried-valley aquifers of the western and central glaciated plains of North America in being large and productive (see Krothe and Kempton, 1988, and Lennox et al., 1988, for general reviews). Individual well fields along its 150-km length yield thousands of liters per second, with total production approaching 1 billion L/d (Shindel et al., 1991).

Facies

Because our focus is on hydrogeologic characteristics of the aquifer system, we divide the sediments into just two facies: a sand and gravel facies inferred to have high permeability, and a mud and diamicton facies inferred to have low permeability. The generalized cross section in Figure 1 is meant to show that these two facies are complexly intermixed.

Outcrops reveal details of the uppermost portions of the sand and gravel facies. Individual beds of sand or sand and gravel predominate and range in thickness from one to several meters. Silty beds, always less than 1 m thick, pinch out within several hundred meters, the width of outcrop, whereas coarser-grained beds usually span the entire exposure. A few bed boundaries are concave upward, providing indications of channel geometries. Paleoflow directions inferred from cross-bedding suggest that flow was dominantly to the southwest, parallel to present drainage. The sand and gravel facies in the subsurface is assumed to have characteristics similar to the outcrops. Terms used on lithologic logs include fine sand, sand, sand and gravel, and gravel.

Because the mud and diamicton facies is poorly exposed at the surface, its characteristics are not known in detail. In the northern Miami Valley, terms used on lithologic logs include clay, silt, gravel with clay, hardpan and till. Norris et al. (1948) and Norris and Spieker (1966) interpreted this facies to be till. Dominic et al. (1996) analyzed a 15-cm-diameter core from the Dayton area and interpreted the 17-m-thick zone of diamictons it penetrated to be predominantly lodgment

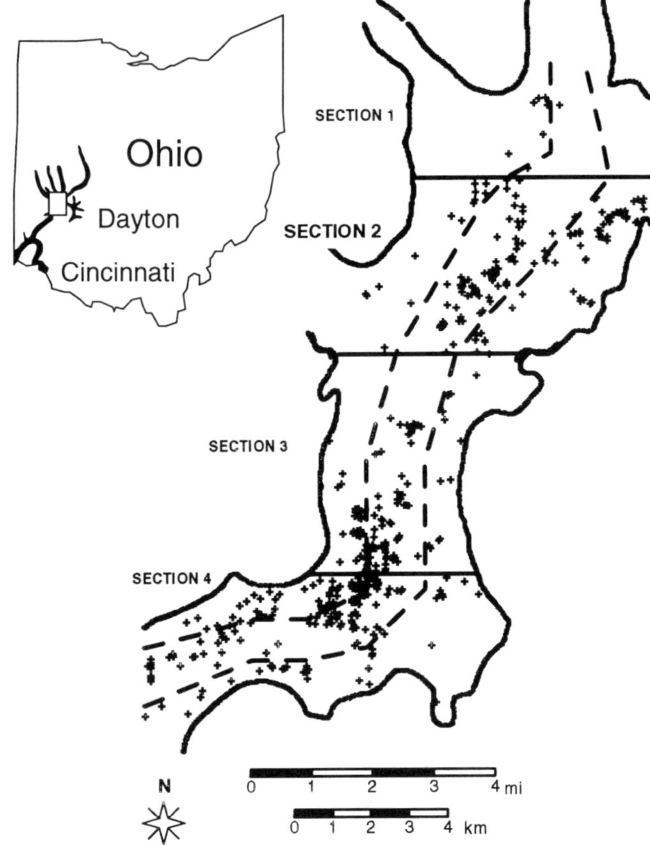

FIG. 2—Location of section 2 and distribution of borehole logs within aquifer system boundaries. Inset shows the location of the Miami Valley aquifer within the state of Ohio and the area of detailed study (rectangle).

till with some evidence of ablation and flow; individual beds of diamicton range from 1 to 5 m in thickness. In the southern Miami Valley (near Cincinnati) this facies is most commonly described as silt or clay. It occurs in beds up to 6 m thick, predominantly at a single elevation. FERMCO (1994) reported rhythmic laminations in some samples. For these reasons, both Brockman (1988) and FERMCO (1994) interpreted the mud facies within the valley near Cincinnati to have been deposited in a proglacial lake. Away from these areas, which have been directly sampled and analyzed, other origins are possible, including ice-contact debris flows (flow till), resedimented colluvium or fine-grained fluvial (vertical accretion) deposits.

Characteristics of the sand and gravel facies can be attributed to deposition by glacial meltwater, predominantly by high-gradient, proglacial streams. In some locations the characteristics of the mud and diamicton facies are known in enough detail to support specific interpretations (e.g., subglacial lodgment near Dayton, proglacial lake(s) near Cincinnati), elsewhere the lithologic descriptions alone do not support specific interpretations. However, these general interpretations can be improved by identifying regions with characteristic proportions and spatial distributions of these two intermixed facies.

Geostatistical Analysis

Figure 2 shows a portion of the aquifer system divided into four sections along its length. Analysis of all these sections, and others, is in progress. To illustrate the importance of identifying facies assemblages, we present results from one of the sections (section 2). For each of 160 boreholes in that section, lithologic descriptions were converted to indicator values. Lithologies of the sand and gravel facies (inferred high permeability) were coded as zero (0), and lithologies of the mud and diamicton facies (inferred low permeability) were coded as one (1). Lithologies were evaluated at each 0.61-m (2-ft) interval of depth, resulting in 11, 293 indicator values in the section.

Analysis by Ritzi et al. (1995, 1996) and Dominic et al. (1996) suggested differences in the distribution of facies from the margins to the middle of the aquifer system. To explore these differences, each section was divided longitudinally into three divisions of roughly equal area. The distribution of facies with depth is shown in Figure 3 for the west, middle and east divisions of the section. Those plots show the proportion of mud and diamicton facies in elevation zones 3.05 m (10 ft) thick. This proportion was estimated for each zone by computing the mean of the declustered indicator data. The cell declustering program, which downweights data locally clustered within one facies, is described by Deutsch and Journel (1992), and its application in this context was discussed by Ritzi et al. (1996).

The mean proportion of mud and diamicton facies in the west division is 0.22, in the middle is 0.17, and in the east is 0.25 (summarized in Table 1). These values were derived by averaging the areally declustered means of elevation zones, which also vertically declusters the data. Standard deviations for the elevation zones are 0.11, 0.09 and 0.13, respectively. Thus, both the mean proportion and the depth variability are similar among the divisions but are slightly lower in the middle.

Vertical transition probabilities (Carle and Fogg, 1996) also were computed for each division. These computations give the probability, t_{km}, of finding facies m at a location distance h away from a location with facies k. Of most interest is the case when k and m represent the same facies, as explained below. As an illustration, Figure 4 shows the results for the middle division of section 2, computed for upward transitions (dip = +90°) with a tolerance angle and bandwidth that ensured pairs were taken from the same borehole. The transition probability plots are fitted with exponential models of the form:

$$t_{kk} = 1-[(1-n_k)(1-exp(-3h/a))] \qquad (1)$$

TABLE 1.—GEOSTATISTICAL RESULTS

	WEST	MIDDLE	EAST
Entire Division			
proportion m&d facies			
mean	0.22	0.17	0.25
standard deviation	0.11	0.09	0.13
mean facies thickness (m)			
mud & diamicton	6.5	2.8	3.6
sand & gravel	12.4	11.0	11.4
Facies Assemblage A			
mean proportion m&d facies	0.42	0.27	0.46
mean facies thickness (m)			
mud & diamicton	7.3	6.7	4.9
sand & gravel	6.6	6.1	8.3
mean facies length (m)			
mud & diamicton			
N-S		230–970	
E-W		140–480	
Facies Assemblage B			
mean proportion m&d facies		0.049	0.032
mean facies thickness (m)			
mud & diamicton		1.6	–
sand & gravel		>10	–

where n_k is the proportion of facies k, h is lag (separation) distance and a is the range. Carle and Fogg (1996) proved that for these autotransition probabilities, the projection to the abscissa of the tangent to this model at $h = 0$ gives the mean length of the facies, l_k. For exponential models, the slope of this line is $-3(1-n_k)/a$, and thus the mean facies length is:

$$l_k = a / (3(1-n_k)) \qquad (2)$$

Because transition probabilities were computed in the vertical direction, the facies lengths derived in this way give the mean thickness of facies; these estimates are compiled in Table 1.

The mean thicknesses of the mud and diamicton facies for the west, middle and east divisions are 6.5 m, 2.8 m and 3.6 m, respectively. For the sand and gravel facies the respective mean thicknesses are 12.4 m, 9.8 m and 11.4 m. Thus, in each division the sand and gravel facies has more than twice the average thickness of the mud and diamicton facies. For each facies, thickness is similar across the divisions but slightly lower in the middle division.

Some points regarding the computation of the vertical transition probability are noted here. According to equation (2), as the proportion of facies k approaches one, its mean length approaches infinity and may exceed the actual thickness of available data. Also, transition probabilities generally were symmetric, i.e., only trivial differences were noted between those computed with +90° and -90° dip. The mean thickness presented in Table 1 is the average of that computed in each of

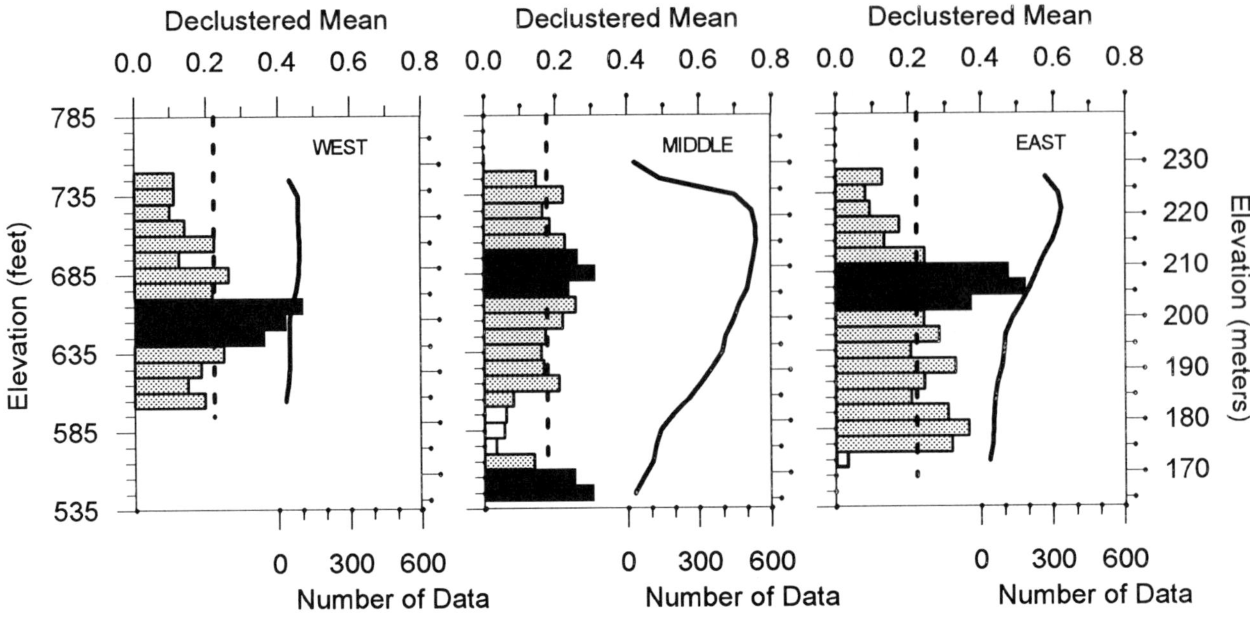

FIG 3.—Depth profiles of the proportion of mud and diamicton facies (given by the declustered mean), shown with the bars and referenced to the upper abscissa. The solid line gives the number of data from which the proportion is computed and is referenced to the lower abscissa. The vertical dashed line indicates the average proportion for the division, as computed from the mean of the bars. Bars shaded black are facies assemblage A, those shaded white are facies assemblage B, and those shaded gray are facies assemblage C.

those directions.

The similarity of these results (mean and standard deviation of facies proportions, facies thicknesses) across the valley suggests that the lateral divisions do not represent strongly different data populations. However, the data can be divided vertically, as follows.

Facies Assemblages

Figure 3 shows that each division contains at least one interval whose proportion of mud and diamicton facies is consistently above the mean. These intervals, referred to as facies assemblage A, were delineated by choosing 9.15-m-thick (30-ft) elevation intervals within which the proportion exceeds the overall mean by more than one standard deviation. In the middle division the two lowermost elevation zones delineate a 6.2-m-thick interval of facies assemblage A. Note that the definition of these zones includes arbitrary characteristics (i.e., that they be approximately 9 m thick and horizontal) that are driven by practical rather than sedimentological considerations. Hydrogeo-logically, these intervals likely act as aquitards. Indeed, Norris and Spieker (1966) presented evidence that locally within sections 1 and 2, a sand and gravel aquifer underlying mud and diamicton exhibits confined conditions. On a valley-wide basis, however, these intervals are composed of more than 50% sand and gravel facies. This suggests that some areas consist almost entirely of mud and diamicton facies but that these cover less than half the area of the valley. Ritzi et al. (1995) showed that

individually these areas are probably are less than 4 km^2 and preferentially located on the valley margin.

Most sections contain intervals with very low proportions of mud and diamicton facies. These intervals define facies assemblage B and were delineated by choosing 9.15-m-thick (30-ft) elevation intervals within which the proportion is less than the overall mean by more than one standard deviation. In section 2 and adjacent sections, facies assemblage B contains from 0 to 5% mud and diamicton facies. In all sections, most depth intervals contain proportions of mud and diamicton facies close to the mean. These intervals define facies assemblage C, which is not restricted to a particular thickness and of which constitutes the bulk of the aquifer system.

Table 1 summarizes estimates of mean proportion and facies thickness for assemblages A and B. Facies thicknesses were estimated from transition probabilities, as described above. These results indicate that although the entire aquifer system is composed of the same two facies, zones exist in which those facies occur in different proportions and thicknesses, supporting their designation as separate facies assemblages. These assemblages were delineated as horizontal intervals within each lateral division. Johnson (1995) and Priscott et al. (1997) described ways of identifying dipping zones within which facies correlation lengths or proportions are maximized. Such methods may be appropriate within the lateral distances over which facies dips are expected to be consistent. In constrast, Figure 3 suggests that facies assemblages occur across

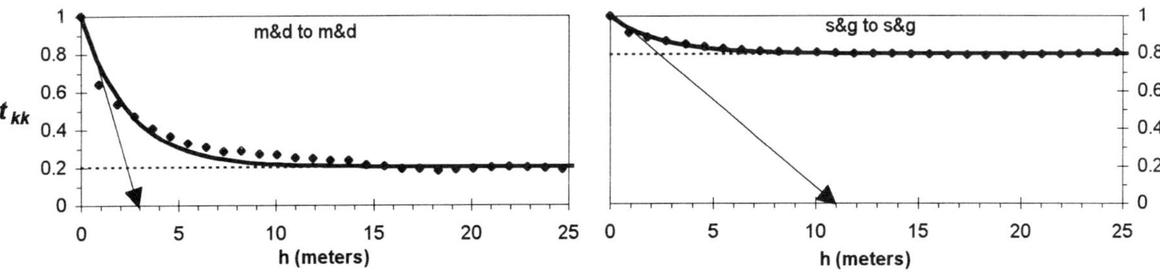

FIG. 4—Vertical autotransition probabilities for the middle division of section 2, with an upward direction of transition. The fitted models are exponential curves. The tangent to the model at $h = 0$ is projected along the arrow to the abscissa to give the mean thickness for each facies.

the valley at different elevations and may not have constant dips. For the following analyses we assume that the facies assemblages are locally horizontal within the middle division.

Facies assemblage A represents the elevation zone(s) over which interconnection among high-permeability sand and gravel facies is more likely to be disrupted. Therefore, it is important to define the spatial distribution of facies within this assemblage.

Spatial Correlation in Facies Assemblage A

Both transition probability and variography provide measures of spatial correlation. Carle and Fogg (1996) showed that the former yields results that are generally more interpretable. Variography can, however, be used where data are less dense; because pairs may begin in either facies, more pairs are included in each lag class. In our study, data were sufficient to support transition probability analysis in the vertical direction only. For measures of lateral spatial correlation, we used variography. To further this illustration we present results for facies assemblage A in the middle division and at 204 to 213 m (670 to 700 ft) in elevation.

We computed directional variograms, corrected for the proportional effect (Johnson and Dreiss, 1989), along azimuths separated by 11.25° of rotation. With 1,542 data in this interval, variograms are not well defined, and directions of maximum and minimum correlation are difficult to determine. Therefore, Figure 5 shows the results for the N-S and E-W directions only. In that the variograms are not well defined, different interpretations are possible. The variograms could be interpreted to have no structure at distances including and beyond the first lag interval. Alternatively interpreting structure to exist, exponential models were fitted as shown in Figure 5; the range in the N-S direction is 500 m and in the E-W direction is 300 m.

Carle and Fogg (1996) showed that the mean length of a facies can be determined from indicator variograms by projecting the slope at $h = 0$ to the proportion of that facies. With exponential models of the form,

$$\gamma = s \, (1\text{-exp}(-3h/a)) \qquad (3)$$

the slope at $h = 0$ equals $3s/a$, where s is the sill (equal to

the population variance). Such slope lines are shown in Figure 5. We consider the slope lines derived from the exponential models to be "maximum" slopes, given that structure does exist in the variograms at this scale. To find "minimum" slopes, we drew lines from the origin through the point at which γ first rises consistently above the sill. In this way, we constrain the range of models that might be defined with more numerous data. Figure 5 shows how estimates of mean lengths of the mud and diamicton facies are derived from the slope lines, given that the proportion of this facies is 0.27. In the N-S direction the estimates range from 230 m to 970 m and in the E-W direction from 140 m to 480 m. Similar estimates could be made for the sand and gravel facies.

Together with the other estimates in Table 1, these results describe the occurrence of mud and diamicton facies within facies assemblage A in the middle division of the valley. Occupying approximately 27% of the volume of this 9.1-m-thick zone, the mud and diamicton facies averages about 7 m thick, <1 km in lateral extent in the N-S direction and <0.5 km E-W. Given its average thickness relative to that of the facies assemblage, the mud and diamicton facies likely are not superposed. This suggests that the facies covers 27% of the area of the middle division, representing the proportion of the aquifer system over which vertical interconnection between sand and gravel facies may be disrupted.

Interpretation of Facies Assemblages

Within this section of the Miami Valley aquifer system, the characteristics of facies assemblage A suggest that it represents till and fine-grained deposits subsequently dissected by glacial meltwater. The average thickness of the mud and diamicton facies (7 m) is similar to but larger than the thickness of till beds identified in a nearby core (1 to 5 m, Dominic et al., 1996). Given that borehole logs generally have lower resolution than is possible with a core, such a difference might be expected. Therefore, we interpret the bulk of mud and diamicton facies in this assemblage to represent till, but other origins not associated with deposition directly from ice are possible. While it is unlikely that the till and other deposits formed a continuous, uniform "sheet," the similarity in facies thicknesses in all divisions across the

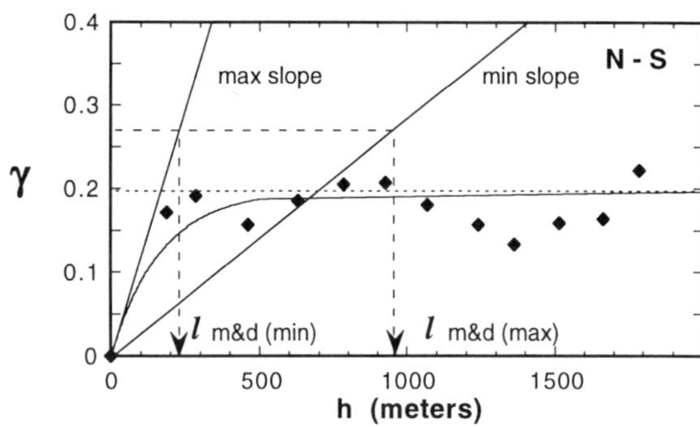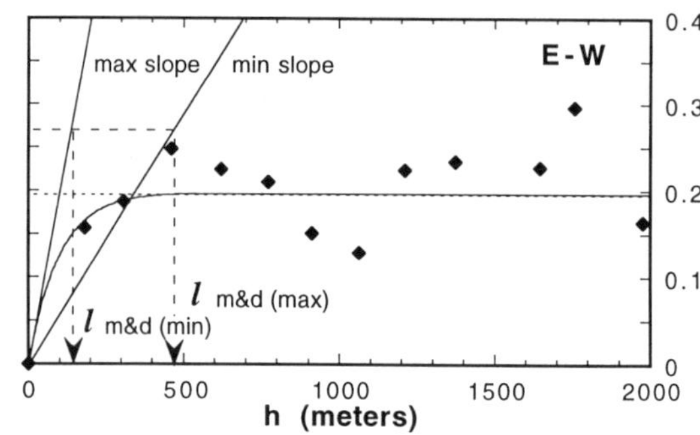

FIG 5.—Variograms computed along N-S and E-W azimuths, fitted with an exponential model. The tangent lines (at $h = 0$) for this and a linear model described in the text are shown; when projected to the proportion of facies k, these give a range of possible mean lengths for the facies. The proportion of mud and diamicton facies (m&d) is 0.27, giving the mean facies lengths indicated by the dashed lines and arrowheads.

valley suggests that these deposits were extensive. As shown in Figure 2, the trend of the bedrock valley in this section is NE-SW, and overall paleoflow would have been similarly aligned. That the mean length of mud and diamicton facies is greater north to south than east to west suggests that it represents remnants uneroded by meltwater flowing parallel to the valley trend. The bulk of the facies assemblage A is composed of sand and gravel, which can be attributed to deposition by high-gradient, multichannel (braided) streams.

The characteristics of facies assemblage C are known only generally. It is dominated by sand and gravel but contains about 20% mud and diamicton facies throughout its depth. Fine-grained sediments accumulate by vertical accretion in fluvial systems, including braided streams. For example, Rust (1978) described gravelly deposits in the Cannes de Roche Formation that contain 15% mud. He interpreted the depositional setting as high-gradient streams on a broad braid plain and attributed the mud to preservation of interchannel areas. A significant amount of interchannel deposits can be preserved on braid plains because active channel belts represent only a small proportion of the overall depositional area, leaving more of the depositional surface unaffected by lateral channel migration. In contrast, relatively less interchannel area should be preserved in valley settings. In the Miami Valley the pervasiveness of mud and diamicton facies suggests that it accumulated with the sand and gravel in fluvial systems, but its relatively high proportion suggests other modes of deposition as well. Mud and diamicton facies may have been deposited in lakes, as debris flows or as till, but these origins cannot be distinguished with the available information. Active channel belts may have occupied the center of the valley more persistently than on the margins, accounting for the lower proportions of mud and diamicton facies there.

The characteristics of facies assemblage B suggest that it represents deposition primarily by glacial meltwater. Unlike facies assemblage C, this assemblage has a relatively small proportion and mean thickness of mud and diamicton facies. Both characteristics suggest that the fine-grained deposits represent vertical accretion associated with fluvial deposition. Our studies in nearby sections indicate that this facies assemblage can be identified more often in the center of the valley, where nonfluvial, fine-grained deposition would be less common than on the margins.

CONCLUSION

This illustration demonstrates that the Miami Valley aquifer system can be divided into meaningful statistical populations. The question of how finely an aquifer system should be divided may best be answered by considering how analyses of the populations are to be used. Determining attributes of populations can aid efforts to represent typical facies distributions in models using such techniques as sequential indicator simulation. In this case dividing more finely into populations will be meaningful if doing so reveals differences in the statistical attributes (proportions, mean lengths and spatial correlation) that lead to differences in the simulation. The attributes might otherwise be used to support conceptual models of the aquifer system and its depositional origin. In this case dividing into populations will be meaningful if it leads to appreciable differences in the interpretation. The populations that are meaningful for these two purposes may be identical but need not be.

In this paper we have focused on defining populations that are meaningful in supporting interpretations of depositional processes. In this case the data populations must be genetically related, and we believe they are best considered as facies assemblages.

Our focus on facies distributions within these assemblages stems from our view that such distributions can be usefully quantified. Together, the statistical attributes of facies distributions and an understanding of their depositional origin provide useful prior estimates for modeling studies and guidance about where those estimates are appropriate. This is the idea embodied in our concept of quantitative facies models. We believe it is useful to develop quantitative facies models at data-rich sites, to compare models from similar depositional settings and to evaluate the usefulness of these models at data-poor sites. Evaluating model usefulness can be done only with numerous, well-studied examples that collectively provide some sense of how transferrable these models might be. All these efforts require, foremost, that meaningful data populations be identified. Once defined, populations can always later be joined, but it probably is best to finely divide aquifer systems at the outset.

ACKNOWLEDGMENTS

This research was supported in part by the National Science Foundation under grants EAR-9305285 and EAR-9628299. Perceptive comments by reviewers Chris Murray and Gary Smith are gratefully acknowledged.

REFERENCES

ANDERSON, M. P., 1989, Hydrogeologic facies models to delineate large-scale spatial trends in glacial and glaciofluvial sediments: Geological Society of America Bulletin, v. 101, p. 501-511.

ANDERSON, M. P., 1990, Aquifer heterogeneity—A geological perspective, *in* Bachu, S., ed., Parameter Identification and Estimation for Aquifer and Reservoir Characterization: Proceedings of the 5th Canadian/American Conference on Hydrogeology, Banff, Alberta, Canada, National Water Well Association, p. 3-22.

ANDERSON, M. P., 1991, Comment on "Universal scaling of hydraulic conductivities and dispersivities in geologic media" by S. P. Neuman: Water Resources Research, v. 27, p. 1381-1382.

BROCKMAN, C. S., 1988, Report of progress on the final phase of glacial geologic mapping in Hamilton County, Ohio: Ohio Geological Survey Open-File Report, 24 p.

CARLE, S. F., and FOGG, G. E., 1996, Transition probability-based indicator geostatistics: Mathematical Geology, v. 28, p. 453-477.

DAVIS, J. M., LOHMAN, R. C., PHILLIPS, F. M., WILSON, J. L., and LOVE, D. W., 1993, Architecture of the Sierra Ladrones Formation, central New Mexico—Depositional controls on the permeability correlation structure: Geological Society of America Bulletin, v. 105, p. 998-1007.

DESBARATS, A. J., 1990, Macrodispersion in sand-shale sequences: Water Resources Research, v. 26, p. 153-163.

DEUTSCH, C. V., and JOURNEL, A. G., 1992, GSLIB, Geostatistical Software Library and User's Guide: New York, Oxford, 340 p.

DOMINIC, D. F., RITZI, R. W., and KAUSCH, K. W., 1996, Aquitard distribution in a northern area of the Miami Valley aquifer, part 2—Interpretation of facies and geostatistical results: Applied Hydrogeology, v. 4, p. 25-35.

FERMCO, 1994, Remedial investigation report for operable unit 5—Volume 1, section 3 of the Fernald Environmental Management Project remedial investigation and feasibility study: Fernald Environmental Information Center, 242 p.

FOGG, G. E., 1986, Groundwater flow and sand body interconnectedness in a thick, multiple-aquifer system: Water Resources Research, v. 22, p. 679-694.

FOGG, G. E., 1989, Stochastic analysis of aquifer interconnectedness—Wilcox Group, Trawick area, east Texas: Texas Bureau of Economic Geology Report of Investigations 189, 68 p.

FOGG, G. E., 1990, Emergence of geologic and stochastic approaches for characterization of heterogeneous aquifers, *in* Moltz, F. J., Melville, J. G., and Guven, O., eds., Proceedings of the Conference on New Field Techniques for Quantifying the Physical and Chemical Properties of Heterogeneous Aquifers: Dublin, Ohio, National Water Well Association, p. 1-17.

FREEZE, R. A., MASSMAN, J., SMITH, L., SPERLING, T., and JAMES, B., 1990, Hydrogeological decision analysis, part 1—A framework: Ground Water, v. 28, p. 738-766.

GOMEZ-HERNANDEZ, J., and SRIVASTAVA, R., 1990, ISIM3D, an ANSI-C three dimensional multiple indicator conditional simulation program: Computers and Geosciences, v. 16, p. 395-440.

JOHNSON, N. M., 1995, Characterization of alluvial hydrostratigraphy with indicator semivariograms: Water Resources Research, v. 31, p. 3217-3227.

JOHNSON, N. M., and DREISS, S. J., 1989, Hydrostratigraphic interpretation using indicator geostatistics: Water Resources Research, v. 25, p. 2501-2510.

JOURNEL, A. G., and ALABERT, F. G., 1989, New method for reservoir mapping: Journal of Petroleum Technology, v. 42, p. 212-218.

KOLTERMANN, C. E., and GORELICK, S. M., 1996, Heterogeneity in sedimentary deposits—A review of structure-imitating, process-imitating, an descriptive approaches: Water Resources Research, v. 32, p. 2617-2658.

KROTHE, N. C., and KEMPTON, J. P., 1988, Region 14, Central glaciated plains, *in* Back, W., Rosenshein, J. S., and Seaber, P. R., eds., Hydrogeology: Geological Society of America, The Geology of North America, v. O-2, p. 129-132

LENNOX, D. H., MAATHUIS, H., and PEDERSON, D., 1988, Region 13, Western glaciated plains, *in* Back, W., Rosenshein, J. S., and Seaber, P. R., eds., Hydrogeology: Geological Society of America, The Geology of North America, v. O-2, p. 115-128.

NORRIS, S. E., CROSS, W. P., and GOLDTHWAIT, R. P., 1948, The water resources of Montgomery County, Ohio: Ohio Division of Water Bulletin 12, 83 p.

NORRIS, S. E. and SPIEKER, A.M., 1966, Ground-water resources of the Dayton area, Ohio: U. S. Geological Survey Water-Supply Paper 1808, 167 p.

PRISCOTT, G. W., RITZI, R. W., and DOMINIC, D. F., 1996, Interpreting abundance, strike, dip, and spatial correlation of till by disaggregating lithologic indicator data (abs.): Geological Society of America, Annual Meeting Abstracts with Programs, p. A-25.

RITZI, R. W., DOMINIC, D. F., BROWN, N. R., KAUSCH, K. W., MCALENNEY, P. J., and BASIAL, M. J., 1995, Hydrofacies distribution and correlation in the Miami Valley aquifer system: Water Resources Research, v. 31, p. 3271-3281.

RITZI, R. W., DOMINIC, D. F., and KAUSCH, K. W., 1996, Aquitard distribution in a northern area of the Miami Valley aquifer, part 1—Three-dimensional geostatistical evaluation of physical heterogeneity: Applied Hydrogeology, v. 4, p. 12-24.

RITZI, R. W., ZARHRADNIK, A., JAYNE, D., and FOGG, G., 1994, Heterogeneity in a glacio-fluvial aquifer: Ground Water, v. 32, p. 666-674.

Rust, B. R., 1978, Depositional models for braided alluvium, *in* Miall, A. D., ed., Fluvial Sedimentology: Canadian Society of Petroleum Geologists Memoir 5, p. 605-625.

Schwartz, F. W., 1977, Macroscopic dispersion in porous media— The controlling factors: Water Resources Research, v. 13, p. 743-752.

Shindel, H. L., Klingler, J. H., Mangus, J. P., and Trimble, L. E., 1991, Water resources data, Ohio, water year 1991: U.S. Geological Survey Water-Data Report OH-91-2, 430 p.

Wingle, W. L., and Poeter, E. P., 1993, Uncertainty associated with semivariograms used for site simulation: Ground Water, v. 31, p. 725-734.

CONDITIONAL SIMULATION OF HYDROFACIES ARCHITECTURE: A TRANSITION PROBABILITY/MARKOV APPROACH

STEVEN F. CARLE

Hydrologic Sciences, University of California, Davis, California 95616
Present address: Lawrence Livermore National Laboratory, L-206, POB 808, Livermore, California 94551

ERIC M. LABOLLE, GARY S. WEISSMANN, AND DAVID VAN BROCKLIN

Hydrologic Sciences, University of California, Davis, California 95616

GRAHAM E. FOGG

Hydrologic Sciences and Dept of Geology, University of California, Davis, California 95616

ABSTRACT: In many hydrogeological investigations of groundwater flow and contaminant transport, considerable uncertainty arises from unknown physical heterogeneity of the aquifer system materials. Indicator geostatistics may offer tools for stochastic simulation of heterogeneity, however, existing methods were not designed with hydrogeological problems in mind. Typically, hydrogeological data are vastly undersampled in lateral directions. Subsequently, geostatistical analyses must rely, in part, on geologic interpretation to yield geologically plausible results. Compared to traditional variogram-based approaches, the transition probability/Markov approach introduced herein more rigorously considers spatial cross-correlations (juxtapositional tendencies), yet is more conducive to integration of geologic interpretation. Three-dimensional (3-D) Markov chains are introduced as a conceptually simple yet theoretically powerful model of spatial variability, supported in theory and practice by numerous prior 1-D geologic applications to vertical sedimentary successions. The geostatistical conditional simulation algorithms of sequential indicator simulation and simulated quenching (zero-temperature annealing) are modified to include consideration for all spatial cross-correlations. "Nonstationarities" related to anisotropy directions, proportions of depositional units and commingling depositional systems also can be considered. Example applications are given for alluvial fan and fluvial depositional systems in California in portions of the Livermore Valley, Kings River alluvial fan and Salinas Valley.

INTRODUCTION

Most researchers agree that realistic modeling of subsurface flow and transport processes requires realistic characterization of system heterogeneities. In particular, transport of fluids and solutes depends strongly on heterogeneity throughout a hierarchy of scales, many of which occur over distances shorter than typical borehole spacing. Difficulty in characterizing such unknown heterogeneity has been a major obstacle to accurate modeling of mass transport in both groundwater and petroleum systems. This difficulty is particularly acute in the lateral directions, where the subsurface typically is undersampled relative to the vertical. For this reason, some researchers (e.g. Anderson, 1989; Fogg, 1989; Phillips and Wilson, 1989; Tetzlaff and Harbaugh, 1989; Journel and Alabert, 1990; Deutsch and Journel, 1992; Koltermann and Gorelick, 1992; Davis et al., 1993; Scheibe and Freyberg, 1995; Webb and Anderson, 1996) have suggested incorporating geologic interpretation to estimate spatial variability. That is, instead of estimating the spatial variability of a hydrologic property, such as hydraulic conductivity (K), one focuses on characterizing the spatial variability of geologic facies. Then, the spatial distribution of geologic facies is simulated to provide a "template" for defining the spatial distribution of K. This approach presumes that a relationship between geologic facies and K can be estimated and that the spatial distribution of the geologic facies adequately represents the spatial distribution of K for purposes of modeling flow and transport processes.

A geologically sound template provides a scientific basis for modeling spatial variability of hydraulic properties. Clearly, subsurface patterns originate from geologic processes; hence, geology should be integral to subsurface characterization whether the approach is deterministic or stochastic. This includes geostatistical approaches, which should intensify, rather than minimize, the integration of geologic interpretation. If possible, the categories chosen for the template should have a geometric context, for example, depositional facies (channel bar, splay, flood plain) as opposed to purely textural descriptors (sand, silt, clay). Then, the interpretational framework of facies models can help guide the construction of geologically plausible hydrogeologic models.

The scale of interest in many hydrogeologic investigations often spans a large portion of a depositional system, for example, an alluvial fan or fluvial sedimentary sequence. At this regional or "aquifer system" scale, geometry and interconnectivity of, for example, fluvial channel and interchannel facies assemblages typically is more important to modeling groundwater flow than are smaller scale sedimentary structures such as cross-bedding. Thus, a model of the spatial distribution of larger scale facies with contrasting hydraulic properties, or "hydrofacies architecture," may be extremely useful as a model of hydraulic heterogeneity at the aquifer system scale.

Geostatistical methods have been applied to characterization of aquifer system heterogeneity, particularly for interpreting spatial variability of hydrofacies (e.g., Johnson and Dreiss, 1989; Wingle and Poeter, 1993; Johnson, 1995; Ritzi et al., 1995) and for conditionally simulating hydrofacies architecture (e.g., Wen and Kung, 1993; Ritzi et al., 1994; Poeter and McKenna, 1995, Bierkens, 1996). These applications have employed the "indicator variogram" to define spatial correlation structures (Deutsch and Journel, 1992). Indicator methods are appealing for hydrogeological numerical modeling applications because of the grid-based geometry, the consideration of conditioning data and the quantitative description of spatial variability. By enabling generation of multiple equiprobable "realizations," conditional simulation provides a means for assessing uncertainty and performing Monte Carlo analyses constrained by borehole data.

The geological community has been hesitant to accept geostatistical simulations of the subsurface, pointing out what they consider to be key shortcomings, such as the stationarity assumption and the inability to represent variable anisotropy directions, for example, radial morphology of an alluvial fan or meandering (Neton et al., 1994). Furthermore, sedimentary facies exhibit strong spatial cross-correlations (juxtapositional tendencies); for example, levee deposits tend to occur vertically and laterally adjacent to channel deposits. Clearly, a geologically plausible characterization of the subsurface ought to account for such juxtapositional relationships. Indeed, an essential aspect of geology is characterization of spatial arrangement of geologic units. Two questions about these geostatistical simulation approaches frequently arise:

1. How can one estimate a three-dimensional (3-D) model of spatial variability given that most geological data sets do not support direct calculation of the variogram in all directions?
2. Are indicator variogram statistics capable of adequately representing the natural configurations of heterogeneity found in the subsurface so that the "realizations" are, in fact, realistic?

This paper addresses both questions and provides an alternative geostatistical approach that facilitates development of models of spatial variability and produces greater geologic realism in conditional simulation results.

NEW GEOSTATISTICAL APPROACH

When a geologic system is modeled by categorizing several geologic units, some fundamental observable attributes are considered:

- volume fraction (proportions),
- lengths (e.g., thickness and lateral extent),
- juxtapositional tendencies,
- directions (e.g., orientation of bedding),
- nonstationarity (spatial trends) in the above.

A geostatistical method should attempt to capture these attributes. In addition, the practitioner needs tools to incorporate this fundamental information into a geostatistical characterization.

The need for a new geostatistical approach emerges in practical hydrogeological applications because existing geostatistical methods do not provide (1) consideration for asymmetric juxtapositional relationships such as fining-upward tendencies; (2) a conceptual framework for incorporating geologic interpretation of proportions, lengths and juxtapositional tendencies into development of cross-correlated spatial variability models; or (3) consideration for locally variable anisotropy directions, for example, radial morphology of alluvial fans, meandering of fluvial depositional units and structure resulting from deformation. Accordingly, these problems have inspired a new transition probability-based approach to indicator geostatistics that simplifies model development and interpretation (Carle and Fogg, 1996). Spatial variability is posed in terms of conditional probabilities of occurrence of the actual categories (geologic units, facies) rather than, for example, variograms of transformed variables such as cumulatively-defined "discretization classes" (Gomez-Hernandez and Srivastava, 1990) or "principal components" (Suro-Perez and Journel, 1991). The transition probability approach is conducive to application of the mathematically simple yet theoretically powerful Markov chain models. Markov chain models can be developed not only by empirical curve-fitting approaches traditionally used in geostatistics but also by conceptual means through interpretation of proportions, mean lengths and juxtapositional tendencies (Carle and Fogg, 1997).

Geologists have long used transition probability measurements in conjunction with Markov chain models as tools for quantitative interpretation of vertical lithologic successions, starting with the pioneering work of Vistelius (1949) followed by many others including Carr et al. (1966), Vistelius (1967), Krumbein (1968), Krumbein and Dacey (1969), Schwarzacher (1969), Doveton (1971) and Miall (1973, 1982). Markov chain models also can be extended to 2-D and 3-D applications by assuming that 1-D Markov chains model spatial variability in any direction (Switzer, 1965; Lin and Harbaugh, 1984; Politis, 1994). A method for developing 2-D or 3-D Markov chains is presented, which enables creation of multi-category models of spatial variability with direction-dependencies in length (anisotropy) as well as juxtapositional relationships.

Although transition probabilities and Markov chains have long been applied in geology and statistics, their application to modern indicator geostatistical approaches is new. Markov chain models of spatial variability can be applied with minor modification to the geostatistical conditional simulation algorithms of sequential indicator simulation (SIS) and simulated quenching (zero-temperature annealing) described by Deutsch and Journel (1992, p. 123-125, 159-160). We find that a two-step of approach of initializing the simulation by SIS and then iteratively improving the simulation by simulated quenching produces better results than either algorithm applied by itself.

The new geostatistical approach has been developed in the context of hydrogeological applications, where data often are sparse yet three-dimensional (3-D) characterizations are needed. Three main steps are involved:

(a) evaluation of spatial variability of geologic units by transition probability measurements (if available) along principal stratigraphic directions of vertical (upward), dip and strike;

(b) development of Markov chain models of spatial variability by quantitative or conceptual means; and

(c) conditional simulation by successive application of the sequential indicator simulation (SIS) and simulated annealing algorithms.

The approach is applied to three Quaternary depositional systems in California (Fig. 1): (1) alluvial fan deposits at Lawrence Livermore National Laboratory (LLNL) in the Livermore Valley of the Coast Range, which contain

FIG. 1.– Map of California showing locations of Lawrence Livermore National Laboratory (LLNL), Kings River alluvial fan, and Salinas Valley.

a large proportion of fine-grained materials; (2) the Kings River alluvial fan in the eastern San Joaquin Valley southeast of Fresno, which contain extensive fluvial deposits; and (3) commingling alluvial fan and fluvial deposits in the Salinas Valley. In these applications we find that the conditional simulation results exhibit geologically plausible stratigraphic relationships consistent with available lithologic data.

TRANSITION PROBABILITY/MARKOV FRAMEWORK

In traditional geostatistical applications, the indicator variogram is incorporated as the measure of spatial variability for categorical variables, such as geologic units. Indicator variogram modeling procedures usually are empirical, involving fitting of mathematical functions to variogram measurements (Deutsch and Journel, 1992, p. 22-25). Alternatively, the transition probability can be used to describe spatial variability as well, particularly where asymmetries such as fining-upward tendencies are present (Carle and Fogg, 1996). Furthermore, the transition probability possesses interpretative advantages for relating fundamental properties of proportions, mean lengths and juxtapositional tendencies to parameters of the spatial variability model. These interpretive advantages are crucial to ensuring development of a geologically plausible model of spatial variability, particularly when data are too sparse to support a purely empirical approach.

In geostatistical conditional simulation applications, the transition probability can be used to formulate cokriging equations used for estimating local conditional probabilities in SIS and objective functions used in simulated annealing (Carle and Fogg, 1996; Carle, 1997). Thus, the transition probability can be used throughout the implementation of indicator geostatistical approaches, whether describing, modeling, or simulating the spatial variability of geologic units.

Definitions

In the application of indicator geostatistics, the indicator variable $I_k(\mathbf{x})$ defines the presence (or absence) of a category k (e.g., a geologic unit) at a location $\mathbf{x} \in \mathbf{D}$ by

$$I_k(\mathbf{x}) = \begin{cases} 1, & \text{if } k \text{ occurs at } \mathbf{x} \\ 0, & \text{otherwise} \end{cases} \qquad k = 1,...,K \quad (1)$$

where K is the number of categories in a region $\mathbf{D}$ (e.g., a stratigraphic sequence). Assuming second-order stationarity, that is, the heterogeneity can be modeled throughout $\mathbf{D}$ by univariate (e.g., mean, proportions) and bivariate spatial statistics (e.g., indicator cross-variogram, indicator cross-covariance, joint probability, transition probability), the *model* of spatial variability will depend only on a separation vector or "lag" h and not on location $\mathbf{x}$. In geostatistics, indicator cross-variograms $\gamma_{jk}(\mathbf{h})$ and, more commonly, indicator variograms $\gamma_{kk}(\mathbf{h})$ (for $j = k$) as defined by

$$2\gamma_{jk}(\mathbf{h}) = E\left\{ (I_j(\mathbf{x}) - I_j(\mathbf{x}+\mathbf{h}))\,(I_k(\mathbf{x}) - I_k(\mathbf{x}+\mathbf{h})) \right\} \quad (2)$$

traditionally are used to describe the spatial variability of indicator variables. Alternatively, the transition probability $t_{jk}(\mathbf{h})$ is simply defined according to a conditional probability:

$$t_{jk}(\mathbf{h}) = \Pr\left\{ k \text{ occurs at } \mathbf{x}+\mathbf{h} \mid j \text{ occurs at } \mathbf{x} \right\}$$

$$= \frac{\Pr\left\{ k \text{ occurs at } \mathbf{x}+\mathbf{h} \text{ } and \text{ } j \text{ occurs at } \mathbf{x} \right\}}{\Pr\left\{ j \text{ occurs at } \mathbf{x} \right\}} \quad (3)$$

where

$$0 \leq t_{jk}(\mathbf{h}) \leq 1 \quad (4)$$

(Ross, 1993, p. 14). Applying (1) to (3), $t_{jk}(\mathbf{h})$ also can be defined in terms of indicator variables as

$$t_{jk}(\mathbf{h}) = \frac{E\left\{ I_j(\mathbf{x})I_k(\mathbf{x}+\mathbf{h}) \right\}}{E\left\{ I_j(\mathbf{x}) \right\}} \quad (5)$$

Under the assumption of stationarity in $\mathbf{D}$, $E\left\{ I_j(\mathbf{x}) \right\}$ is equivalent to the volumetric proportions p_k of category k,

$$E\left\{ I_k(\mathbf{x}) \right\} = p_k \qquad \forall k \quad (6)$$

Applying (5) and (6) to (2), $\gamma_{jk}(\mathbf{h})$ relates to $t_{jk}(\mathbf{h})$ by

$$2\gamma_{jk}(\mathbf{h}) = p_j \left[2t_{jk}(0) - t_{jk}(\mathbf{h}) - t_{jk}(-\mathbf{h}) \right] \quad (7)$$

(Carle and Fogg, 1996).

Thus, transition probabilities can be converted to indicator cross-variograms, if desired. Although indicator cross-variograms can be converted into transition probabilities in a similar manner, any information on asymmetry (e.g., a fining-upward tendency) would be lost from the outset because indicator cross-variograms assume symmetry by definition in (2). Additionally, existing indicator conditional simulation methods do not directly incorporate cross-correlation information because modeling of the full matrix of cross-covariances is considered tedious and impractical (Deutsch and Journel, 1992, p. 82, 149). Alternatively, indicator cokriging can be implemented with the transition probability (Carle and Fogg, 1996), inherently pointing to the Markov chain as a model of spatial variability for categorical (indicator) variables that fully considers cross-correlations (Carle and Fogg, 1997).

Continuous-Lag Markov Chains

Markov chain models applied to time series assume, in theory, that future occurrences depend on the present and not on the past. This simple stochastic model can address spatial applications by replacing the time lag with a spatial lag h_ϕ in a direction ϕ. The resulting mathematical expression of the continuous-lag formulation of a Markov chain model is

$$\mathbf{T}(h_\phi) = \exp\left[\mathbf{R}_\phi h_\phi \right] \quad (8)$$

(Krumbein, 1968; Agterberg, 1974, p. 457; Ross, 1993, p.

290), where $\mathbf{T}(h_\phi)$ denotes a $K \times K$ matrix of transition probabilities,

$$\mathbf{T}(h_\phi) = \begin{bmatrix} t_{11}(h_\phi) & \cdots & t_{1K}(h_\phi) \\ \vdots & \ddots & \vdots \\ t_{K1}(h_\phi) & \cdots & t_{KK}(h_\phi) \end{bmatrix}$$

and $\mathbf{R}_\phi$ denotes a $K \times K$ matrix of transition rates,

$$\mathbf{R}_\phi = \begin{bmatrix} r_{11,\phi} & \cdots & r_{1K,\phi} \\ \vdots & \ddots & \vdots \\ r_{K1,\phi} & \cdots & r_{KK,\phi} \end{bmatrix}$$

with entries $r_{jk,\phi}$ describing a conditional rate of change from category j to category k per unit length in the direction ϕ. By differentiation of (8) with respect to h_ϕ at $h_\phi = 0$, the transition rates are related to transition probabilities by

$$r_{jk,\phi} = \frac{\partial t_{jk}(0)}{\partial h_\phi} \tag{9}$$

(Ross, 1993, p. 290). Despite its theoretical simplicity, the Markov chain model has shown remarkable applicability to vertical lithological successions (Vistelius, 1949; Krumbein, 1968; Krumbein and Dacey, 1969; Schwarzacher, 1969; Harbaugh and Bonham-Carter, 1970; Doveton, 1994).

Comparison to Discrete-Lag Markov Chains

Geologic applications more commonly have employed the discrete-lag Markov chain, which is formulated from a transition probability matrix $\mathbf{T}(\Delta h_\phi)$ obtained from measurements at a discrete lag interval Δh_ϕ, usually chosen rather arbitrarily as a sampling interval. For the discrete-lag Markov chain model, $\mathbf{T}(n\Delta h_\phi)$ for $n = 0, 1, 2, ..., \infty$ is computed by successively applying the Chapman-Kolmogorov equation:

$$\begin{aligned} \mathbf{T}(0) &= \mathbf{I} \\ \mathbf{T}(\Delta h_\phi) &= \mathbf{I}\,\mathbf{T}(\Delta h_\phi) \\ \mathbf{T}(2\Delta h_\phi) &= \mathbf{T}(\Delta h_\phi)\,\mathbf{T}(\Delta h_\phi) \\ &\;\;\vdots \\ \mathbf{T}(n\Delta h_\phi) &= \mathbf{T}\left[(n-1)\Delta h_\phi\right]\,\mathbf{T}(\Delta h_\phi) \end{aligned} \tag{10}$$

where $\mathbf{I}$ denotes the identity matrix (Agterberg, 1974, p. 420-421). However, any discrete-lag Markov chain model can be formulated as a continuous-lag Markov chain model by applying (8) with $\mathbf{R}_\phi$ determined by

$$\mathbf{R}_\phi = \frac{\ln\left[\mathbf{T}(\Delta h_\phi)\right]}{\Delta h_\phi} \tag{11}$$

Advantages of the continuous approach are threefold. First, (8) represents the spatial variability model as a continuous function instead of discrete outputs only at fixed lag intervals as in (10). Second, the continuous-lag formulation of a Markov chain model is not dependent on issues of data spacing or sampling intervals because transition rates can be established by applying (11) to transition probabilities at any lag interval. Third, the ability to formulate the Markov chain as a continuous function enables the application to geostatistical simulation methods of SIS and simulated annealing.

Categorization – Facies Context

The first requirement toward applying the transition probability/Markov approach described is *categorization* or definition of the hydrofacies. In hydrogeologic applications at the aquifer system scale, categorization according to depositional facies is a desirable, if not mandatory, approach because it provides a scientific (geologic) basis for predicting geometry and spatial arrangement of hydraulic heterogeneity (Anderson, 1989; Anderson and Woessner, 1992, p. 29-37). At the LLNL site, comparison of well test results with detailed sedimentological analysis of core (Noyes, 1991) provide sufficient basis for establishing hydrofacies based on depositional facies. Admittedly, any facies-based categorization is only approximate. We would nevertheless argue that any attempt to place the subsurface characterization into a depositional facies context is an improvement over prevailing approaches in subsurface hydrology, such as the Unified Soil Classification System (Casagrande, 1948; ASTM, 1996) or Gaussian random fields, because it initiates and reinforces a geologically-informed characterization.

In many hydrogeologic applications, such as the Kings River fan and Salinas Valley sites, data consist primarily of descriptions of sediment textures, with little facies interpretation or hydraulic testing information. For the Kings River fan, facies interpretations were augmented by core samples, geophysics, soils surveys and analogy with modern systems. In the Salinas Valley case, the facies categories are only inferred from textural information. Regardless of data quality, the facies interpretation is needed to develop the three-dimensional geometric inter-

TABLE 1.– BASIS FOR DEFINITION OF HYDROFACIES FROM LLNL CORE DATA.

#	facies	texture	%
1	*debris flow*	poorly-sorted clay-gravel	7
2	*flood plain*	clay and silt	56
3	*levee*	silty or clayey fine sand	19
4	*channel*	sand and gravel (rounded)	18

pretive context that facies models provide (Miall, 1992). Nonetheless, the categorizations presented should be viewed as "operationally defined" to various stages of completion. This paper emphasizes the spatial variability modeling and conditional simulation of such characterizations rather than the methods used in facies interpretation.

Application to LLNL – Vertical Direction

Groundwater contaminants, primarily volatile organic compounds, are present in alluvial fan deposits underlying LLNL (Fig. 1) over a regional extent of several km^2 to depths as great as 100 m (Thorpe et al., 1990). The aquifer system materials consist of a network of ancient stream channel and levee (including crevasse-splay) deposits and debris flows embedded in fine-grained flood plain deposits (Noyes, 1991). The lateral and vertical spatial extent of contaminants may be exacerbated by natural hydraulic interconnectivity of the subsurface channel/levee network (Carle, 1996). Working models of the hydrofacies architecture are needed to realistically assess contaminant fate and potential effectiveness of remediation efforts.

As summarized in Table 1, four hydrofacies were categorized as *debris flow, flood plain, levee* and *channel* facies based on detailed interpretations (Noyes, 1991) of portions of 5,500 m of core from vertical boreholes located near the southwest portion of LLNL and vicinity. The facies categorization was extended to the remainder of the core by interpretation of geophysical logs and textural descriptions noted by site geologists (Qualheim, 1988).

Figure 2 shows vertical (z)-direction transition probability measurements for the core data and a continuous-lag Markov chain model (8) developed by applying (11) to obtain a vertical transition rate matrix $\mathbf{R}_z$ for the $\mathbf{T}\,(\Delta h_z = 0.9\ \mathrm{m})$ measurement. The Markov chain model provides an excellent overall fit to the transition probability measurements, including the off-diagonal cross-correlations. As demonstrated by this four-category example, the Markov chain efficiently generates sixteen model structures $t_{jk}(h_z)$ from sixteen parameters $r_{jk,z}$ for $j, k =$

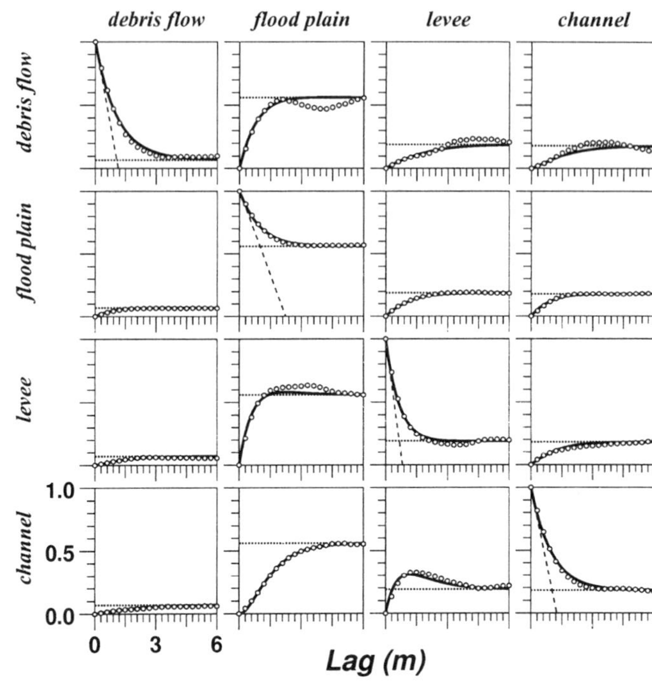

FIG. 2.– Matrix of vertical-direction transition probabilities for LLNL core data: measurements (dots) and Markov chain model (solid lines) obtained by applying equation (17) to the matrix of transition probability measurements at 0.9 m lag. Intersection of dashed line (tangent) with lag axis indicates mean length; dotted line indicates proportions.

1, ..., 4. Mathematically, the matrix exponential form of the Markov chain (8) is composed of linear combinations of exponential functions by (28), as described in the Appendix, where evaluation of the matrix logarithm (11) also is detailed. Nonetheless, a Markov chain can produce non-exponential-looking structures, for example, the "Gaussian"-like *channel →flood plain* and "hole-effect" *channel →levee* transition probability models in Figure 2.

MODEL CONCEPTUALIZATION

In the previous example the LLNL core data were relatively abundant for characterizing spatial variability of facies in the vertical direction and, thus, were conducive to application of the direct method of (11) to establish $\mathbf{R}_z$. However, spatial variability measurements may contain considerable uncertainty in lateral directions and, thus, may not be amenable to empirical modeling procedures. Only by understanding how model parameters relate to spatial variability can one address the important practical question, "Does this model make sense, geologically and probabilistically?" In practice, the parameters of a transition probability model can be directly related to fundamental interpretable properties of proportions, mean length, asymmetry and juxtapositional tendencies (Carle and Fogg, 1996; 1997).

Proportions

Assuming stationarity, the "sill" (asymptotic limit) of the transition probability is related to proportions by

$$\lim_{h_\phi \to \infty} t_{jk}(h_\phi) = p_k \qquad \forall j, k \qquad (12)$$

If the proportions of the geologic units are known a priori, then (12) can be used to establish sills for the transition probability matrix. Conversely, the sill of the spatial variability model implies the assumed proportions. Equation (12) shows that the sills of each entry in the transition probability matrix will correspond to the proportions of the column category. For example, in Figure 2, the horizontal dotted lines indicate the sill of the vertical (z)-direction transition probability models for the kth (column) category in accordance with (12). As $h_z \to \infty$, the transition probability models clearly approach $p_1 \approx 0.07$, $p_2 \approx 0.56$, $p_3 \approx 0.19$ and $p_4 \approx 0.18$, corresponding to the percentages given in Table 1.

The point here is not to evaluate proportions from the transition probability sills, but rather to emphasize the straightforward relationship between proportions and the transition probability sill, which is useful for both data and model interpretation and, more importantly, for selection of plausible parameters for the model of spatial variability.

Mean Length

Although the "range" (the lag at which the sill is reached) commonly is used in geostatistics to describe spatial continuity and anisotropy, the "mean length" may function in a similar role. Let an "embedded occurrence" denote a distinct occurrence of a single category along a line in a particular direction, for example, a bed of gravel bounded by silt below and sand above. The mean length $\overline{L}_{k,\phi}$ of the category k in the direction ϕ is defined as the total length of category k along lines in the direction ϕ divided by the number of embedded occurrences of k along lines in the

direction ϕ. For example, the vertical (z)-direction mean length $\overline{L}_{k,z}$ would correspond to a "mean thickness." Thus, mean length represents a general term for "mean thickness" in any (stratigraphic) direction.

Mathematically, mean length $\overline{L}_{k,\phi}$ relates to a diagonal transition probability $t_{kk}(h_\phi)$ by

$$-\frac{\partial t_{kk}(0)}{\partial h_\phi} = \frac{1}{\overline{L}_{k,\phi}} \qquad (13)$$

(Carle and Fogg, 1996). Application of (9) to (13) shows that a diagonal transition rate $r_{kk,\phi}$ directly relates to mean length by

$$r_{kk,\phi} = -\frac{1}{\overline{L}_{k,\phi}} \qquad (14)$$

According to (13), the mean length will be indicated on a plot of the autotransition probability by the intersection of the tangent at the origin with the ordinate axis, as depicted by the dashed lines in the diagonal entries of Figure 2. Equation (13) can be used during model development either to establish the slope at the origin given knowledge or interpretations of mean length or, conversely, to interpret mean length implied by a diagonal transition probability model. Furthermore, (14) implies that knowledge or estimation of mean length can be used to establish diagonal transition rates for Markov chain models.

In the two-category ($K = 2$) case of a Markov chain, the mean length $\overline{L}_{k,\phi}$ in a direction ϕ relates to the corresponding "effective range" parameter of "$3a_\phi$" traditionally used in geostatistics for an exponential structure (Deutsch and Journel, 1992, p. 23) by

$$\overline{L}_{k,\phi} = \frac{a_\phi}{1 - p_k} \qquad k = 1, 2$$

In situations of three or more categories ($K \geq 3$), however, each entry of the Markov chain becomes a sum of two or more exponential structures, which may have complex coefficients. Thus, mean length provides a more direct parameter than the "range" for development of Markov chain models with three or more categories.

Asymmetry

The term "asymmetry" is used here to denote dependence of the bivariate statistics on whether the lag is positive or negative. An important distinction between $\gamma_{jk}(\mathbf{h})$ and $t_{jk}(\mathbf{h})$ evident in (7) is that $\gamma_{jk}(\mathbf{h})$ is intrinsically symmetric because $\gamma_{jk}(\mathbf{h}) = \gamma_{jk}(-\mathbf{h})$, whereas (3) allows for asymmetry or the possibility that $t_{jk}(\mathbf{h}) \neq t_{jk}(-\mathbf{h})$ or, equivalently, $p_j t_{jk}(\mathbf{h}) \neq p_k t_{kj}(\mathbf{h})$ for $j \neq k$. This is an important consideration for modeling spatial variability of stratigraphic units in fluvial deposits, where vertically asymmetric juxtapositional relationships occur as a result of fining-upward tendencies (e.g., Allen, 1970). In Figure 2, a *channel* →*levee* $(4 \rightarrow 3)$ fining-upward tendency is indicated because $p_4 t_{43}(h_z) \gg p_3 t_{34}(h_z)$ for small h_z.

If juxtapositional relationships are indeed symmetric between two categories j and k in a direction ϕ, then the relation

$$t_{jk}(h_\phi) = \left(\frac{p_k}{p_j}\right) t_{kj}(h_\phi) \qquad (15)$$

holds, where h_ϕ is a lag in a direction ϕ. Differentiation of (15) with respect to h_ϕ at $h_\phi = 0$ and application of (9) yields the relation

$$r_{jk,\phi} = \left(\frac{p_k}{p_j}\right) r_{kj,\phi} \qquad (16)$$

for symmetric juxtapositional relationships between categories j and k in a direction ϕ. Equations (15) and (16) are useful during interpretation for determining whether juxtapositional relationships are symmetric and in model development for establishing symmetrical juxtapositional tendencies in the transition probability or transition rate matrices, if desired.

Juxtapositional Tendencies

When developing either a geological or a geostatistical interpretation, one might want to evaluate the extent to which measured or modeled transition probabilities indicate preferential juxtapositional tendencies (e.g., levee deposits laterally or vertically adjacent to channel deposits) versus relatively disordered facies successions. Indeed, the main application of transition probability/Markov models by geologists has been to quantitatively analyze how juxtapositional tendencies compare relative to various states of disorder, either relative to conditional probabilities of embedded occurrences (Miall, 1973, 1982), entropy of transition frequencies (Hattori, 1976), independence of transition frequencies (Turk, 1979, 1982) and transition probabilities with respect to proportions (Carle and Fogg, 1996). These quantitative interpretational frameworks also can be posed with respect to transition rates for continuous-lag Markov chains (Carle, 1996; Carle and Fogg, 1997).

In practice, the interpretation of transition rates with respect to proportions may be the simplest to implement because it requires the fewest assumptions. In this interpretive framework, one considers occurrences of facies successions relative to volumetric proportions p_k, that is, whether an occurrence of k adjacent to j occurs greater or lesser than indicated by the proportions p_k. Consider that $t_{jj}(h_\phi)$ is the conditional probability of a unit j "transitioning" to itself over a lag h_ϕ, so that the conditional probability of transitioning to any other unit $\neq j$ is $[1 - t_{jj}(h_\phi)]$. If the transition probability $\widehat{t}_{jk}(h_\phi)$ for $k \neq j$ depends on proportions p_k, then

$$\widehat{t}_{jk}(h_\phi) = [1 - t_{jj}(h_\phi)]\frac{p_k}{1 - p_j} \qquad \text{for } k \neq j$$
$$(17)$$

Differentiating (17) by h_ϕ at $h_\phi = 0$ and applying (14) yields a corresponding transition rate $\widehat{r}_{jk,\phi}$ referenced with respect to category proportions:

$$\widehat{r}_{jk,\phi} = \frac{p_k}{\overline{L}_{j,\phi}(1 - p_j)} \qquad \text{for } k \neq j \qquad (18)$$

Thus, a measured or modeled transition rate $r_{jk,\phi}$ can be compared to $\widehat{r}_{jk,\phi}$ to judge whether k occurs adjacent to j in the direction ϕ to a lesser or greater degree relative to the proportion of k.

Application to LLNL – Strike Direction

Borehole data usually are insufficient to quantify spatial variability in lateral directions, not only because of typically sparse lateral spacing but also unknown variations in depositional dip and strike. Thus, the development of spatial variability models can obviously benefit from integration of geologic interpretation. Recognizing that a Markov chain analysis can be used as an interpretive tool then, conversely, geologic interpretation can be used to help establish geologically plausible transition rates.

For example, the transition rate matrix developed for the vertical (z)-direction model of the LLNL data (Fig. 2) viewed in strictly quantitative terms is:

$$\mathbf{R}_z = \begin{bmatrix} -0.875 & 0.706 & 0.104 & 0.064 \\ 0.088 & -0.447 & 0.150 & 0.209 \\ 0.024 & 1.080 & -1.227 & 0.123 \\ 0.040 & 0.000 & 0.766 & -0.806 \end{bmatrix} \mathrm{m}^{-1}$$

Alternatively, one could apply the concepts of mean length (to the diagonal entries) and juxtapositional tendencies relative to proportions (to the off-diagonal entries) to obtain an equivalent, but more conceptual, expression of the transition rate matrix:

$$
\mathbf{R}_z = \begin{bmatrix} -\dfrac{1}{\overline{L}=1.14} & 1.34\widehat{r} & 0.58\widehat{r} & 0.38\widehat{r} \\ 1.24\widehat{r} & -\dfrac{1}{\overline{L}=2.24} & 0.77\widehat{r} & 1.14\widehat{r} \\ 0.23\widehat{r} & 1.27\widehat{r} & -\dfrac{1}{\overline{L}=0.82} & 0.43\widehat{r} \\ 0.58\widehat{r} & 0.00\widehat{r} & 4.1\widehat{r} & -\dfrac{1}{\overline{L}=1.24} \end{bmatrix} \text{m}^{-1}
\tag{19}
$$

where, for simplicity, subscript notation is dropped such that $\overline{L}$ denotes the mean length as defined in (14), and $\widehat{r}$ denotes a transition rate dependent on proportions by (18). Insights into mean length and juxtapositional tendencies obtained from outcrops, geophysical interpretation, or facies models could be used to establish geologically plausible diagonal and off-diagonal transition rates.

Some additional tools from the laws of probability can simplify the conceptual development of a transition rate matrix. The row sums of $\mathbf{R}_\phi$ must obey

$$
\sum_{k=1}^{K} r_{jk,\phi} = 0 \qquad \forall j
\tag{20}
$$

and the column sums must obey

$$
\sum_{j=1}^{K} p_j r_{jk,\phi} = 0 \qquad \forall k
\tag{21}
$$

(Ross, 1993, p. 273). Application of (20) and (21) eliminates the need to specify row and column entries in $\mathbf{R}_\phi$ involving one category β, herein referred to as the "background category." Conceptually the background may be viewed as the category that "fills in the space" not occupied by the other categories. Thus, for a four-category system, only $3 \times 3 = 9$ of the $4 \times 4 = 16$ entries in the transition rate matrix need direct specification; the remaining entries can be determined by (20) and (21) according to the laws of probability. In any application, however, the entries in the transition rate matrix, including row and column entries involving the background category, should obey

$$
r_{jj,\phi} < 0 \qquad \forall j
$$

$$
0 \leq r_{jk,\phi} \leq -r_{jj,\phi} \qquad \forall j, k \neq j
$$

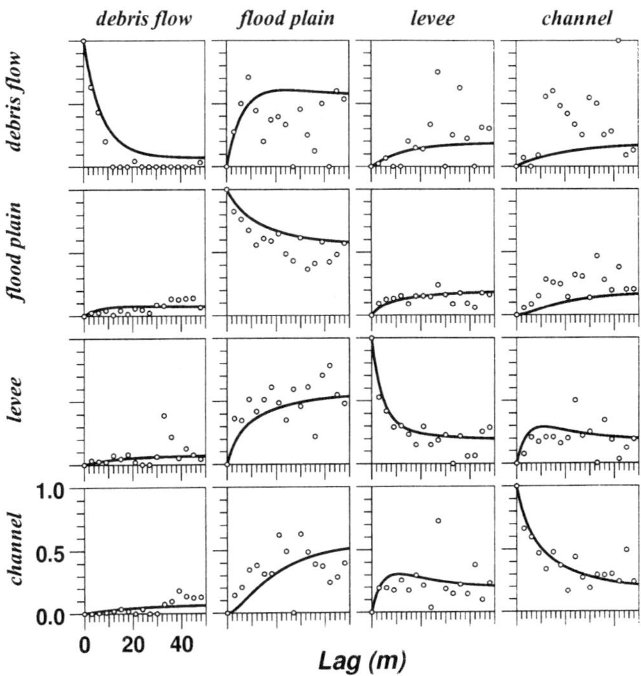

FIG. 3.– Matrix of strike-direction transition probabilities for LLNL core data: measurements (dots) and a Markov chain model (solid lines) based on conceptual interpretation of proportions, mean lengths, and juxtapositional tendencies.

and

$$
0 \leq r_{jk,\phi} \leq -\left(\frac{p_k}{p_j}\right) r_{kk,\phi} \qquad \forall j, k \neq j
$$

in order to satisfy (4).

In this example, *flood plain* (category 2) was chosen as background because, in general, it occupies space not occupied by other facies deposited by higher energy depositional processes. Mathematically, however, any category can be chosen as background. If symmetry of juxtapositional tendencies between two categories j and k in the direction ϕ is assumed, then relation (16) can be applied. For example, in the four-category case, if all juxtapositional relationships are assumed symmetric, then only *three* off-diagonal transition rates need to be specified once mean lengths, proportions and the background category are selected.

Strike (x)- and dip (y)-direction transition rate matrices $\mathbf{R}_x$ and $\mathbf{R}_y$ can be developed conceptually by applying estimates of mean length and proportions in conjunction with interpretation of lateral juxtapositional tendencies implied by Walther's Law, that a vertical facies succession corresponds to a lateral sequence of depositional environments (Leeder, 1982, p. 140). The mean lengths $\overline{L}_{k,x}$ for the strike direction can be established for the uncertain data in Figure 3 by considering plausible strike:vertical length ratios. Obviously, what constitutes a "plausible" mean length or elongation ratio is an interpretive matter that depends on the depositional system. Empirical geomorphic models of fluvial channel width:depth ratios (e.g., Etheridge and Schumm, 1978) could be used to guide the interpretation. In this example, assumed strike (x)-direction mean lengths of 8 m, 6 m and 10 m correspond to strike:vertical elongations ratios of 7.0:1, 7.4:1 and 8.1:1 for *debris flow*, *levee* and *channel*, respectively. The mean length for *flood plain*, the assumed background category, need not be specified. If the strike (x)-direction juxtapositional tendencies for *channel* $\rightarrow$*levee*, *channel* $\rightarrow$ *debris flow* and *levee* $\rightarrow$*debris flow* are assumed similar with respect to proportions as those for the vertical direction (19) in accordance with Walther's Law, the full strike-direction transition rate matrix $\mathbf{R}_x$ could be estimated by

$$
\mathbf{R}_x = \begin{bmatrix} -\frac{1}{\overline{L}=8} & c_1 & s & s \\ c_2 & c_2 & c_2 & c_2 \\ 0.23\widehat{r} & c_1 & -\frac{1}{\overline{L}=6} & s \\ 0.58\widehat{r} & c_1 & 4.1\widehat{r} & -\frac{1}{\overline{L}=10} \end{bmatrix} \mathrm{m}^{-1}
$$

where c_1 and c_2 denote successive application of (20) and (21) to respective row and column sums of $\mathbf{R}_x$, and s denotes an imposition of symmetric juxtapositional tendencies by applying (16). Note that only six of the sixteen entries in $\mathbf{R}_x$ need direct specification once proportions, symmetry and a background category are assumed. The solid lines in Figure 3 show the resulting strike-direction Markov chain model. A dip (y)-direction model could be developed in a similar manner.

Clearly, other interpretations could be made to estimate strike-direction transition rates for this model of facies architecture. These could be accomplished, for example, by tuning the off-diagonal transition rates to accommodate different preservation potentials for facies successions in different directions. Trial-and-error procedures also can be applied by generating stochastic simulations with an initial rate matrix, then adjusting the transition rates to obtain desired juxtapositional tendencies. The Markov chain framework, although simple in theory, provides considerable flexibility and versatility for modeling 1-D spatial variability.

MULTIDIMENSIONAL MARKOV CHAINS

2-D or 3-D Markov chain models can be developed by assuming that spatial variability in any direction can be characterized by a 1-D Markov chain (Switzer, 1965; Lin and Harbaugh, 1984; Politis, 1994). Although this may seem like a tenuous theoretical leap, the assumption here is merely that Markov chains might characterize spatial variability not only in the vertical but in other stratigraphic directions such as dip or strike. In a typical hydrogeological application, data coverage usually is inadequate to directly develop a 1-D Markov chain model for each of the infinity of directions. Alternatively, model development can focus on the principal directions, say the strike (x), dip (y) and vertical (z). Then 1-D Markov chain models for any direction can be interpolated from the principal direction models.

Considering that the transition probability matrix $\mathbf{T}(h_\phi)$ for an arbitrary direction ϕ depends entirely on $\mathbf{R}_\phi$, the interpolation of Markov chain models can be accomplished by ellipsoidally interpolating entries in the transition rate matrices for the principal x, y and z directions by

$$
|r_{jk,\phi}| = \sqrt{\left(\frac{h_x}{h_\phi}r_{jk,x}\right)^2 + \left(\frac{h_y}{h_\phi}r_{jk,y}\right)^2 + \left(\frac{h_z}{h_\phi}r_{jk,z}\right)^2}
$$

$$(22)$$

for all j and $k \neq \beta$, where h_x, h_y and h_z are the x, y and z direction components of $h_\phi = \sqrt{h_x^2 + h_y^2 + h_z^2}$. The remaining entries in $\mathbf{R}_\phi$ involving j or $k = \beta$ can be determined by applying (20) and (21). For the negative lag vector components, say h_{-x}, entries from the rate matrix $\mathbf{R}_{-x}$ corresponding to the opposite direction $-x$ are defined by

$$
r_{jk,-x} = \left(\frac{p_k}{p_j}\right) r_{kj,x}
$$

and used in (22) in place of entries for $\mathbf{R}_x$, in accordance with the backward Kolmogorov differential equation (Agterberg, 1974, p. 455-456). Figure 4 shows a two-dimensional transition probability model for the strike-vertical $(x - z)$ plane as interpolated from the vertical (z)- and strike (x)-direction models previously developed and shown in Figures 2 and 3. The prescribed strike-direction symmetry is evident in that $t_{jk}(h_x, h_z) = t_{jk}(-h_x, h_z)$ for every entry. However, asymmetries are evident in the vertical direction, particularly for $t_{34}(h_x, h_z)$ and $t_{43}(h_x, h_z)$ because of a strong fining-upward tendency of *channel* $\rightarrow$*levee*.

Transition Probability - Strike-Vertical (x-z) Plane

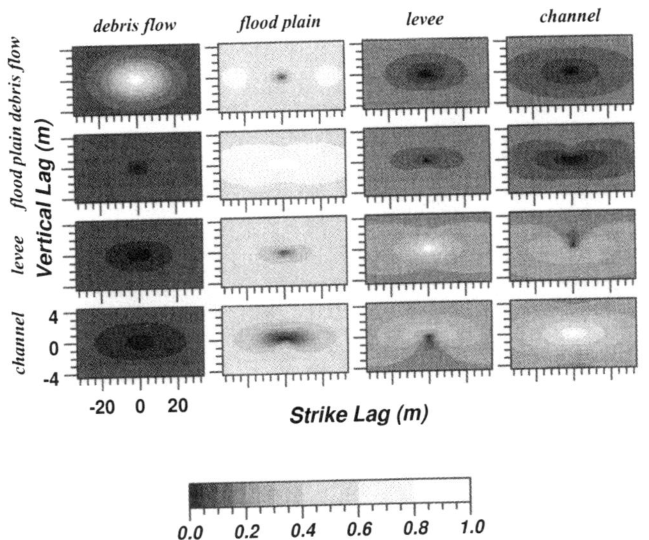

FIG. 4.– Two-dimensional Markov chain model for strike-vertical plane established by interpolating strike- and vertical-direction 1-D Markov chain models.

CONDITIONAL SIMULATION

Considering that true subsurface heterogeneity or "the truth" cannot be exactly determined, conditional simulation algorithms attempt to produce multiple images that possess *patterns* of heterogeneity or "spatial variability" characteristic of the truth while honoring available data. The so-called "realizations" may be useful for implementing realistic models of groundwater flow and contaminant transport, uncertainty analyses, or Monte Carlo inversions.

The conditional simulation technique applied in this paper involves two steps: (1) establishment of an "initial configuration" by the sequential indicator simulation (SIS) algorithm (Deutsch and Journel, 1992, p. 123-125), and (2) iterative improvement of the SIS-generated initial configuration by the simulated quenching (zero-temperature annealing) algorithm (Deutsch and Journel, 1992, p. 159-160). The two steps are mutually dependent because SIS alone will not yield stochastic simulations that adequately honor the model of spatial variability, and quenching will not succeed without a rudimentary initial configuration (Carle, 1997). Both the SIS and quenching steps may rely on the same Markov chain model of spatial variability and, consequently, are conducive to implementation in succession.

Sequential Indicator Simulation

The initialization step follows the sequential indicator simulation (SIS) algorithm described by Deutsch and Journel (1992, p. 123-125, 148), except that a transition probability-based indicator cokriging estimate $[i_k(\mathbf{x}_0)]^*_{coK}$ is used to approximate the local conditional probabilities from data at N locations by

$$\Pr\{k \text{ occurs at } \mathbf{x}_0 \mid i_j(\mathbf{x}_\alpha);\ \alpha = 1, ..., N;\ j = 1, ..., K\}$$

$$\approx [i_k(\mathbf{x}_0)]^*_{coK} = \sum_{i=1}^{N}\sum_{j=1}^{K} i_j(\mathbf{x}_\alpha)w_{jk,\alpha} \qquad (23)$$

where $i_j(\mathbf{x}_\alpha)$ represents the value of an indicator variable at a location $\mathbf{x}_\alpha$ as defined in (1). The weighting coefficients $w_{jk,\alpha}$ are computed from a transition probability-based cokriging system of equations:

$$\begin{bmatrix} \mathbf{T}(\mathbf{x}_1 - \mathbf{x}_1) & \cdots & \mathbf{T}(\mathbf{x}_N - \mathbf{x}_1) \\ \vdots & \ddots & \vdots \\ \mathbf{T}(\mathbf{x}_1 - \mathbf{x}_N) & \cdots & \mathbf{T}(\mathbf{x}_N - \mathbf{x}_N) \end{bmatrix}\begin{bmatrix} \mathbf{W}_1 \\ \vdots \\ \mathbf{W}_N \end{bmatrix}$$

$$= \begin{bmatrix} \mathbf{T}(\mathbf{x}_0 - \mathbf{x}_1) \\ \vdots \\ \mathbf{T}(\mathbf{x}_0 - \mathbf{x}_N) \end{bmatrix} \qquad (24)$$

where

$$\mathbf{W}_i = \begin{bmatrix} w_{11,i} & \cdots & w_{1K,i} \\ \vdots & \ddots & \vdots \\ w_{K1,i} & \cdots & w_{KK,i} \end{bmatrix}$$

(Carle and Fogg, 1996). Using transition probability-based indicator cokriging instead of the traditional indicator kriging approach improves consideration of spatial cross-correlations in estimating local conditional probabilities for SIS. Figure 5 (top) shows the initial configuration state of a conditional simulation of strike-vertical architecture after applying the SIS algorithm using (23) and (24) with the strike-vertical 2-D Markov chain model shown in Figure 4. In Figure 6, vertical transition probabilities measured from this initial configuration seriously depart from the transition probability model, indicating that the SIS algorithm, by itself, does not generate a pattern of spatial variability consistent with the 2-D Markov chain model.

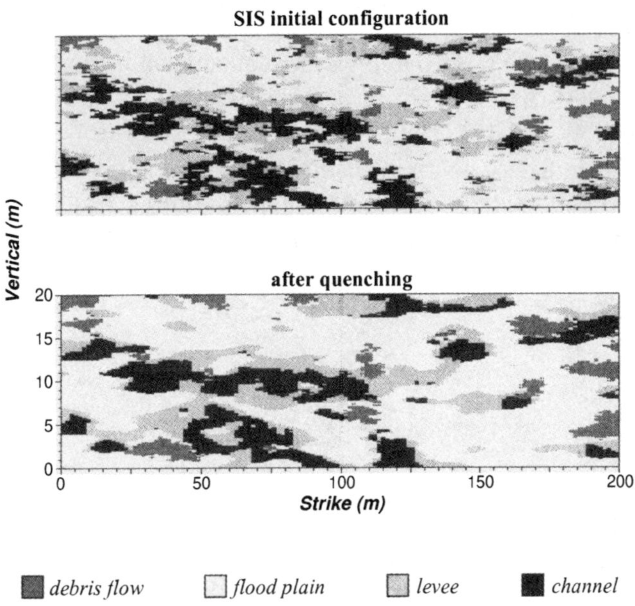

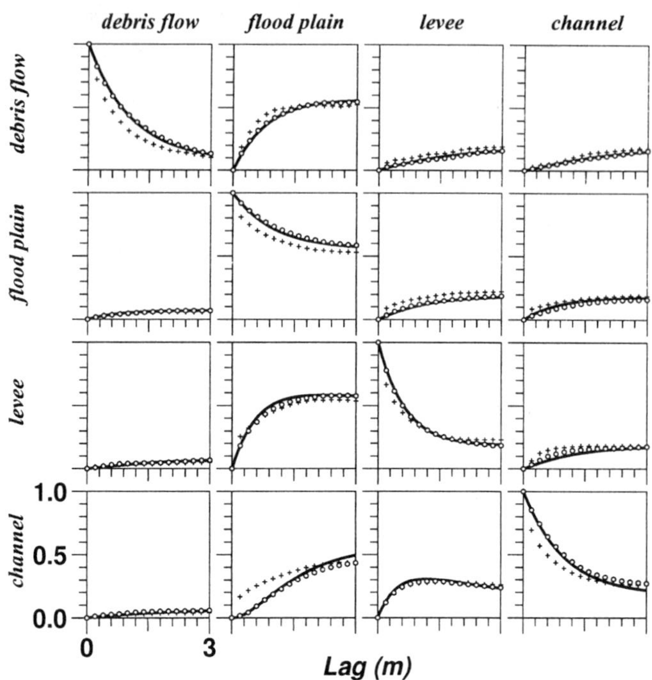

FIG. 5.– At top, SIS-based initial configuration of a 2-D conditional simulation of strike section hydrofacies architecture at LLNL. At bottom, final version of conditional simulation after implementing simulated quenching on initial configuration. Shaded center locations indicate conditioning data.

FIG. 6.– Comparison of the Markov chain model (solid lines) with vertical transition probabilities computed from SIS-generated initial configuration (crosses) and final conditional simulation in Figure 5 after applying simulated quenching (dots).

Simulated Quenching

The simulated quenching step is implemented to improve agreement between measured and modeled transition probabilities, starting from a SIS-generated initial configuration. The quenching step attempts to solve the optimization problem of

$$\min\left\{ \mathbf{O} = \sum_{l=1}^{M}\sum_{j=1}^{K}\sum_{k=1}^{K} \left(t_{jk}(\mathbf{h}_l)_{MEAS} - t_{jk}(\mathbf{h}_l)_{MOD}\right)^2 \right\} \quad (25)$$

where $\mathbf{O}$ denotes an objective function, $\mathbf{h}_l$ denote $l = 1, ..., M$ specified lag vectors, and "$MEAS$" and "MOD" distinguish measured and modeled transition probabilities, respectively (Deutsch and Journel, 1992, p. 159-160). The simulated quenching algorithm is implemented by repeatedly cycling through each nodal location of the conditional simulation and inquiring whether a change to another category will reduce $\mathbf{O}$; if so, the change is accepted. This iterative improvement procedure continues until $\mathbf{O}$ is minimized, or a limit on the number of iterations is reached. Conditioning is maintained by not allowing changes of categories at conditioning locations. "Artifact disconti-

nuity" problems (Deutsch and Cockerham, 1994), where simulated patterns do not jibe with conditioning data, can be avoided by initializing the simulation with a SIS step instead of random values, choosing a minimal number of the closest lag vectors $\mathbf{h}_l$ for (25) and positioning the lag vectors in a configuration that considers the anisotropy of the spatial variability (Carle, 1997). Figure 5 (bottom) shows the result of applying simulated quenching to the SIS-generated initial configuration. After quenching, measured and modeled transition probabilities show excellent agreement (Fig. 6).

Note in Figure 5 that asymmetries of the fining-upward tendencies of *channel* →*levee* are clearly apparent in the quenched conditional simulation. Such asymmetric juxtapositional relationships, which are strongly evident in the vertical transition probability measurements, cannot be simulated by traditional indicator variogram-based geostatistical methods.

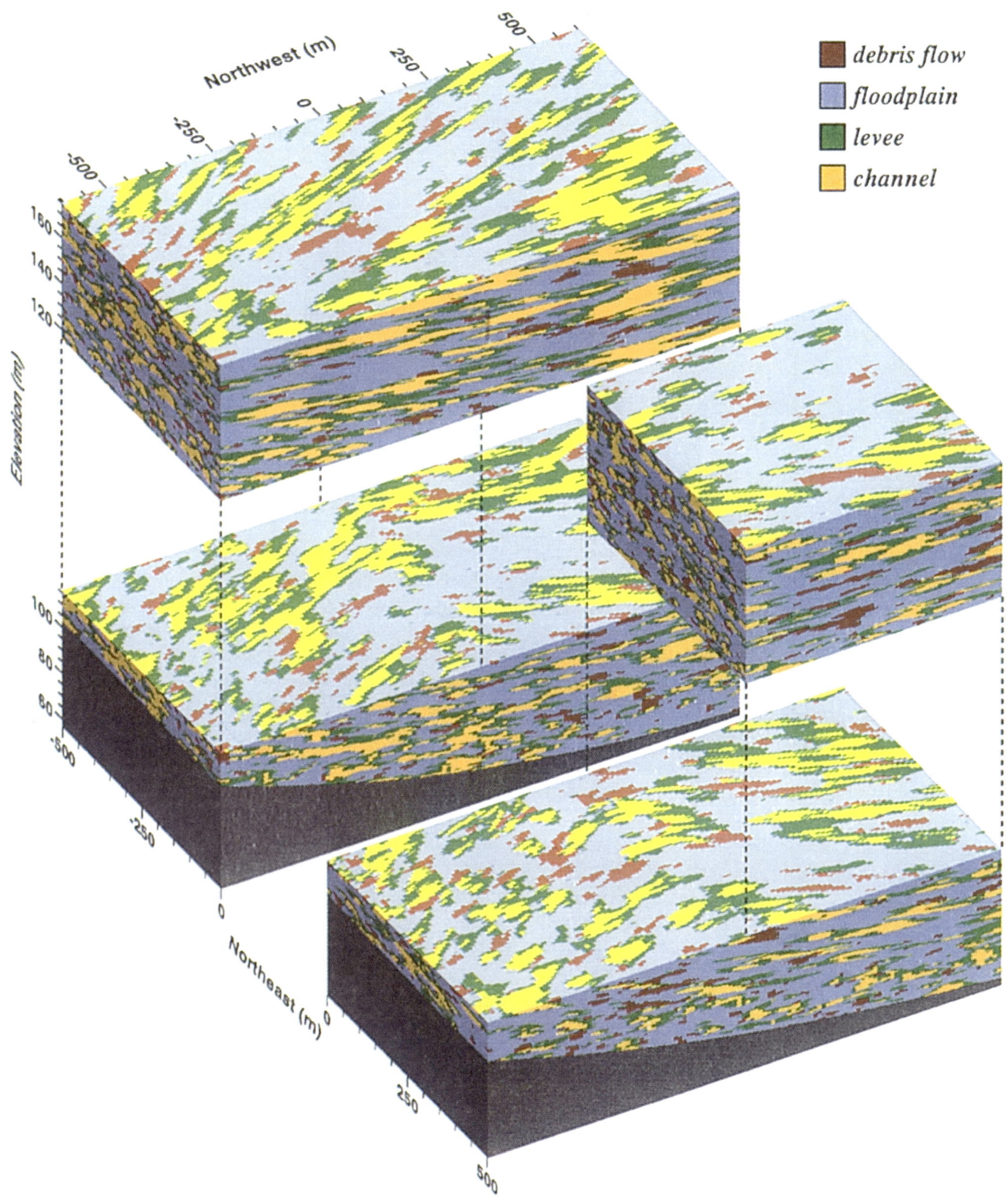

FIG. 7.– Three-dimensional simulation of hydrofacies architecture at LLNL conditioned by 125 borehole logs on a $334 \times 120 \times 400$ node grid with a discretization of 3 m $\times$10 m $\times$0.3 m ($x \times y \times z$).

Stationarity– Limitations and Possibilities

The stochastic methods described in this paper rely on an assumption of stationarity in the models of spatial variability. From the standpoint of geologic interpretation of alluvial fans, Neton et al. (1994) have criticized stochastic approaches for not considering directional variations, trends in lithofacies proportions and fan commingling. Granted, stationarity assumptions can constrain applicability of stochastic methods to limited regions. Nevertheless, it is possible to project nonstationary characteristics into the stochastic simulations, even though a stationary model of spatial variability is assumed. In the SIS step, conditioning data can enforce a nonstationarity in proportions, such as the coarsening-upward or fining-outward of an alluvial fan, through the "unbiasedness" property (consideration of *local* proportions) intrinsic to the transition probability-based indicator cokriging estimate. The quenching step generally preserves major features of the initial configuration, so that trends in proportions may persist to the end result of the conditional simulation (Carle, 1996). Nonstationarities in direction can be included in the conditional simulation if an a priori model of local directional variations is conceived, as shown in Figure 7 for the LLNL site, where seven parasequences associated with episodes of Quaternary alluvial fan deposition have been mapped (Blake et al., 1995). In this example, stratigraphic dip directions radiate from different source areas, such that anisotropy directions vary laterally and with depth corresponding to each depositional package.

If several depositional systems occur within a study region, different models of spatial variability can be ascribed to each depositional system. An example of this is given later in application to commingling braided-river and alluvial fan deposits in the Salinas River Valley. It also is possible to develop location-dependent spatial variability models.

KINGS RIVER ALLUVIAL FAN

The Kings River alluvial fan is located southeast of Fresno, where the Kings River enters the San Joaquin Valley at the western base of the Sierra Nevada (Fig. 1). Agricultural use of herbicides, pesticides and fertilizer has led to extensive non-point source contamination of the aquifer system (Nightingale, 1970; Kloos, 1983; Domagalski and Dubrovsky, 1991). Realistic models of hydrofacies architecture of the Kings River alluvial fan are needed to assess contaminant fate over a large regional extent.

TABLE 2.– BASIS FOR DEFINITION OF HYDROFACIES FROM KINGS RIVER ALLUVIAL FAN CORE DATA.

#	facies	c-horizon texture	%
1	channel lag *gravel*	-	3
2	channel bar *sand*	sand, loamy sand	46
3	levee/bar *fine sand*	(fine) sandy loam	26
4	overbank *fines*	clay, silt, clay loam	25

In an effort to quantitatively characterize aquifer system heterogeneity, a detailed study area was established on a medial portion of the fan. Data available include 1:24,000-scale soil survey maps of c-horizon textures (U.S. Department of Agriculture, 1971), geophysical logs for seven test wells ranging in depth from 36 to 81 m, approximately 150 m of continuous core from three of the test wells, and 3.9 km of shallow seismic reflection and ground-penetrating radar (GPR) data. The core, logs and GPR and seismic data were obtained by the U.S Geological Survey in 1994 and 1995.

Geology

The Kings River alluvial fan deposits consist primarily of fluvially derived elongate gravel and sand bodies surrounded by overbank fine-grained material. Deposition primarily occurred during glacial episodes of the Sierra Nevada, with periods of nondeposition or erosion during the interglacials (Huntington, 1980). The fan sediments observed in the core and c-horizon (2 m depth) soil mapping were categorized into four hydrofacies based on sedimentalogical interpretations for the Kings River fluvial system (Table 2). Proportions of the hydrofacies are given according to the core data. Vertical juxtapositional relationships indicative of the fluvial setting, such as fining-upward tendencies (e.g., where *gravel* is overlain by *sand*, and *sand* is overlain by *fine sand*) were observed in the well data.

Sedimentologic categories similar to those seen in the core were recognized on the soil survey maps, as given by the c-horizon (2 m depth) textures shown in Table 2. Notably, *gravel* tends to be buried deeper than the 2-m depth of the soil samples; thus, corresponding soils generally are not evident on the map. The c-horizon map indicates relatively straight channel bodies, as well as lateral fining-outward (relative to channel axes) tendencies of *sand →fine sand →fines*. Compared to the core data, the proportions of fine-grained deposits are much greater in the c-horizon. Much of this difference probably relates

Transition Probability - Vertical (z) Direction

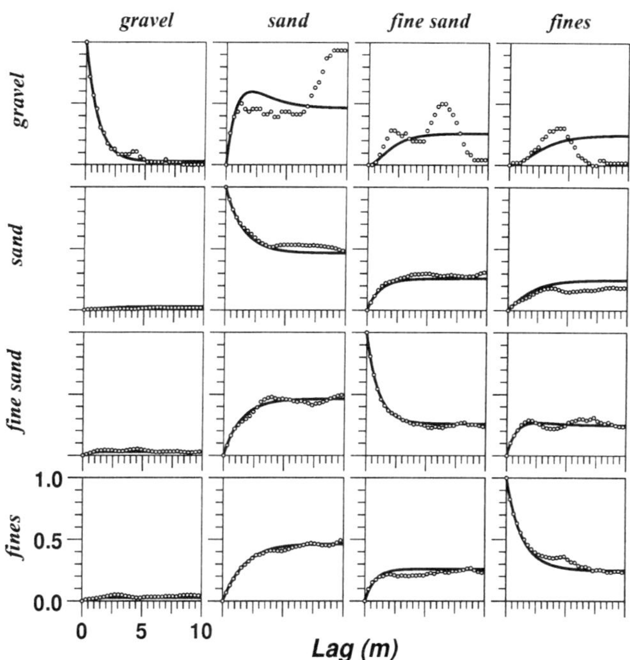

FIG. 8.– Matrix of vertical-direction transition probabilities for Kings River alluvial fan core data: measurements (dots) and Markov chain model (solid lines).

Transition Probability - Dip (y) Direction

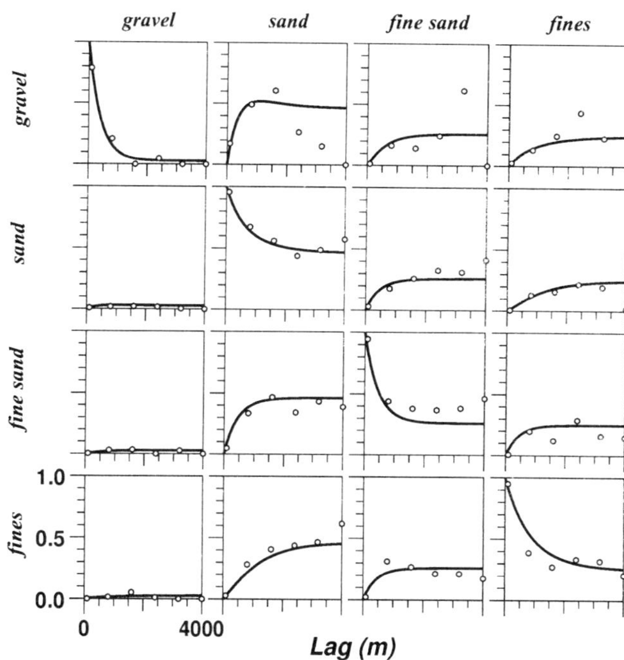

FIG. 9.– Matrix of dip-direction transition probabilities based on Kings River core data: measurements (dots) and Markov chain model (solid lines).

to preservation potentials of the sediment types, where coarse-grained sediment tends to be preferentially preserved, and fine-grained material tends to be eroded and transported to distal portions of the fan.

Markov Chain Modeling

Vertical (z)-direction transition probability measurements obtained from the core data, shown in Figure 8, were modeled by a Markov chain with a transition rate matrix of

$$\mathbf{R}_z = \begin{bmatrix} -\frac{1}{\bar{L}=1.06} & 2.10\hat{r} & 0.00\hat{r} & 0.00\hat{r} \\ 0.57\hat{r} & -\frac{1}{\bar{L}=2.73} & 1.44\hat{r}_{23,z} & 0.59\hat{r} \\ 1.64\hat{r} & 0.68\hat{r} & -\frac{1}{\bar{L}=1.33} & 1.53\hat{r} \\ 1.69\hat{r} & 0.67\hat{r} & 1.51\hat{r} & -\frac{1}{\bar{L}=1.70} \end{bmatrix} \text{ m}^{-1}$$

where the symbols are analogous to those in (19), $\bar{L}$ denoting mean lengths by (14) and $\hat{r}$ denoting a transition rate relative to proportions by (18). Transition rates for the last row and column involving the background category (fines) were calculated by applying (20) and (21). Fining-upward tendencies are indicated by the transition rates of

2.10$\hat{r}$ and 1.44$\hat{r}$ for *gravel* →*sand* and *sand* →*fine sand*, versus 0.57$\hat{r}$ and 0.68$\hat{r}$ for the opposing *sand* →*gravel* and *fine sand* →*sand* transitions. A tendency for *gravel* to occur as channel lag deposits beneath *sand* is strongly evident in the zero vertical transition rates for *gravel* →*fine sand* and *gravel* →*fines*. The relatively high transition rates of 1.53$\hat{r}$ for *fines* →*silty sand* and 1.51$\hat{r}$ for *silty sand* →*fines* indicate that *fines* and *silty sand* tend to be associated with each other. The relatively high transition rates of 1.64$\hat{r}$ for *silty sand* →*gravel* and 1.69$\hat{r}$ for *fines* →*gravel* indicate a tendency of the fluvial cycles to initialize in finer grained deposits.

The test wells were aligned in the general dip direction, which enabled an estimation of dip (y)-direction transition probabilities (Fig. 9), although considerable uncertainty results from the limited data and large well spacing. Markov chain modeling for the dip direction was facilitated by assuming the proportions given in Table 2, the overbank fines as the background category and symmetrical lateral juxtapositional relationships, as indicated by the soil survey. As a result, only six parameters needed to be specified for the dip-direction: mean lengths of the

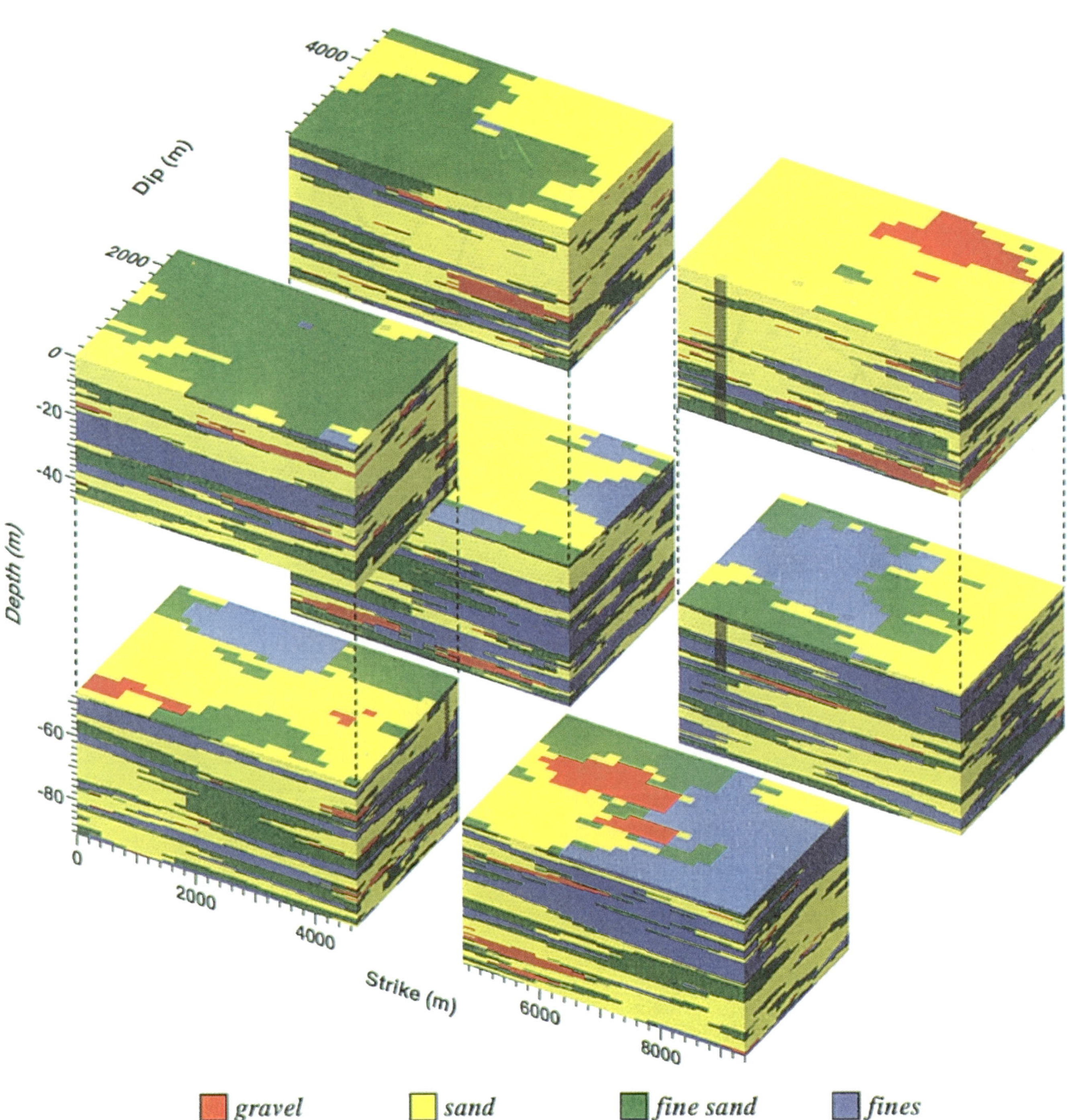

FIG. 10.– Conditional simulation of hydrofacies architecture in a portion of the Kings River alluvial fan. Conditioning data (shaded) consist of core and geophysical logs from seven boreholes.

gravel, *sand* and *fine sand* categories and juxtapositional tendencies for *gravel → sand*, *sand → fine sand* and *gravel → fine sand*. Mean lengths of 400 m for *gravel*, 1,200 m for *sand* and 600 m for *fine sand* were assumed, which appeared consistent with spatial continuity indicated by the c-horizon maps and seismic and GPR surveys. The off-diagonal transition rates were tuned by trial-and-error to fit the data after first trying values measured from the soil survey and analogous to those applied to the vertical transition rates (a Walther's Law-style interpretation). The resulting dip (y)-direction transition rate matrix was prescribed by

$$\mathbf{R}_y = \begin{bmatrix} -\dfrac{1}{\overline{L}=400} & 1.50\widehat{r} & 0.55\widehat{r} & c_1 \\ s & -\dfrac{1}{\overline{L}=1200} & 1.44\widehat{r} & c_1 \\ s & s & -\dfrac{1}{\overline{L}=600} & c_1 \\ c_2 & c_2 & c_2 & c_2 \end{bmatrix} \mathrm{m}^{-1}$$

where s, c_1 and c_2 are prescribed by (16), (20) and (21), respectively.

Seismic and GPR data also indicated lateral continuity of fine-grained units for the strike (x) direction. Assumed mean lengths of 325 m for *sand* and 215 m for *fine sand* were checked for consistency with the soil map. A strike (x)-direction transition rate matrix was developed in a manner similar to the dip (y)-direction rate matrix by

$$\mathbf{R}_x = \begin{bmatrix} -\dfrac{1}{\overline{L}=150} & 1.50\widehat{r} & 0.55\widehat{r} & c_1 \\ s & -\dfrac{1}{\overline{L}=325} & 1.44\widehat{r} & c_1 \\ s & s & -\dfrac{1}{\overline{L}=215} & c_1 \\ c_2 & c_2 & c_2 & c_2 \end{bmatrix} \mathrm{m}^{-1}$$

The resulting transition dip (y)- and strike (x)-direction transition rate matrices yielded fining-outward tendencies that are, in general, less strict than the fining-upward tendencies of the vertical transition rates. This interpretation allows for the possibility that *gravel* may occur laterally adjacent to all categories, although primarily adjacent to *sand*.

Conditional Simulation

The conditional simulation of a portion of the Kings River alluvial fan shown in Figure 10 was generated on a 51 × 51 × 301-node grid with a discretization of 50 m × 100 m × 0.3048 m ($x \times y \times z$). Conditioning was maintained by treating both the core data and facies interpretations of the geophysical logs as hard data. The simulation shows juxtapositional relationships that are consistent with a con-

ceptual geologic model of the Kings River system. The *gravel* and *sand* categories tend to be strongly associated as elongate channel deposits with strong fining-upward tendencies of *gravel → sand* and *sand → fine sand*, as expected for a fluvial depositional system. The *fine sand* and *fines* categories tend to be associated with each other, as would be expected for upper channel, levee and overbank facies.

SALINAS VALLEY

In the Salinas Valley, elevated levels of nitrate in ground water are the result of historical land uses, which include irrigated agriculture, feed lots and on-site sewage disposal. Determining the impacts of such land-use practices on groundwater quality requires careful analysis of nitrate transport from land surface to water table. In turn, the impact of physical heterogeneity in the vadose zone is crucial to development of realistic nitrate transport models. A detailed study area was established at Wing Ranch, about 15 km southeast of Salinas (Fig. 1), in an effort to stochastically characterize three-dimensional vadose zone heterogeneity using the geostatistical methods described herein. Interestingly, the vadose zone at Wing Ranch consists of alluvial deposits from two depositional systems. This section describes an approach to including such a "nonstationarity" into geostatistical simulation.

Geology

The vadose zone at the Wing Ranch study area consists of fluvial deposits of the Salinas River and alluvial fan deposits originating from granitic Salinian basement terrains in the Gabilan Range to the northeast (Durham, 1974; Tinsley, 1975). Coarse-grained material from the Salinas River deposits can be distinguished from the alluvial fan deposits by the presence of Tertiary marine shale and chert derived from Monterey Formation rocks in the Sierra de Salinas southwest of Salinas Valley.

Seven boreholes were drilled to a depth of approximately 30 m. Cores were obtained continuously and categorized according to provenance and texture (Maserjian, 1993) as:

Salinas River deposits:	alluvial fan deposits:
1. *sand* (53%)	1. *sand* (43%)
2. *silty sand* (47%)	2. *silty sand* (46%)
	3. *fines* (11%)

Transition Probability - Vertical (z) Direction

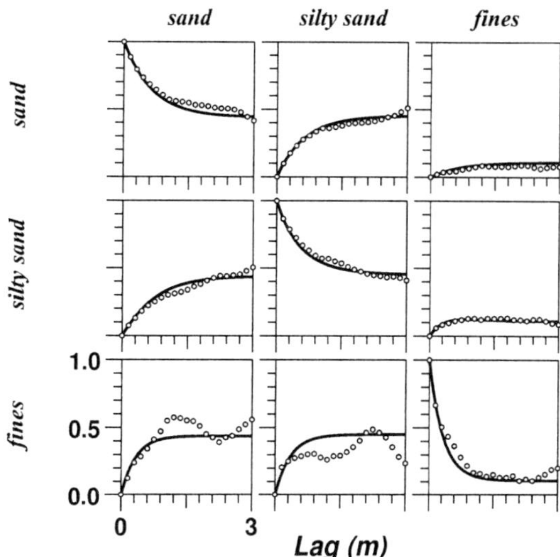

Transition Probability - Vertical (z) Direction

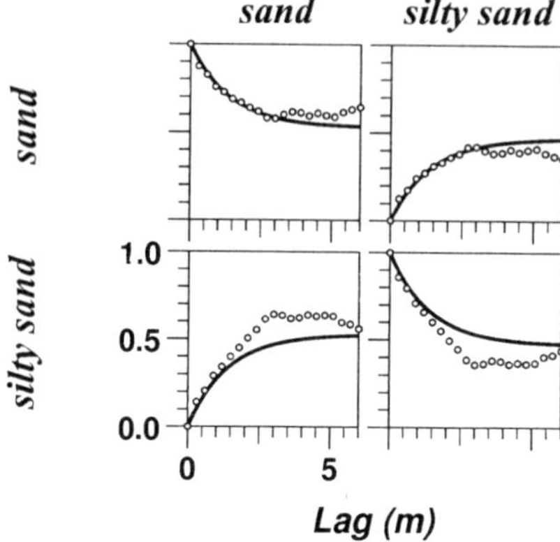

FIG. 12.—Matrix of vertical transition probability measurements based on core data for alluvial fan deposits at Wing Ranch study area: measurements (dots) and Markov chain model (solid lines).

FIG. 11.— Matrix of vertical-direction transition probabilities based on core data for Salinas River deposits at Wing Ranch study area: measurements (dots) and Markov chain model (solid lines).

where the percentages indicate the proportions of the textures observed in each depositional system. Based on lithologies observed in the study area, the alluvial fan deposits generally overlap the Salinas River deposits, although some interfingering of the depositional systems occurs.

Markov Chain Modeling

The local patterns of heterogeneity in the Salinas River and alluvial fan deposits were assumed to be independent of each other. Vertical (z)-direction transition rate matrices for the two depositional systems were developed by fitting the vertical-direction transition probability measurements (Figs. 11 and 12) obtained from the core data, yielding

$$\mathbf{R}_{z,\text{SALINAS}} = \begin{bmatrix} -\frac{1}{\overline{L}=2.95} & c_1 \\ c_2 & c_2 \end{bmatrix} \text{m}^{-1}$$

and

$$\mathbf{R}_{z,\text{ALLUVIAL}} = \begin{bmatrix} -\frac{1}{\overline{L}=1.15} & 1.04\hat{r} & c_1 \\ 0.63\hat{r} & -\frac{1}{\overline{L}=0.98} & c_1 \\ c_2 & c_2 & c_2 \end{bmatrix} \text{m}^{-1}$$

where background categories were assumed as *silty sand*

for the Salinas River system and *fines* for the alluvial fan system; diagonal transition rates are defined according to the mean length $\overline{L}$ as in (14); c_1 and c_2 denote application of (20) and (21), respectively; and $\hat{r}$ denotes a transition rate dependent on proportions by (18). Note that no off-diagonal transition rates need to be specified in the two-category Salinas River case.

The lateral spacing of the boreholes is too large to establish a 3-D model of spatial variability directly from the data. Alternatively, transition rate matrices for the strike (x) and dip (y) directions were established interpretatively by assuming:

1. proportions, as given previously,
2. mean lengths corresponding to geologically plausible elongation ratios,
3. similar juxtapositional tendencies of *sand* →*silty sand* ($1{\to}2$) as observed in the vertical, and
4. symmetric juxtapositional relationships in the strike (x) and (y) directions.

As a result, strike (x)- and dip (y)-direction transition rate matrices were obtained by

$$\mathbf{R}_{x,\text{SALINAS}} = \begin{bmatrix} -\frac{1}{\overline{L}=50} & c_1 \\ c_2 & c_2 \end{bmatrix} \text{m}^{-1}$$

$$\mathbf{R}_{y,\text{SALINAS}} = \begin{bmatrix} -\frac{1}{\overline{L}=200} & c_1 \\ c_2 & c_2 \end{bmatrix} \text{m}^{-1}$$

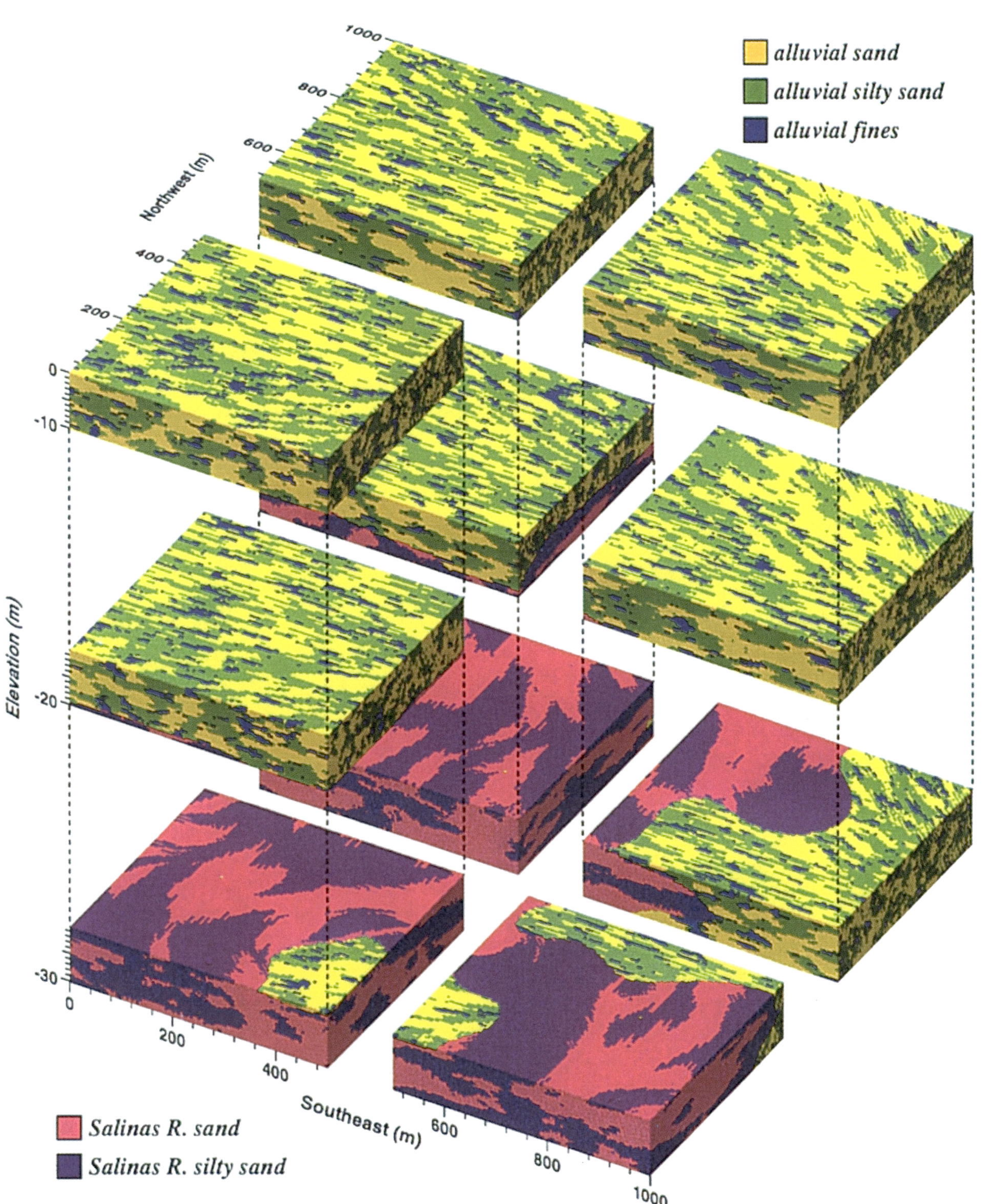

FIG. 13.– Conditional simulation of hydrofacies architecture at Wing Ranch study area, Salinas Valley, incorporating commingling alluvial fan and Salinas River deposits. Conditioning data consist of core descriptions from seven boreholes.

$$R_{x,\text{ALLUVIAL}} = \begin{bmatrix} -\frac{1}{\overline{L}=10} & 1.04\hat{r} & c_1 \\ s & -\frac{1}{\overline{L}=8} & c_1 \\ c_2 & c_2 & c_2 \end{bmatrix} \text{m}^{-1}$$

$$R_{y,\text{ALLUVIAL}} = \begin{bmatrix} -\frac{1}{\overline{L}=40} & 1.04\hat{r} & c_1 \\ s & -\frac{1}{\overline{L}=30} & c_1 \\ c_2 & c_2 & c_2 \end{bmatrix} \text{m}^{-1}$$

where s denotes application of (16), and c_1 and c_2 denote application of (20) and (21).

Conditional Simulation

A conditional simulation of hydrostratigraphy underlying the Wing Ranch study area (Fig. 13) was generated on a $101 \times 101 \times 101$-node grid with a discretization of 15 m $\times$ 15 m $\times$ 0.3 m ($x \times y \times z$). The simulation was produced in a similar manner as given in previous examples, except that another simulation step was added to distinguish the two depositional systems. First, conditional simulations of each depositional system were generated independently, as if only one depositional system existed in the vadose zone. Then, an additional conditional simulation was generated having two categories representing the two depositional systems. Using the proportions of the two systems, conditioning data and large assumed mean lengths, a geologically reasonable interface between the fluvial and alluvial fan systems was produced by a conditional simulation, which was then used to splice together the two independent conditional simulations of the Salinas River and alluvial fan hydrofacies architecture.

Nonstationarities in anisotropy directions also were considered. Meandering features were assimilated into the Salinas River deposits by assuming that anisotropy directions vary in the horizontal plane according to a Gaussian random field (Deutsch and Journel, 1992, p. 136-137). Stratigraphic dip directions of the alluvial fan deposits were fixed according to a radial morphology. The simulation illustrates two "nonstationarities" that can be incorporated in the geostatistical simulation of hydrostratigraphy: (1) nonstationarity attributed to different patterns of heterogeneity in each of two interfingering depositional systems, and (2) nonstationarity in anisotropy directions attributed to morphological features, such as meandering or radial alluvial fan morphology. These nonstationarities obviously would impact field-scale transport of nitrates. Conditional simulations such as this have been used for field-scale mod-

els of flow and transport to estimate contaminant arrival times from the land surface to the water table (Fogg et al., 1995).

SUMMARY AND CONCLUSIONS

A new transition probability-based geostatistical approach to conditional simulation has been presented with example applications to alluvial and fluvial hydrofacies architecture. The approach improves the ability to incorporate geologic interpretation into geostatistical methods of conditional simulation, which is crucial for hydrogeological applications. Hydrogeological data typically are too sparse in lateral directions to apply the traditional geostatistical approach of empirical curve-fitting to variogram measurements. Furthermore, existing geostatistical approaches do not fully account for all spatial cross-correlations and, thus, cannot account for certain juxtapositional relationships such as fining-upward tendencies, which commonly occur in alluvial depositional systems.

The foundation of the approach is quantitative description of spatial variability of categorical variables (hydrofacies) in terms of the transition probability. The transition probability offers advantages over the more prevalent "indicator variogram" where integration of indirect, subjective, or conceptual information is desired. By simple graphical observation, features of the transition probability can be related precisely to geological attributes of proportions, mean length and juxtapositional tendencies. Such interpretability not only improves understanding of spatial variability measurements, but facilitates construction of geologically plausible spatial variability models.

The adoption of the transition probability as a measure of spatial variability points to employment of the Markov chain as a mathematically and conceptually simple yet theoretically powerful model of spatial variability. Markov chains have a long-standing track record in geology for quantitative analysis of vertical stratigraphic sequences. In the examples given, Markov chains consistently provide excellent models for measured vertical-direction transition probabilities of hydrofacies categories at three sites of alluvial and fluvial deposits in California.

Markov chain models can be extended to 3-D applications by modeling spatial variability in principal stratigraphic directions of vertical, strike and dip using either quantitative or conceptual means, then interpolating those 1-D models to all directions. Traditional geostatistical approaches rely on abundant data or an exhaustive "reference image" to empirically obtain a model of spatial variability. Alternatively, the Markov chain approach provides a conceptual framework for developing models of spatial variability through integration of fundamental information on proportions, mean length and juxtapositional tendencies. This approach is particularly useful in lateral directions, where data are typically sparse relative to the vertical. Sylvester's theorem (see Appendix) provides the mathematical foundations for calculating transition rate matrices and transition probabilities of 3-D, continuous-lag Markov chain models.

The 3-D Markov chain models of spatial variability then can be applied to both sequential indicator simulation (SIS) and simulated annealing geostatistical conditional simulation algorithms. In our experience, excellent conditional simulation results can be obtained by first applying SIS to generate an "initial configuration," then iteratively improving the SIS result by simulated quenching (zero-temperature annealing). As applied to fluvial deposits, the conditional simulations display geologically plausible juxtapositional tendencies. For example, at Lawrence Livermore National Laboratory, levee deposits tend to occur above and laterally adjacent to channel deposits; in channel bodies of the Kings River alluvial fan, gravel lag deposits tend to occur at the base of sand deposits. These fining-upward tendencies are strongly supported by core data but could not be properly simulated by prevalent indicator kriging-based SIS approaches. It is also possible to incorporate "nonstationarities" into the stochastic simulation results, even though models of spatial variability fundamentally assume stationarity. Nonstationarities in direction, such as meandering or the radial morphology of an alluvial fan, can be incorporated in the SIS and simulated quenching algorithms with an a priori map (deterministic or stochastic) of anisotropy directions.

In the techniques and examples given, we have emphasized integration of geologic interpretation into geostatistical methods. Granted, any geostatistical approach can be applied from a purely quantitative standpoint, such that categories defined according to "levee" or "channel lag" or "overbank fines" would assume no geologic significance. However, such geologic facies have geometric and locational significance in geologic interpretation, providing insight to characterizing true hydraulic heterogeneity patterns that otherwise would be difficult to characterize solely by statistics derived from typical hydrogeological data sets. Thus, we have attempted to present a geostatistical modeling approach that not only maximally utilizes univariate (proportions) and bivariate (transition probability) spatial statistics, but also brings out the "geo" aspect of "geostatistics."

APPENDIX: MATHEMATICAL FOUNDATIONS

In order to develop continuous-lag Markov chains as geostatistical models of spatial variability, one must be able to perform the following:

- evaluate the *matrix exponential* form of Markov chain given by (8),
- evaluate the *matrix logarithm* of a transition probability matrix given by (11), and
- convert a discrete-lag Markov chain to a continuous-lag Markov chain by combining (8) and (11).

However, (8) and (11) cannot be computed directly from the matrix entries. In either case, the key step is to find the eigenvalues of $\mathbf{R}_\phi$ or $\mathbf{T}(h_\phi)$, which can be computed using codes for real general matrices given by Smith et al. (1976) or Press et al. (1992).

For notational simplicity, let the lag $h = h_\phi$ and $\mathbf{R} = \mathbf{R}_\phi$. A square $(K \times K)$ matrix such as $\mathbf{R}$ can be expressed in diagonal form with respect to its eigenvalues by

$$\mathbf{R} = \sum_{l=1}^{K} \lambda_l \mathbf{Z}_l \qquad (26)$$

where the λ_l for $l = 1, ..., K$ denote the eigenvalues of $\mathbf{R}$, and $\mathbf{Z}_l$ denotes a spectral component matrix associated with each eigenvalue λ_l. The spectral component matrices $\mathbf{Z}_l$ can be determined directly from the eigenvalues and matrix $\mathbf{R}$ by

$$\mathbf{Z}_l = \frac{\prod_{m \neq l} (\lambda_m \mathbf{I} - \mathbf{R})}{\prod_{m \neq l} (\lambda_m - \lambda_l)} \qquad l = 1, ..., K \qquad (27)$$

where $\mathbf{I}$ denotes the identity matrix. The continuous-lag Markov chain (8) then can be computed from

$$\mathbf{T}(h) = \sum_{l=1}^{K} \exp(\lambda_l h) \mathbf{Z}_l \qquad (28)$$

through application of Sylvester's theorem (Agterberg, 1974, p. 406-412). Recognizing that (28) represents a canonical form of $\mathbf{T}(h)$, two useful conclusions can be drawn for the Markov chain model:

1. The eigenvalues $\theta_l(h)$ of $\mathbf{T}(h)$ relate to the eigenvalues

λ_l of $\mathbf{R}$ by

$$\theta_l(h) = \exp\left(\lambda_l h\right) \quad \text{or} \quad \lambda_l = \frac{\ln \theta_l(h)}{h} \qquad \forall l = 1, ..., K \tag{29}$$

2. Both $\mathbf{R}$ and $\mathbf{T}(h)$ have identical spectral component matrices $\mathbf{Z}_l$.

As a result, if a Markov chain model is assumed, a transition probability matrix $\mathbf{T}(\Delta h)$ for a discrete lag Δh can be used to compute $\mathbf{R}$ by applying (29) to (26) to obtain

$$\mathbf{R} = \sum_{l=1}^{K} \frac{\ln \theta_l(\Delta h)}{\Delta h} \mathbf{Z}_l \tag{30}$$

where $\theta_l(\Delta h)$ and $\mathbf{Z}_l$ are the eigenvalues and spectral component matrices, respectively, corresponding to $\mathbf{T}(\Delta h)$. Application of (30) to (8) yields

$$\mathbf{T}(h) = \sum_{l=1}^{K} \theta_l(\Delta h)^{h/\Delta h} \mathbf{Z}_l \tag{31}$$

which represents a continuous-lag version of the more commonly used discrete-lag Markov chain model (10). We emphasize that the advantage of (31) over (10) is the continuous functional representation of the model, that is, the ability to calculate $\mathbf{T}(h)$ at any h, not just integer multiples of Δh. Expression (28) shows that a Markov chain model corresponds to a linear combination of exponential functions. Nonetheless, rather nonexponential looking structures can be obtained from a Markov chain model, as evident in some of the off-diagonal transition probabilities for the examples given.

ACKNOWLEDGMENTS

The authors thank J. M. Davis and J. H. Doveton for commentary that substantially improved the paper. This work was supported by Lawrence Livermore National Laboratory, U.S. Geological Survey Water Resources Research Grant 14-18-001-61909, the Monterey Water Resources Agency, U.S. Environmental Protection Agency Grant R819658, Center for Ecological Health Research at University of California at Davis, N.I.E.H.S. Superfund Grant ES-04699, Occidental Chemical Company, University of California Toxic Substances Teaching and Research Program and U.S. Army Corps of Engineers Waterways Experimentation Station. We are also grateful to Lisa Maserjian for collecting and describing cores from the Salinas Valley site and to Karen Burow of the U.S. Geological Survey for providing cores from the Kings River alluvial fan. Although the information in this document has been funded in part by the U.S. Environmental Protection Agency, it does not necessarily reflect the views of the Agency, and no official endorsement should be inferred.

REFERENCES

AGTERBERG, F. P., 1974, Geomathematics: New York, Elsevier Scientific Publishing Co., 596 p.

ALLEN, J. R. L., 1970, Studies in fluviatile sediments, a comparison of fining-upward cyclothems with special reference to coarse-member composition and interpretation: Journal of Sedimentary Petrology, v. 40, p. 298-323.

ANDERSON, M. P., 1989, Hydrogeologic facies models to delineate large-scale spatial trends in glacial and glaciofluvial sediments: Geological Society of America Bulletin, v. 101, p. 501-511.

ANDERSON, M. P., AND WOESSNER, W. W., 1992, Applied Groundwater Modeling: San Diego, Academic Press Inc., 391 p.

AMERICAN SOCIETY FOR TESTING AND MATERIALS, 1996, Committee D-18 on soil and rock: American Society for Testing and Materials, 257 p.

BIERKENS, M. F. P., 1996, Modeling hydraulic conductivity of a complex confining layer at various spatial scales: Water Resources Research, v. 32, p. 2369-2382.

BLAKE, R. G., NOYES, C. M., AND MALEY, M. P., 1995, Hydrostratigraphic analysis–The key to cost-effective ground water cleanup at Lawrence Livermore National Laboratory: Lawrence Livermore National Laboratory Report UCRL-JC-120614, 13 p.

CARLE, S. F., 1996, A transition probability-based approach to geostatistical characterization of hydrostratigraphic architecture: Unpubl. Ph.D. Dissertation, University of California, Davis, 233 p.

CARLE, S. F., 1997, Implementation schemes for avoiding artifact discontinuities in simulated annealing: Mathematical Geology, v. 29., p. 231-244.

CARLE, S. F., AND FOGG, G. E., 1996, Transition probability-based indicator geostatistics: Mathematical Geology, v. 28, p. 453-476.

CARLE, S. F., AND FOGG, G. E., 1997, Modeling spatial variability with one- and multidimensional continuous-lag Markov chains: Mathematical Geology, v. 29, p.

CARR, D. D., HOROWITZ, A., HRARBAR, S. V., RIDGE, K. F., ROONEY, R., STRAW, W. T., WEBB, W., AND POTTER, P. E., 1966, Stratigraphic sections, bedding sequences and random processes: Science, v. 28, p. 89-110.

CASAGRANDE, A., 1948, Classification and identification of soils: Transactions of the American Society of Civil Engineers, v. 113, p. 901-992.

DAVIS, M. J., LOHMANN, R. C., PHILLIPS, F. M., WILSON, J. L., AND LOVE, D. W., 1993, Architecture of the Sierra Ladrones Formation, central New Mexico–Depositional controls on the permeability correlation structure: Geological Society of America Bulletin, v. 105, p. 998-1007.

DEUTSCH, C. V., AND COCKERHAM, P. W., 1994, Practical

considerations in the application of simulated annealing in stochastic simulation: Mathematical Geology, v. 26, p. 67-82.

DEUTSCH, C. V., AND JOURNEL, A. G., 1992, GSLIB, Geostatistical Software Library and User's Guide: New York, Oxford University Press, 340 p.

DOMAGALSKI, J. L., AND DUBROVSKY, D. M., 1991, Regional assessment on nonpoint-source pesticide residues in ground water, San Joaquin Valley, California: U.S. Geological Survey Water-Resources Investigations Report 91-4027, 64 p.

DOVETON, J. H., 1971, An application of Markov chain analysis to the Ayrshire Coal Measures succession: Scottish Journal of Geology, v. 7, p. 11-27.

DOVETON, J. H., 1994, Theory and applications of vertical variability measures from Markov chain analysis, in Yarus, J. M., and R. L. Chambers eds., Stochastic Modeling and Geostatistics–Principles, Methods, and Case Studies: American Association of Petroleum Geologists Computer Applications in Geology, no. 3, p. 55-64.

DURHAM, D. L., 1974, Geology of the southern Salinas Valley area, California: U.S. Geological Survey Professional Paper 819, 111 p.

ETHERIDGE, F. G., AND SCHUMM, S. A., 1978, Reconstructing paleochannel morphologic and flow characteristics–Methodology, limitations, and assessment, in Miall, A. D., ed., Fluvial Sedimentology: Canadian Society of Petroleum Geologists Memoir 5, p. 703-721.

FOGG, G. E., 1989, Emergence of geologic and stochastic approaches for characterization of heterogeneous aquifers, in Proceedings, New Field Techniques for Quantifying the Physical and Chemical Properties of Heterogeneous Aquifers, Dallas, Texas, March 20-23, 1989: National Water Well Association, Dublin, Ohio, p. 1-17.

FOGG, G. E., ROLSTON, D. E., LABOLLE, E. M., BUROW, K. R., MASERJIAN, L. A., DECKER, D. A., AND CARLE, S. F., 1995, Matrix diffusion and contaminant transport in granular geologic materials, with case study of nitrate contamination in Salinas Valley, California: Final Technical Report submitted to Monterey County Water Research Agency and U.S. Geological Survey under Water Resources Research Award No. 14-08-0001-G1909, 65 p. and appendices.

GOMEZ-HERNANDEZ, J. J., AND SRIVASTAVA, R. M., 1990, ISIM3D, an ANSI-C three-dimensional multiple indicator conditional simulation program: Computers and Geosciences, v. 16, p. 395-440.

HARBAUGH, J. W., AND BONHAM-CARTER, G. F., 1970, Computer Simulation in Geology: New York, Wiley Interscience, 575 p.

HATTORI, I., 1976, Entropy in Markov chains and discrimination of cyclic patterns in lithologic successions: Mathematical Geology, v. 8, p. 477-497.

HUNTINGTON, G.L., 1980, Soil-land form relationships of portions of the San Joaquin River and Kings River alluvial depositional systems in the Great Valley of California: Unpubl. Ph.D. Dissertation, University of California, Davis, 147 p.

JOHNSON, N. M., 1995, Characterization of alluvial hydrostratigraphy with indicator semivariograms: Water Resources Research, v. 31, p. 3205-3216.

JOHNSON, N. M., AND DREISS, S. J., 1989, Hydrostratigraphic interpretation using indicator geostatistics: Water Resources Research, v. 25, p. 2501-2510.

JOURNEL, A. G., AND ALABERT, F., 1990, New method for reservoir mapping: Journal of Petroleum Technology, v. 42, p. 212-218.

KLOOS, H., 1983, DBCP pesticide in drinking water wells in Fresno and other communities in the Central Valley of California: Ecology of Disease, v. 2, p. 363-367.

KOLTERMANN, C. E., AND GORELICK, S. M., 1992, Paleoclimatic signature in terrestial flood deposits: Science, v. 256, p. 1775-1782.

KRUMBEIN, W. C., 1968, Fortran IV computer program for simulation of transgression and regression with continuous-time Markov models: Computer Contribution 26, Kansas State Geological Survey, 38 p.

KRUMBEIN, W. C., AND DACEY, M. F., 1969, Markov chains and embedded Markov chains in geology: Mathematical Geology, v. 1, p. 79-96.

LEEDER, M. R., 1982, Sedimentology: London, George Allen & Unwin, 344 p.

LIN, C., AND HARBAUGH, J. W., 1984, Graphic Display of Two- and Three- Dimensional Markov Computer Models in Geology: New York, Van Nostrand Reinhold, 180 p.

MASERJIAN, L. A., 1993, Hydrogeologic analysis of the unsaturated zone, northern Salinas Valley, California: Unpubl. M.S. Thesis, University of California, Davis, p.

MIALL, A. D., 1973, Markov chain analysis applied to an ancient alluvial plain succession: Sedimentology, v. 20, p. 347-365.

MIALL, A. D., 1982, Analysis of fluvial depositional systems: American Association of Petroleum Geologists, Education course note series # 20, 75 p.

MIALL, A. D., 1992, Alluvial deposits, in Walker, R. C., and James, N. P., eds., Facies Models: Geological Association of Canada, p. 119-142.

NETON, M. J., DORSCH, J., OLSON, C. D., AND YOUNG, S. C., 1994, Architecture and directional scales of heterogeneity in alluvial fan aquifers: Journal of Sedimentary Research, v. B64, p. 245-247.

NIGHTINGALE, H. I., 1970, Statistical evaluation of salinity and nitrate content and trends beneath urban and agricultural areas - Fresno, California: Ground Water, v. 8, p. 22-28.

NOYES, C. N., 1991, Hydrostratigraphic analysis of the Pilot Remediation Test Area, LLNL, Livermore, California: Unpubl. M.S. Thesis, University of California, Davis, 165 p.

PHILLIPS, F. M., AND WILSON, J. L., 1989, An approach to estimating hydraulic conductivity spatial correlation scales using geologic characteristics: Water Resources Research, v. 25, p. 141-143.

POETER, E. P., AND MCKENNA, S. A., 1995, Reducing uncertainty associated with ground-water flow and transport predictions: Ground Water, v. 33, p. 899-904.

POLITIS, D. N., 1994, Markov chains in many dimensions: Advances in Applied Probability, v. 26, p. 756-774.

PRESS, W. H., TEUKOLSKY, S. A., VETTERLING, W. T., AND FLANNERY, B. P., 1992, Numerical Recipes in Fortran: New York, Cambridge University Press, 963 p.

QUALHEIM, B. J., 1988, Well log report for the LLNL ground water project 1984-1987: Lawrence Livermore National Laboratory Report UCID-21342 (updated 1989, 1993), 12 p. and 304 oversize sheets.

RITZI, R. W., DOMINIC, D. F., BROWN, N. R., KAUSCH, K. W., MCALENNEY, P. J., AND BASIAL, M. J., 1995, Hydrofacies distribution and correlation in the Miami Valley aquifer system: Water Resources Research, v. 31, p. 3271-3281.

RITZI, R. W., JAYNE, D. F., ZAHRADNIK, A. J., FIELD, A. A., AND FOGG, G. E.,, 1994, Geostatistical modeling of heterogeneity in glaciofluvial, buried-valley aquifers: Ground Water, v. 32, p. 666-674.

ROSS, S., 1993, Introduction to Probability Models, 5th ed.: San Diego, Academic Press, 556 p.

SCHEIBE, T. D., AND FREYBERG, D. L., 1995, Use of sedimentological information for geometric simulation of natural porous media structure: Water Resources Research, v. 31., p. 3259-3270.

SCHWARZACHER, W., 1969, The use of Markov chains in the study of sedimentary cycles: Mathematical Geology, v. 12, p. 213-234.

SMITH, B. T., BOYLE, J. M., DONGARRA, J. J., GARBOW, B. S., IKEBE, Y., KLEMA, V. C., AND MOLER, C. B., 1976, Matrix eigensystem routines–EISPACK guide, *in* Lecture Notes in Computer Science, 2d ed.: New York, Springer-Verlag, v. 6, 551 p.

SURO-PEREZ, V., AND JOURNEL, A. G., 1991, Indicator principal component kriging: Mathematical Geology, v. 23, p. 759-788.

SWITZER, P., 1965, A random set process in the plane with a Markovian property (note): Annals of Mathematical Statistics, v. 36, p. 1859-1863.

TETZLAFF, D. M., AND HARBAUGH, J. W., 1989, Simulating Clastic Sedimentation: New York, Van Nostrand Reinhold, 202 p.

THORPE, R. K., ISHERWOOD, W. F., DRESEN, M. D., AND WEBSTER-SCHOLTEN, C. 1990, CERCLA Remedial Investigations Report for the LLNL Livermore site: Lawrence Livermore National Laboratory Report UCAR-10299.

TINSLEY, J. C., III, 1975, Quaternary geology of the northern Salinas Valley, Monterey County, California: Unpubl. Ph.D. Dissertation, Stanford University, Stanford, California, 201 p.

TURK, G., 1979, Transition analysis of structural sequences– Discussion: Geological Society of Americal Bulletin, v. 90, p. 989-991.

TURK, G., 1982, Markov chains–Letters to the Editor: Mathematical Geology, v. 14, p. 539-542.

U.S. DEPARTMENT OF AGRICULTURE, 1971, Soil survey, eastern Fresno area, California: U.S. Department of Agriculture Soil Conservation Service, Washington, D.C., 323 p.

VISTELIUS, A. B., 1949, On the question of the mechanism of formation of strata: Doklady Akademii Nauk, SSSR, v. 65, p. 191-194.

VISTELIUS, A. B., 1967, Studies in Mathematical Geology: New York, Consultants Bureau, p. 252-258.

WEBB, E. K., AND ANDERSON, M. P., 1996, Simulation of preferential flow in three-dimensional heterogeneous conductivity fields with realistic internal architecture: Water Resources Research, v. 32, p. 533-545.

WEN, X. H., AND KUNG, C. S., 1993, Stochastic simulation of solute transport in heterogeneous formations–A comparison of parametric and nonparametric geostatistical approaches: Ground Water, v. 31, p. 953-965.

WINGLE, W. L., AND POETER, E. P., 1993, Uncertainty associated with semivariograms used for site simulation: Ground Water, v. 31, p. 725-734.

COMBINING GEOLOGIC INFORMATION AND INVERSE PARAMETER ESTIMATION TO IMPROVE GROUNDWATER MODELS

EILEEN P. POETER AND SEAN A. MCKENNA

Department of Geology and Geological Engineering, Colorado School of Mines, Golden, Colorado 80401
Current address for McKenna:
Geohydrology Department, Sandia National Laboratories, Albuquerque, New Mexico 87185

ABSTRACT: Two synthetic examples and one field example demonstrate how improper definition of the spatial distribution of geohydrologic units in conceptual models of groundwater flow leads to erroneous parameter estimations, resulting in poor predictions of flow system behavior. Efficient and objective parameter estimation is obtained through inverse modeling techniques. The examples demonstrate the sensitivity of the estimated geohydrologic parameter values within the geohydrologic units to variations in spatial distribution of those units. In all three examples, geological information in the form of stratigraphic opinion or geophysical interpretation regarding unit distributions, as a well as geologic rules ranking hydraulic conductivity of units, improves conceptual model development.

INTRODUCTION

Development of conceptual models is the first step in a groundwater modeling project. A conceptual model qualitatively describes the geometry, hydraulic properties and boundary conditions of the groundwater system. It provides an overview of the location and volume of water recharging and discharging the system, as well as the patterns of flow within the system. Given the nonunique character of field data, multiple conceptual models are developed at the beginning of a project and, depending on the character of the incoming data and results of ongoing analyses, each conceptual model is either retained for further consideration or eliminated during the modeling process. Spatial distribution of hydraulic properties is the aspect of conceptual models of greatest interest to sedimentologists who contribute information to groundwater modelers. A groundwater modeler may specify this distribution either deterministically or stochastically (Fogg, 1989). The distribution of units that are either homogeneous or exhibit functionally related geohydrologic properties is defined using geostatistical estimation or simulation, a sedimentary process model or, most commonly, the "crayola" method (Rizzo and Dougherty, 1994) in which the modeler colors maps and cross sections based on observations, judgment, experience and intuition. Sedimentologists can interact with groundwater modelers by better defining geohydrologic units. This introduction explains why such interaction is valuable, and the body of this article provides examples.

Regardless of the method used to define the units, the next step is to calibrate the model. Calibration is adjustment of hydraulic parameters (e.g., hydraulic conductivity [K], storage coefficient, recharge rate) within units of the model until the results of the model closely match field observations (e.g., hydraulic heads in wells, flow rates of groundwater discharging to springs, streams or wells). This calibration often is accomplished by trial and error but can be accomplished using an "inverse" modeling algorithm (parameter estimation). Calibration through inverse modeling involves minimizing the differences between field observations and model results by using a computer algorithm to efficiently determine the optimal set of parameter values (Poeter and Hill, 1997). However, more often the trial-and-error approach is used, in which the modeler tries to select the best parameter values based on experience, intuition and knowledge of the reasonable range of parameters. The modeler enters these values, runs the computer model, compares the model results to field observations, decides how to adjust the parameters to improve the match of results and observations, and repeats the process until satisfied with the fit.

If the units of the conceptual model are not properly delineated, the final parameter values estimated for the units will not reflect the hydraulic properties in the field. Predictions made using the model likely will be erroneous, and decisions based on those predictions may not result in the best, or even a desirable, action. Unreasonable parameter values are more likely to be obtained using the inverse modeling approach because modelers using the trial-and-error approach will not select unreasonable parameter values (e.g., a higher hydraulic conductivity for a silt deposit than an adjacent sandy gravel deposit). Instead, the modeler tends to accept greater discrepancies in the match between field observations and model results (i.e., sacrifices the best fit between the model and the data) to maintain reasonable model parameters. The inverse calibration procedure will determine the parameters that provide the best fit even though their absolute or relative magnitude may be unreasonable. Also, when the trial-and-error approach is used, the modeler often forces the desired conditions at the boundaries of the model. For example, if inflow is expected on a particular model boundary, the modeler may specify inflow on that boundary. In contrast, the inverse algorithm can estimate flow rates at the boundaries but produce results contrary to the expected pattern in order to obtain the best fit to the field observations (Anderman et al., 1995). Such a discrepancy can occur because the inferred flow direction is uncertain when determined using sparse field data (e.g., the angle at which equipotential lines meet the boundary determines whether flow is inward or outward at the boundary). This output of unreasonable values often is disillusioning to new users of inverse models but actually is a great benefit, as illustrated later in this article. At first glance, it may seem that trial and error provides the more reasonable result and so would be the preferred method. However, when the modeler sacrifices honoring data to maintain "reasonableness," important information about the system is ignored and errors in the conceptual model are disregarded.

It is our experience that unreasonable results produced by the inverse model indicate a problem with the conceptual model. Often that problem is a result of poorly defined distribution of units. Unit definition is the area in which sedimentologists can best collaborate with groundwater modelers. Sedimentologists can address three items to improve the groundwater modeling process:

1. identifying the location and character (sharp/gradational) of boundaries of geohydrologic units;
2. providing expert opinion on and measures of the continuity and directional trends of geohydrologic units; and
3. defining geologic "rules" regarding the relative values of hydraulic properties of geohydrologic units.

This paper illustrates the problems faced by groundwater modelers that can be alleviated through sedimentological and stratigraphic knowledge. Examples include both deterministic and stochastic development of conceptual models for synthetic problems and stochastic development of a conceptual model for a field problem.

Geological input to groundwater modelers should be quantified and should "fill in" information between the typical one-dimensional data obtained from boreholes. Although a general method does not exist for incorporat-

ing descriptive insight into definition of units for flow models, this presentation illustrates some approaches to incorporating sedimentological and stratigraphic information into groundwater models so that sedimentologists and stratigraphers can identify appropriate methods of transferring their knowledge to groundwater modelers.

Groundwater professionals have called for increased quantification of geologic information. As Konikow and Papadopulos (1988) noted, "In these times of computer models and simulations, it should be reemphasized that the real world is complex, three-dimensional, heterogeneous, and commonly anisotropic. This reality can only be described accurately through careful geohydrologic research in the field." Similarly, Stephenson et al. (1991) pointed out, "New field techniques must be developed to provide spatially distributed data rather than point data for geohydrologic analysis." Farvolden and Cherry (1991) suggested a new, more quantitative perspective to geologic work. This paper is one contribution toward the long-term goal of quantifying sedimentologic and stratigraphic information in a manner that enhances geohydrologic achievement.

IMPACT OF THE CONCEPTUAL MODEL ON PARAMETER ESTIMATION: SYNTHETIC EXAMPLES

Synthetic problems are useful for illustrating the impact of a poor conceptual model on estimated parameter values because the "true" subsurface conditions are known and can be compared to the outcome of the modeling process.

Synthetic Example: Deterministic Conceptual Model

A simple basin configuration, roughly 30 by 40 km, is presented in Figure 1A. Flow is generally northward with groundwater discharges to a body of surface water at the north end of the basin and to the stream. The perimeter of the basin is defined by the groundwater divide. Saturated thickness of the unconfined aquifer varies from 30 m to 150 m within the basin. This synthetic basin is hydraulically two-dimensional; that is, hydraulic character does not vary with depth. Consequently, the hydraulic conductivity distribution (analogous to the distribution of hydrologically similar units, known as hydrofacies) shown on the map prevails to the depth at which bedrock occurs. Four units are delineated with increasing mean grain size and hydraulic conductivity (K) (Fig. 1B). Recharge is in the form of infiltration, distributed as illustrated in Figure 1C. Recharge is zero in the regional discharge area at the north end of the basin where vegetative consumption prevents infiltration to the water table. Hydraulic conductivity observations available at fourteen locations were used to constrain the zonation of hydrologic units in the groundwater model (Fig. 2A). Groundwater flow was simulated in the synthetic basin to provide accurate values of hydraulic heads (Fig. 1A). The simulation also yielded one groundwater discharge measurement of 0.2 m³/s base flow over the entire length of the northeastern tributary from the headwaters to the confluence with the trunk stream also. These head and streamflow observations were used to estimate hydraulic conductivities and recharge rate throughout the area where they have not been measured.

In calibrating this simple basin model to the observed heads and flows, we found that neither errors in field observations or in the initial estimates of parameters nor minor variations in distribution of geohydrologic units causes problems in obtaining a reasonable parameter estimation solution. Problems arise when a poor conceptual model (e.g., distribution of units that significantly differs from the true distribution) is utilized, even when error-free observations and true starting values are used for the parameters. Although perceived as a problem, it is actually beneficial because it renders inverse modeling an excellent tool for differentiating reasonable and unreasonable conceptual models.

For this simple basin model, the hydraulic conductivity values of the four units (K1–K4) and the magnitude of the recharge, R (given its relative spatial distribution), were estimated. We used MODFLOWP (Hill, 1992) exclusively to estimate parameters and found it to be robust. Hydraulic conductivity distributions and estimated parameter values for the cases considered on this study are presented in Figures 2 and 3, respectively. In the following descriptions the term "observations" refers to the observations of hydraulic head at the fourteen observation locations (Fig. 2A) and of stream base flow in the northeast tributary.

Case 1, base case.—

The base case includes error-free observations, a perfect conceptual model and correct "true" values of parameters as a starting point. The parameter estimation process quickly converges on the true parameter values (Figs. 2B, 3).

Case 2, incorrect initial estimates.—

Case 2 utilizes error-free observations, a perfect conceptual model and incorrect initial values of the param-

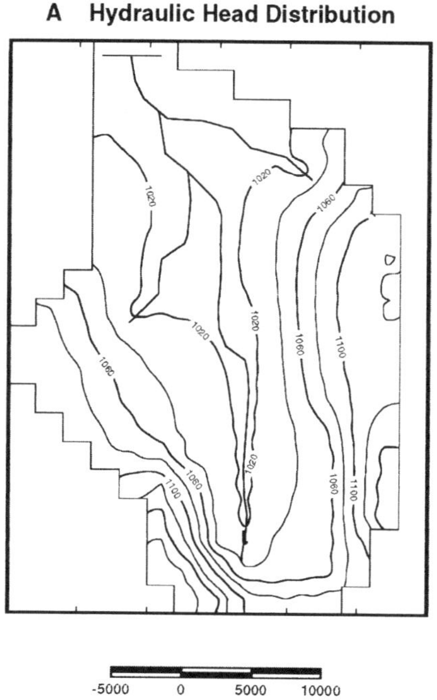

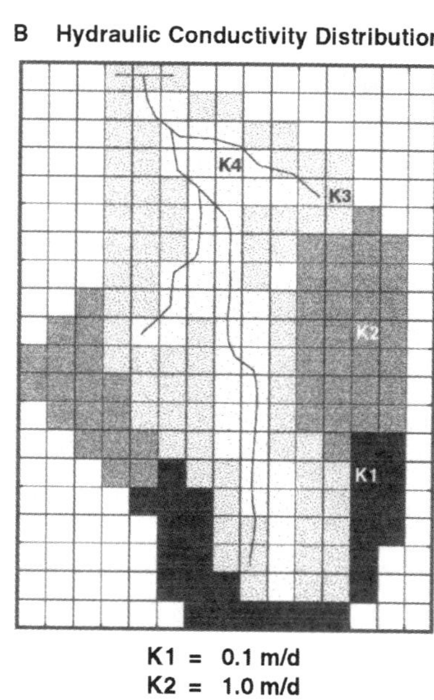

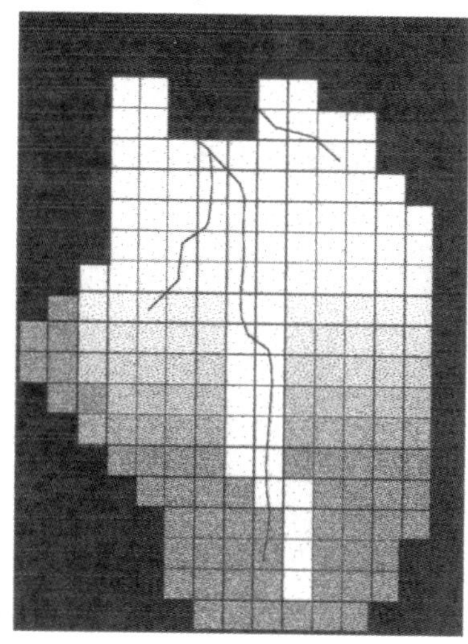

FIG. 1.—Simple deterministic synthetic basin configuration. (A) Hydraulic head distribution. (B) Hydraulic conductivity increases from 0.1, to 1.5 and 10 m/d, respectively, from black to lightest gray. (c) Recharge increases linearly with decreasing gray tone from 1.5×10^{-5} m/d in the south to 3×10^{-4} m/d in the northern area; black is zero recharge or outside of the basin boundary. Black lines denote streams.

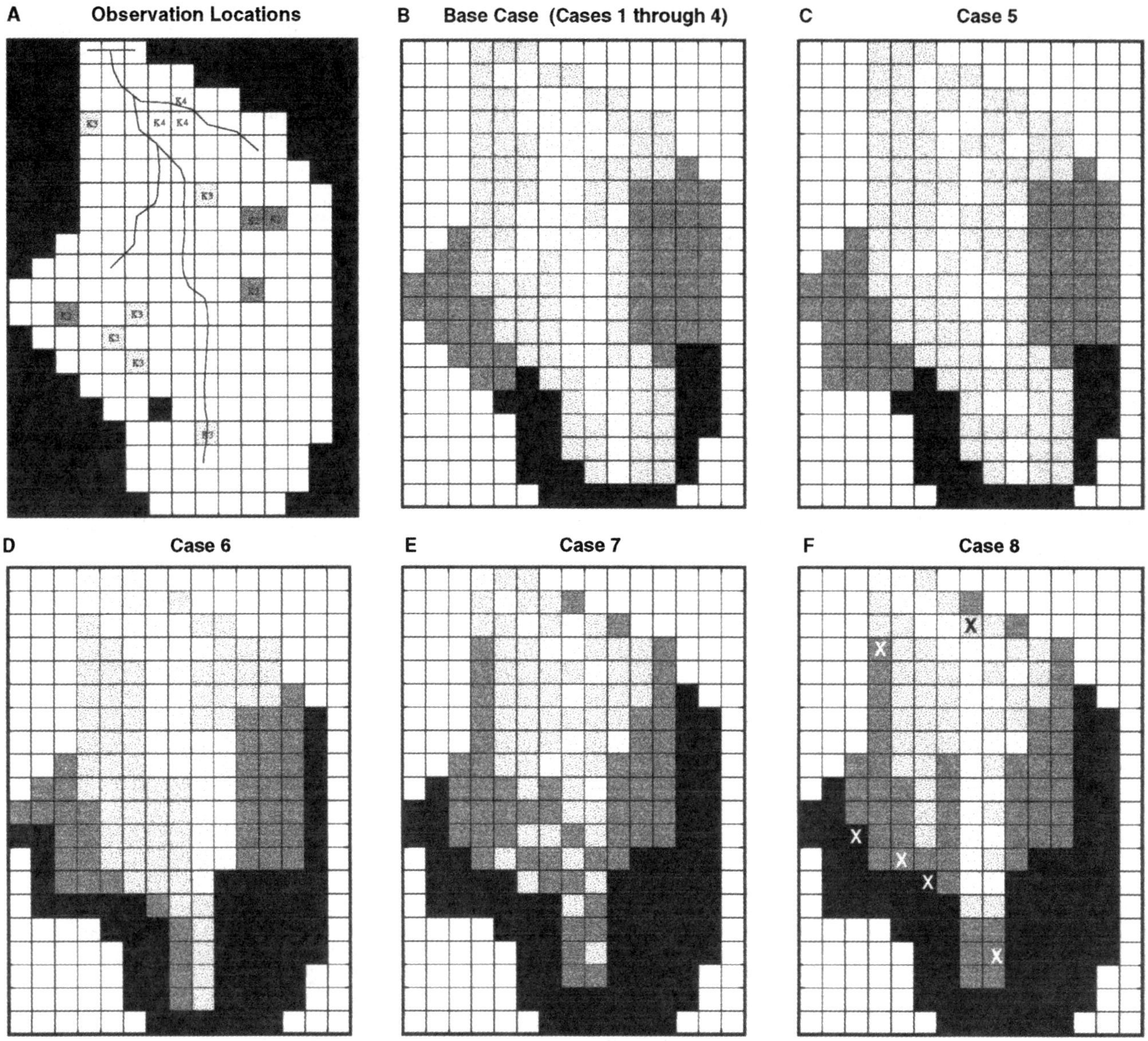

A   Observation Locations

B   Base Case (Cases 1 through 4)

C   Case 5

D   Case 6

E   Case 7

F   Case 8

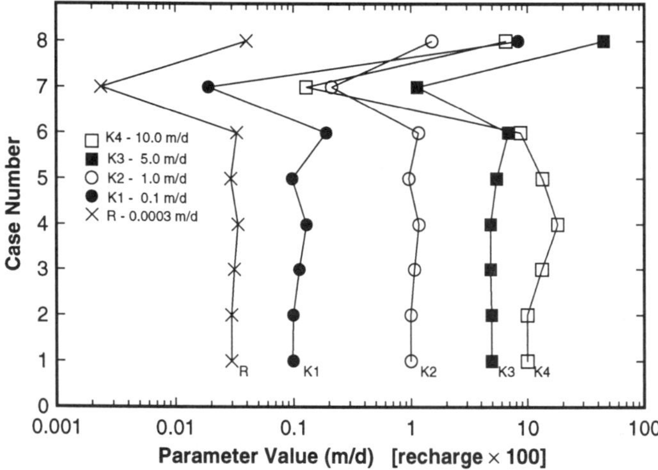

eters, including various combinations of parameters defined as two of orders of magnitude too small or large. The parameter estimation process quickly converges on the true parameter values (Figs. 2B, 3).

FIG. 2. (above)—Observation locations where information on hydraulic head and lithology (as represented by units 1–4) are available and (B–F) alternative conceptual model units for hydraulic conductivity. X's in (F) represent locations where unit contacts were shifted such that the dominant lithology at the observation location was not assigned to the grid block.

FIG. 3. (left)—Estimated parameter values for alternative models presented in Figure 2. True parameter values are indicated in the legend.

Case 3, error in observations.—

Case 3 includes error-laden observations having normally distributed noise with a standard deviation of 2 m on head observations and flow underestimated by 4 percent, a perfect conceptual model and true parameter values for initial estimates. The code converges readily to values near the true values. Estimated parameter values are slightly different than the true values, reflecting values that better fit the erroneous observations (Figs. 2B, 3).

Case 4, larger error in observations.—

Case 4 is identical to case 3 but with a standard deviation of 4 m on the head observations and flow measurement underestimated by 7.5 percent (Figs. 2B, 3). As found in case 3, but to a greater extent, estimated parameter values are somewhat different than true values, reflecting parameter values that better fit the erroneous observations (Figs. 2B, 3).

Case 5, minor unit distribution error.—

Case 5 includes error-free observations and accurate initial values for the parameters but utilizes a more discontinuous definition of the coarsest grained (high hydraulic conductivity, K4) unit than exists in the base case. Again this results in rapid convergence to parameter values close to the correct values (Figs. 2C, 3).

Case 6, minor unit distribution error.—

In case 6 the percentage of the finest grained unit is overestimated, and the coarsest grained unit is connected to the hydraulic boundary on the north end of the domain. In this case, errors in parameter estimations begin to arise. In order to better match the observed heads, hydraulic conductivity of the fine-grained unit (K1) is overestimated, compensating for the overabundance of the fine-grained unit in the model. Hydraulic conductivity is underestimated for the coarse-grained unit (K4), compensating for the hydraulic connection that does not actually exist (Figs. 2D, 3).

The cases presented to this point pose formidable problems to the parameter estimation algorithm; however, no difficulty exists in obtaining reasonable estimates. The remaining two cases show how larger errors in the unit distribution pattern can create large errors in the estimates of parameter values.

Case 7, major unit distribution error.—

In case 7 the percentage of the fine-grained unit (K1) is substantially overestimated, and the percentage of coarse-grained unit (K4), which is much more discontinuous than in the base case, is underestimated. In this case an unsatisfactory set of parameter values is estimated. All the parameters, including recharge rate, are estimated to be an order of magnitude or more lower than base case values. Estimated hydraulic conductivity of the coarse-grained unit is lower than for the medium-grained unit (Figs. 2E, 3).

Case 8, major unit distribution error.—

Sometimes the predominant geology in a small borehole is not thought to represent that of a 4-million-m² flow model-grid block. Consequently, case 8 includes six locations where geologic boundaries are shifted slightly so that lithologic observations at the grid block center are not represented as the lithology of the flow-model grid block. This conceptual model yields a more reasonable recharge rate but overestimates hydraulic conductivity of the three finest grained units. The relative order of hydraulic conductivity between units again is incorrect, with the finest grained unit exhibiting a higher hydraulic conductivity than the coarsest grained unit and a reversal in expected hydraulic conductivities for the two coarsest grained units (Figs. 2F, 3).

The problem with these model configurations may seem obvious for this simplistic problem in which the truth is known, but it is obscure when working with field data. The information that sedimentologists and stratigraphers can provide regarding proportion of various types of materials in the field, their continuity or lack thereof, and their relative hydraulic conductivity is invaluable to groundwater modelers.

Synthetic Example: Stochastic Conceptual Model

The synthetic problem discussed here was presented in greater detail by Poeter and McKenna (1995) (Fig. 4). This problem demonstrates conceptual models with unit distributions that honor the available geologic data but were not compatible with geohydrologic data. Unit distributions were improved through addition of soft information, yet a portion of the improved conceptual models were not compatible with geohydrologic data. Models producing parameter values in conflict with geologic "rules" and those yielding a poor fit to the hydraulic heads after optimization of the parameter values were eliminated. The use of soft information coupled with the process of model elimination significantly reduced the uncertainty associated with predictions of concentration at a specified location and time.

The synthetic system is a portion of an alluvial valley aquifer with a north-to-south hydraulic head gradient

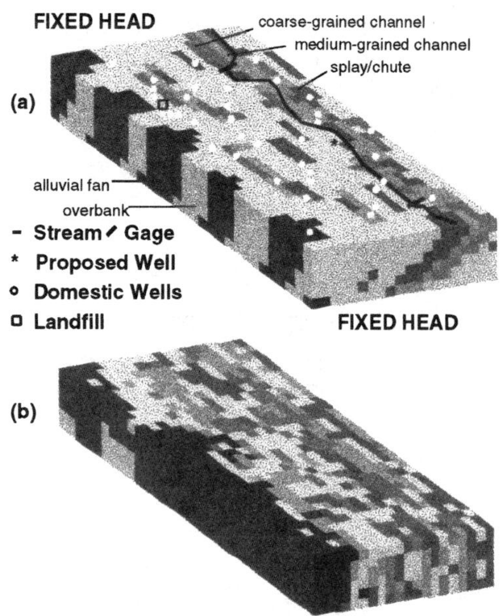

FIG. 4.—Synthetic stochastic model. Facies patterns (dark to light): alluvial fans, splay/chute deposits, coarse-grained channel deposits, medium-grained channel deposits, overbank deposits.

(yielding inflow at the north end and outflow at the south end), no-flow conditions at the eastern, western and underlying boundaries, recharge on the surface and discharge to the stream. An average recharge of 2×10^{-4} m/d is applied over most of the area, with three times that rate applied on a 200-m-wide strip along the east and west margins margin of the valley and no recharge near the stream which is a discharge area. The 31 domestic wells penetrate to various depths and pump at low rates (Fig. 4A). The model domain is 6,000 m long, 2,000 m wide, 60 m to 80 m thick (depending on elevation of the water table in the uppermost layer) and is represented by a grid of 30 by 20 by 6 cells. A landfill and stream are located as shown in Figure 4A. Of ultimate interest is the probable contaminant concentration at the proposed well location 50 years after a leak commences at the landfill (Fig. 4A).

The alluvial fan unit is limited to the western edge of the alluvial valley and is discontinuous. The main channel unit is oriented north-south and is further east in successively younger layers. The discontinuous coarse-grained channel unit occurs on the outside bends of meanders. The medium-grained channel unit is more continuous. Fine- to medium-grained channel units along the main channels represent splay deposits. Other medium-grained units located within the flood plain represent chute cutoff oxbow lakes. Fine- to medium-grained units of similar distribution represent neck cutoff oxbow lakes. This unit definition and spatial distribution is a gross generalization of the classic fluvial lithofacies as presented by Allen (1965), Bridge and Leeder (1979) and Walker and Cant (1984). Five units, all with a horizontal-to-vertical conductivity anisotropy ratio of 2, are used in the synthetic model with initial estimates of hydraulic conductivity given in Table 1.

Groundwater flow was simulated in the synthetic problem to obtain 37 composite head observations and one flow observation for input to inverse modeling using MODFLOWP (Hill, 1992). A breached landfill was simulated using the MT3D code (Zheng, 1991), yielding a known, normalized concentration of $C/Co = 0.0012$ at a proposed well location 50 years after the landfill began leaking. Due to their similarity of average hydraulic properties, the alluvial fan and medium-grained channel deposits (units 1 and 4) were estimated as one value in the inverse modeling exercise. Parameter estimation was performed on five parameters: overall vertical anisotropy of hydraulic conductivity throughout the model (i.e., a constant ratio of vertical to horizontal hydraulic conductivity

TABLE 1.—HYDRAULIC CONDUCTIVITIES AND VOLUME PERCENTAGES FOR UNITS IN THE STOCHASTIC SYNTHETIC EXAMPLE.

Units	Mean* Hydraulic Conductivity (m/d)	Percent of Sediment Volume
1 Heterogeneous alluvial fan	50	6
2 Fine-grained overbank deposits	0.05	61
3 Fine/medium-grained splay/chute deposits	5	11
4 Medium-grained channel deposits	50	14
5 Coarse-grained channel deposits	500	8

* All units have a constant value of effective porosity and log-normally distributed (but no spatially correlated) hydraulic conductivities with a standard deviation of ⅙ in natural log space (thus, hydraulic conductivity varies by a factor of three within any one unit, and there is no significant overlap of conductivities between units).

that applies at all locations and in all units) and four hydraulic conductivities.

Prior information on the estimated parameters was defined as the mean of the true distribution with (1) a standard deviation of 0.5 of the ln of hydraulic conductivity to constrain the estimates of hydraulic conductivity to reasonable values, and (2) a standard deviation of 0.2 to constrain estimates of the vertical anisotropy factor. Sedimentologists can provide insight to reasonable magnitudes and standard deviations for prior information by making numerous small-scale measurements of air permeability at outcrops (e.g., Davis et al., 1993) and/or providing some other quantitative measure of hydraulic heterogeneity based on grain size, sorting, packing and cementation.

Using semivariograms developed with "hard" data from the 37 wells at the observation locations, stochastic, geostatistical simulations were conducted using a modified version (McKenna, 1994) of ISIM3D (Gomez-Hernandez and Srivastava, 1990) in the public domain software collection UNCERT (Wingle et al., 1997). The simulations yielded 100 multiple realizations of unit distributions that honor both the location of the units and the statistics of their distribution as known from well logs. An overview of the simulation process is included in the appendix to this paper. Hard data are those with negligible uncertainty, such as direct measurements of hydraulic conductivity or observations of lithology. Soft data are information with non-negligible uncertainty, such as indirect measurements gathered in a geophysical survey and expert opinion regarding geologic fabric or structure.

Inverse groundwater flow modeling was undertaken within the numerous subsurface interpretations to estimate hydraulic conductivities of the units for each of the realizations. Only 8 percent of the first 100 realizations gave parameter estimates exhibiting the proper relative order of hydraulic conductivity for the units. Consistently, the estimated conductivity of the alluvial fan/medium-grained channel sand unit was lower than that of the overbank unit. This underestimation led us to suspect that the stochastic models included too much continuity in the alluvial fan/-medium-grained channel deposits. Thus, the inverse modeling procedure was estimating low hydraulic conductivity for the alluvial fan unit in order to fit the "measured" heads, which exhibit higher gradients produced by the "true" discontinuous units of high hydraulic conductivity.

One can conclude that knowledge of the relative order of hydraulic conductivity of units is useful in eliminating unreasonable conceptual models. Once relative order is established through measurement or indirect relationships,

as mentioned above, it becomes a valuable addition to measurements that sedimentologists now make in the field.

Examination of the geostatistical realizations confirmed that the simulated alluvial fan unit was too well connected in the initial 100 realizations. Furthermore, the unit was found to be either completely continuous along the left boundary of the domain or connected to the high-conductivity unit of the continuous channel unit near the centerline of the domain (Fig. 4B). Providing quantitative measures of continuity is another item sedimentologists can contribute to conceptual groundwater model development. This may be accomplished through outcrop observation or through a combination of core description and depositional analysis to determine the ratio of sediment supply to accumulation space (Cross, 1991). However, as Cross cautioned, a "catalogue" of the shape and size of units in a given depositional environment is not likely to prove fruitful because many factors that control continuity are independent of depositional environment. Incorporating geophysical data from outcrop investigations also contributes to assessment of unit continuity. Outcrop logging, as discussed by Slatt et al. (1992), might be feasible not only on vertical sections but also along horizontal transects exposed in outcrop.

When we recognized that the alluvial fan unit was too continuous in the realizations, we added "soft" data to the simulation process by specifying a decreasing probability of occurrence of alluvial fan material with distance from the surface manifestation of each fan. Although probability values were determined in an ad hoc manner, their impact on the result was striking. Sensitivity to the selected probability values generally was minimal with the exception of a noticeable impact when probability values were above or below the threshold values at which the units were simulated either as continuous or discontinuous. For each 10 m of depth directly below the alluvial fan, the probability of occurrence of alluvial fan unit was specified as 95%, 80%, 50%, 10% and 0 to represent propagation of the fan across the valley through time. On the fan's periphery, a 95% probability of occurrence 100 m to the east and 200 m to the north or south was specified (distances are limited by grid spacing in the model) with a zero probability at greater distances. A 10% probability of occurrence was specified at locations underlying these peripheral units. This information is expert opinion that can be provided by sedimentologists based on their field observations in surrounding areas and experience with similar geologic settings.

The same semivariograms then were used with both the hard and soft data to geostatistically simulate 400 realizations of hydrofacies arrangement. Further reduction of uncertainty was attained by eliminating individual realizations using the results of inverse flow modeling; that is, realizations were eliminated if the flow model did not converge, if the relative order of conductivities was not as expected, or if the converged parameter estimation yielded a relatively poor fit to the field measurements as indicated by a large objective function value. Only 2.5 percent of the realizations remained after those eliminations. Uncertainty in the contaminant concentration at the proposed well site substantially decreased. Its mean concentration approached the true value as the project progressed (Fig. 5) from the first 100 realizations with the unrealistically continuous alluvial fan unit (Fig. 5A), through the improved characterization by incorporating expert opinion on the alluvial fan unit (Fig. 5B), and finally to the elimination of realizations based on inverse modeling results (Fig. 5C). This progression resulted in a better characterized system, as indicated by a decrease in the predicted range of possible contaminant concentrations at the point of interest.

It would be beneficial to identify a measure that does not require as much computation as that required by the flow and transport modeling process. After examining the percentage distribution of units, the semivariograms of the realizations and connectivity between the landfill and the proposed well location, we found no apparent correlation between these parameters and the character of either the retained or eliminated models. Some of the retained models exhibited a similar proportion of units and similar variograms as the true configuration of units, while others did not. Similarly, some retained models exhibited a well connected path of the coarsest two units between the landfill and the proposed well location, while others did not. The "true" situation does have a continuous path of the two highest units between the landfill and the proposed well site. Subtle differences in the patterns of the units contribute to the movement of contaminants to the proposed well location. Apparently, a few fine-grained units can obstruct the path (resulting in a poor connectivity by our measure) but still yield concentrations similar to the true concentrations as long as no alternative well-connected path of high-conductivity unit is present to capture and divert the plume. Thus, our measure of continuity was not a useful measure. This does not imply that accurate representation of the units is not important. Indeed, many of the retained realizations appeared to be more similar to the true unit distribution than the discarded realizations (Fig. 6), suggesting that soft data did improve the character of unit distributions. However, we have not identified an appropriate measure of representative units other than matching

heads, flow rates and concentrations through the computationally intensive modeling process.

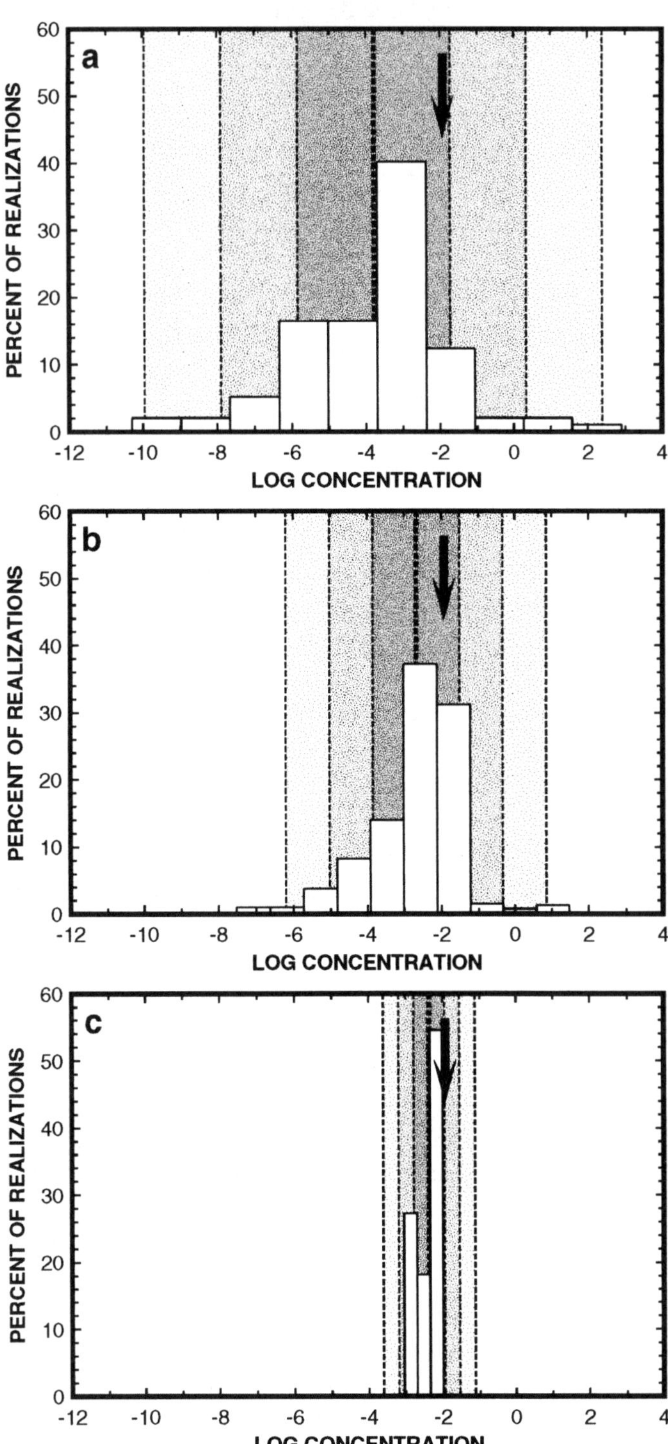

FIG. 5.—Histograms of predicted concentrations at the proposed well location 50 years after breach of the landfill. (A) based on hard data and field-estimated hydraulic parameters; (B) based on hard and soft geologic data and field-estimated hydraulic parameters; and (C) after eliminating configurations from (B) and estimating parameters using inverse modeling.

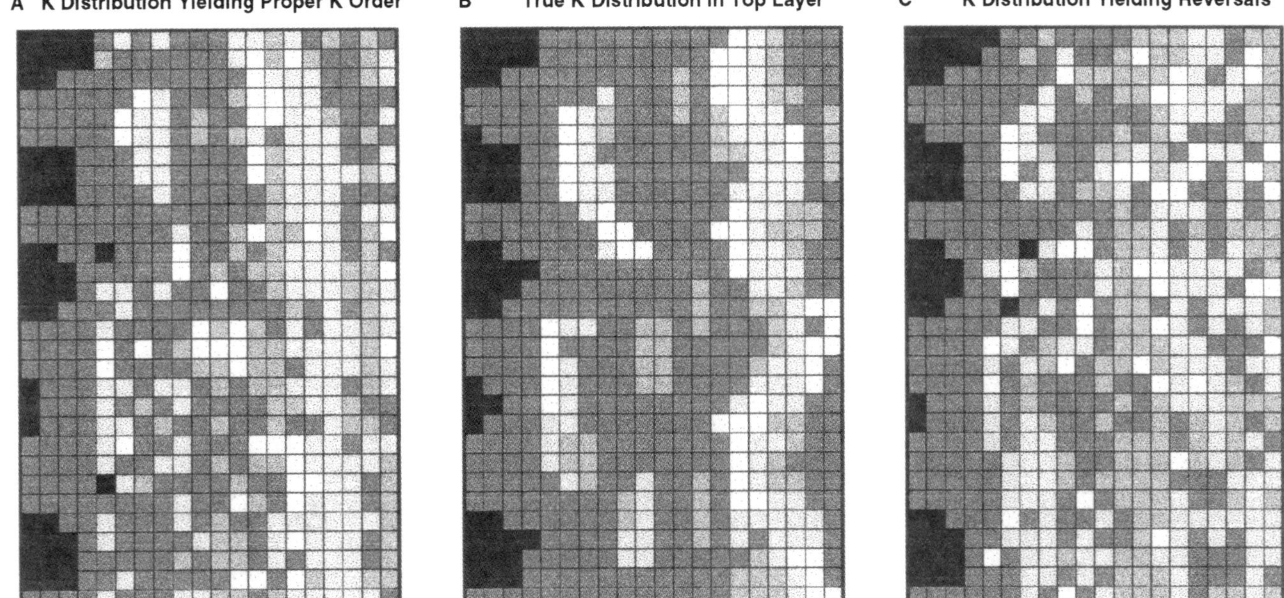

FIG. 6.—Unit distribution in the top layer of (A) a typical retained realization, is more similar (notice the trend of the channel unit) to (B) the true system, than (C) a typical eliminated realization.

IMPACT OF THE CONCEPTUAL MODEL ON PARAMETER ESTIMATION: FIELD EXAMPLE

Stochastic geostatistical modeling of unit locations conditioned to both hard and soft information was undertaken at a field site near Golden, Colorado (McKenna, 1994). The realizations resulting from the geostatistical simulations were used as input to a groundwater flow model. Both forward groundwater flow modeling, using field estimates of hydraulic conductivity, and inverse parameter estimation were undertaken (McKenna and Poeter, 1995).

Site Description

The site overlies two sedimentary formations in the Denver Basin—the Fountain Formation and the Lyons Formation. The units strike N-S and dip to the east. Their contact lies in the eastern half of the site. The depositional environment of the Fountain Formation is interpreted as a braided stream system of three depositional units: stream channel, flood-plain mudstones and deltaic-lacustrine sandstones (Howard, 1966; Hubert, 1960; Weimer and Land, 1972). Their distribution is a complex function of depositional processes and postdepositional diagenesis. The Lyons Formation is a fluvial deposit in this locale and consists of fine- to medium-grained sandstone containing large amounts of conglomerate and minor lenticular shale and mudstone (Weimer and Land, 1972).

Sixteen wells (Fig. 7), ranging from 30 to 50 m deep (and including 130 m of core), exhibit water levels 3 to 4 m below ground elevation and reveal bedding planes dipping an average 33° to the east (McKenna, 1994). Geophysical logs include neutron-epithermal neutron, natural

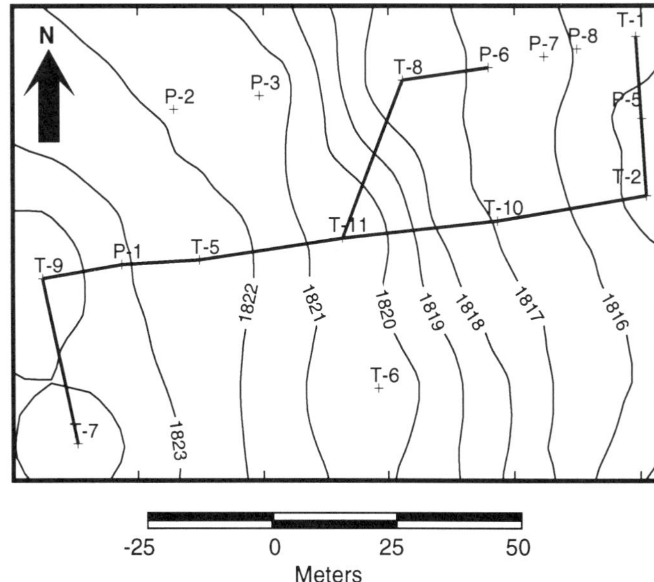

FIG. 7.—Site map of well locations, tomography lines and groundwater table elevation of the CSM survey field site (after McKenna and Poeter, 1995).

gamma, electrical conductivity and sonic functions. Other available data include small-scale air-permeability measurements from core samples, larger scale hydraulic conductivity measurements from downhole packer tests and cross-borehole seismic tomograms from eight well pairs (Fig. 7). Hydraulic heads were measured in 28 locations—8 from open wells and 20 from 1.5- to 3.0-m-long screens in hydraulically sealed units.

The recovered cores were logged to determine the depositional units. Geophysical log measurements were used to better determine diagenetic effects on the depositional units and to estimate unit locations in wells without core (McKenna and Poeter, 1995). Small-scale air-permeability measurements obtained on the core were used to define the hydrologic similarity of different units. By combining geological, geophysical and hydrological information, eight units were defined with distinct hydrologic properties distinguished by sonic velocity differences (Fig. 8).

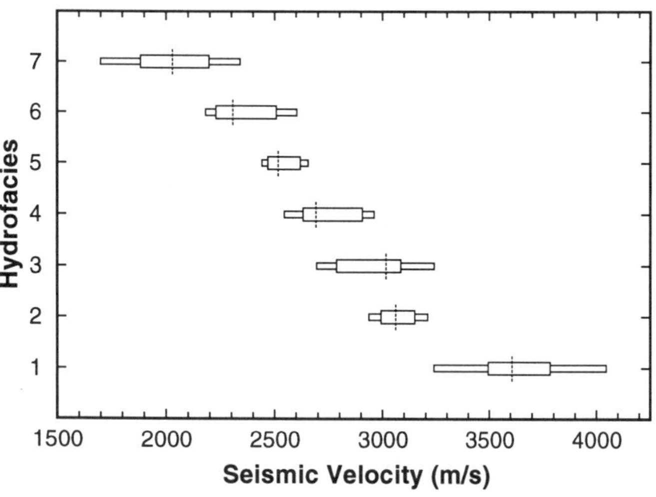

FIG. 8.—Box plots showing distribution of seismic velocity within each unit. The dashed line is the median of each distribution, and boxes show the 10th, 25th, 75th and 90th percentiles of the distribution (after McKenna and Poeter, 1995).

Cross-Borehole Seismic Tomography

Cross-borehole seismic tomography provides high-resolution images of aquifer heterogeneities without the limitations of traditional surface seismic surveys. Eight cross-borehole seismic-transmission tomography surveys were conducted (Fig. 7) to obtain high-resolution images (maximum resolution 1.3 ± 0.4 m) of seismic velocity between boreholes (McKenna and Poeter, 1995). The tomograms were used to derive soft data estimates of the distribution of units between wells.

Geostatistical Simulation

A total of 200 multiple indicator geostatistical realizations was created with a version of ISIM3D (Gomez-Hernandez and Srivastava, 1990) modified to incorporate three types of soft information (McKenna, 1994) and publicly available as UNCERT software (Wingle et al., 1997; http://uncert.mines.edu). Each realization was 123 m (E-W) by 93 m (N-S) by 44 m (405 by 305 by 144 ft) and discretized into 81 by 61 by 72 elements. Hard information derived from the geologic interpretation and the logs was used to constrain an ensemble of 100 realizations (hard-data ensemble). An additional 100 realizations were created using the hard information and two types of soft information (hard/soft-data ensemble). The first type of soft information was the result of the tomographic inversions used to give an imprecise estimate of units at locations between wells (McKenna and Poeter, 1995; McKenna, 1994). The imprecision was accounted for during the simulation process according to techniques developed by Alabert (1987). The second type of soft information was the estimated probability of occurrence of a given unit at a location, as provided by the geophysical logs (McKenna and Poeter, 1995).

Flow Modeling

Forward flow modeling was undertaken to examine reduction of uncertainty in head prediction due solely to incorporation of soft information into the subsurface characterization. Inverse flow modeling was conducted to determine the influence of additional soft data on estimation of hydraulic conductivities within the geostatistically simulated units.

The groundwater gradient is essentially west to east across the site (Fig. 7). Thus, fixed head boundaries were imposed at the west and east site limits with zero-flux boundaries along the north and south limits to simulate easterly flow. The types of conditions that could be applied on the upper domain boundary were limited by the inverse groundwater flow model, MODFLOWP, which, at the time, could not accommodate convertible layers (i.e., layers that are both confined and unconfined in space and/or time). Recharge around the site (Kiusalaas, 1992) is essentially zero. For these reasons the top boundary of the aquifer was modeled as a confined system with zero recharge. Rubin and Dagan (1988) had demonstrated that fixed-head boundary conditions could have an adverse effect on predicted heads in heterogeneous media. To surmount this problem, we added "buffer units" to the west and east ends of the model. Many of the full-scale geostatistical realizations of 355,752 elements were too large to solve for flow with available facilities and within a reasonable time frame. Thus, the realizations were upscaled to cell dimensions of 6.1 m by 6.1 m by 0.6 m (x, y, z). Incorporation of 16 of the fine-scale cells into one coarse cell was accomplished by assigning the most common units to the larger cell.

Field estimates of hydraulic conductivity and measurements of head were used to model groundwater flow and determine head residuals. The only difference between the hard- and hard/soft-data forward models was distribution of the units. This difference, due to incorporation of soft information, reduced the mean average absolute error on the predicted heads by 23 percent (from 3.8 to 2.9 m) (Figs. 9A, 9B).

Average absolute head deviation obtained from forward flow modeling was large, even after incorporating soft information into the definition of unit distribution (Figs. 9A, 9B). However, head deviations were minimized through automated calibration using MODFLOWP (Hill, 1992). Estimates of hydraulic conductivities made with available field data did not provide the optimum fit to the measured heads. Through trial and error calibration of several realizations, updated initial conductivity estimates were obtained for inverse modeling. Field estimates and initial estimates of hydraulic conductivities for both hard- and hard/soft-data realizations are given in Table 2. Hydraulic conductivities of units 6 and 8 were believed to be known with greater confidence from borehole packer tests relative to other units; therefore, they were not estimated.

Three rules were developed for acceptable realizations:

1. Unit 1 (well-cemented conglomerates) had lower hydraulic conductivity than unit 3 (less consolidated sandstones).
2. Unit 8 (fracture units) had the highest estimated hydraulic conductivity of any unit.
3. After applying criteria 1 and 2, the half of the remaining solutions with the smallest head residuals was retained.

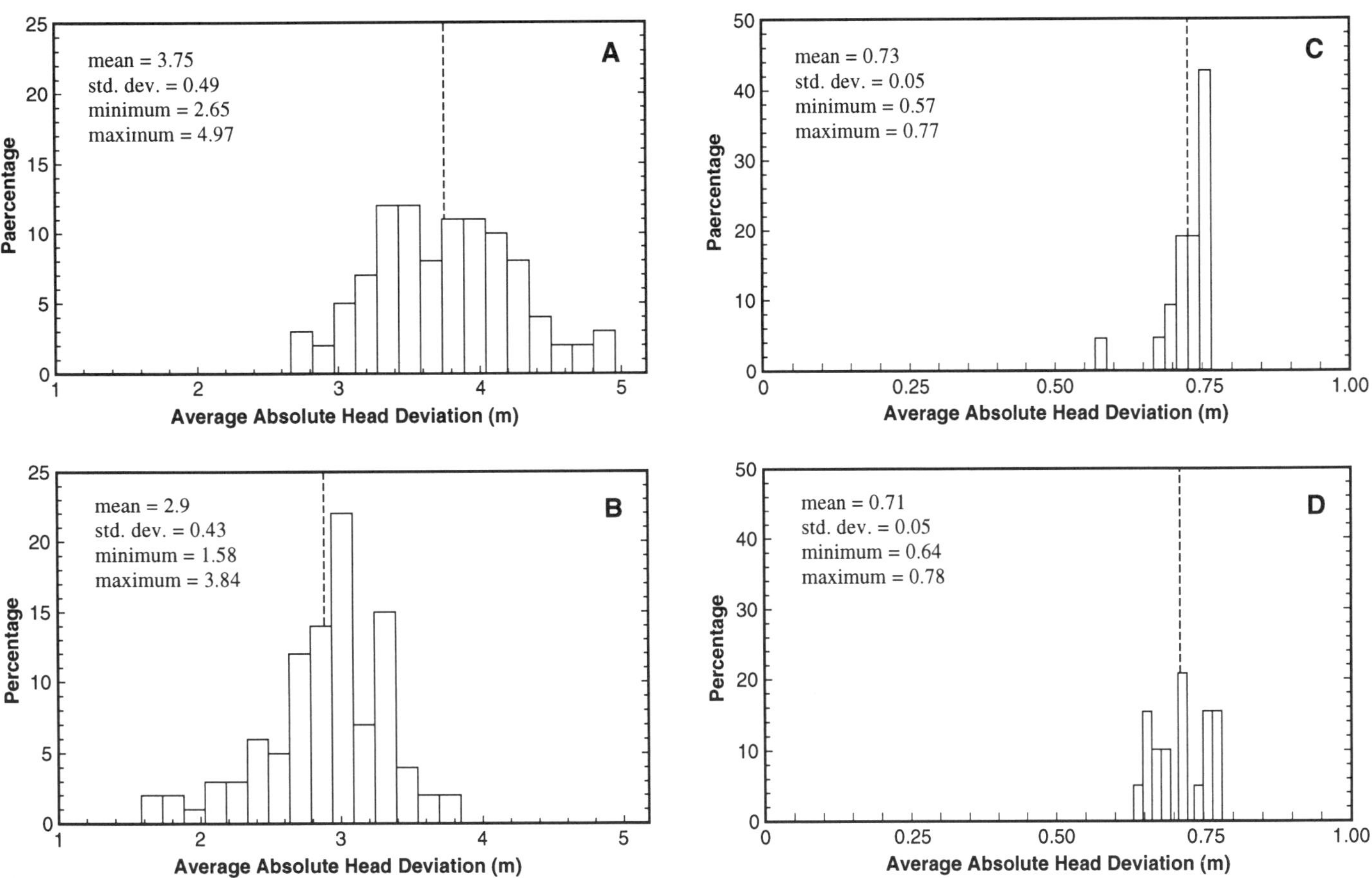

FIG. 9.—Distributions of average absolute head deviations for the hard-data (A) and the hard/soft-data (B) realizations resulting from forward groundwater flow modeling using field estimates of hydraulic conductivity. Distributions of average absolute head deviations after inverse parameter estimation are shown for the hard data in (C) and the hard/soft data in (D). Note the change in the scale of the axes from (A) and (B) to (C) and (D). Dashed lines indicate the mean value (after McKenna and Poeter, 1995).

After applying these elimination criteria, the hard-data ensemble was reduced from 100 to 21 plausible realizations, and the hard/soft-data ensemble from 100 to 19. After parameter estimation, residual head distributions were much smaller (Figs. 9C, 9D) than those produced by forward modeling (Figs. 9A, 9B). Average absolute head deviations were similar for both the hard (mean 0.73 m, standard deviation 0.05 m) and hard/soft (mean 0.71 m, standard deviation 0.05 m) simulations. Because of this similarity, one could argue that the hard data alone produced the same quality realizations. However, estimates of hydraulic conductivity calculated from hard-data realizations had greater standard deviations (except for unit 4, where the standard deviations were roughly equal), suggesting less certainty in the result when only hard data were used (Fig. 10). The lower standard deviations of estimated conductivity associated with the hard/soft-data simulations is attributed to reduction in geologic uncertainty resulting from use of soft data. The median value of hydraulic conductivity for every unit is higher for the hard-data versus hard/soft-data realizations. One factor contributing to this difference is the absence of the highest conductivity unit (unit 8) in the hard-data realizations. No overlap occurs in distributions of hydraulic conductivity estimates in units 4 and 5 from the hard and hard/soft realization ensembles (Fig. 10), indicating that one of the approaches (hard data versus hard/soft) was inaccurate for these two units.

Discussion of the Field Example

Hydraulic conductivity field data were insufficient to determine which assessment is inaccurate. Better characterization of conductivity, or an appropriate surrogate, would provide more rigorous rules that may differentiate the representativeness of the hard versus hard/soft simulations. Available information allowed definition of units and general rules for the relative hydraulic properties of units that, through the elimination process, limited the alterna-

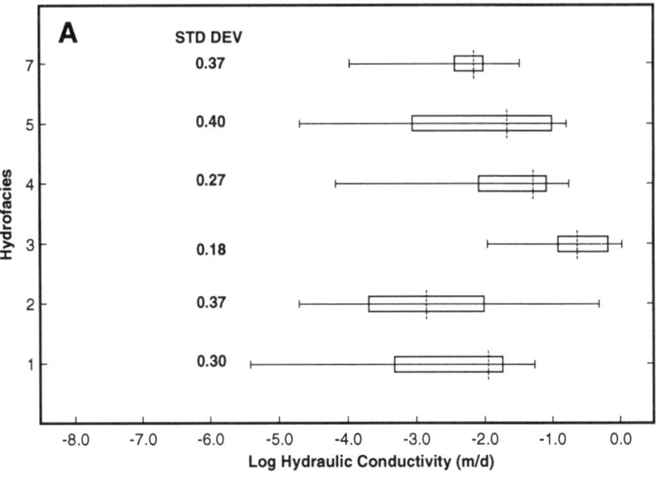
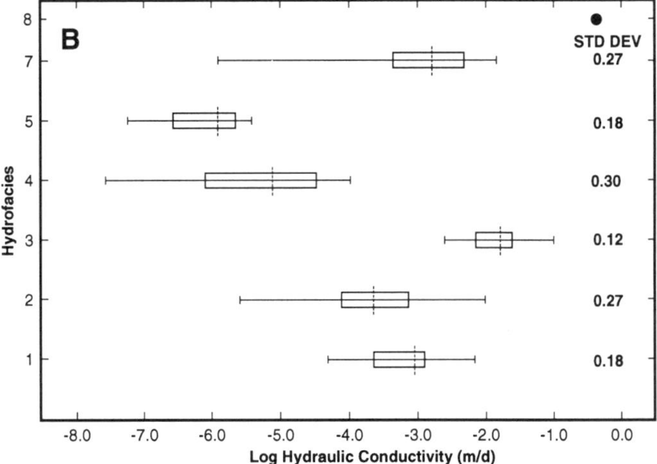

FIG. 10.—Box and whisker plots of hydraulic conductivity estimates for the 21 hard-data (A) and 19 hard/soft-data (B) realizations. Dashed lines indicate the median; the ends of the box are the 25th and 75th quartiles; and the ends of the whiskers are the minimum and maximum of the distribution. Values next to the whiskers are standard deviations. The hydraulic conductivity value of unit 8, which was not estimated, is shown as a filled circle in (B) (after McKenna and Poeter, 1995).

tive subsurface geohydrologic interpretations produced by coupled geostatistical simulation and inverse flow modeling. However, data were insufficient to determine whether

TABLE 2.—FIELD AND INITIAL ESTIMATES OF HYDRAULIC CONDUCTIVITY FOR
INVERSE MODELING IN THE FIELD EXAMPLE.*

	Unit Conductivity (log of m/d)							
Data Set	1	2	3	4	5	6	7	8
Field	−3.5	−3.5	−3.0	−2.9	−1.9	−2.9	−0.9	0.4
Hard	−1.8	−1.9	−1.1	−1.9	−1.9	−2.9	−0.5	—
Hard/Soft	−2.8	−2.9	−2.1	−4.9	−5.9	−2.9	−3.1	0.4

* After McKenna and Poeter (1995)

the relative values produced by the hard or those produced by the hard/soft simulations were more representative of field conditions. It is unlikely that an even distribution of hydraulic conductivity tests between units can be obtained during the first phase of a field investigation because the initial data are needed to define the units. Once units are delineated, subsequent field investigations can be designed to test all the units. Of the 18 field-scale hydraulic conductivity tests completed at the field site, one is in unit 4 and none are available for unit 5. If this study were extended, further testing would be conducted in each unit in order to develop stricter rules for selection of plausible realizations.

To obtain unique estimates of hydraulic conductivities (as opposed to obtaining linear combinations such as ratios), head observations and either flow observations or prior estimates of parameters are required. In this study 28 head observations but no flow observations were available. However, known values of the conductivity of units 6 and 8 were available from prior hydraulic testing, and this information was used to obtain unique estimates of hydraulic conductivities (to the extent that the known values were accurate). These field-measured values were used as prior knowledge in the inversion, and no significant correlation of hydraulic conductivity existed between any of the estimated conductivity values.

This field study demonstrates how quantitative techniques for incorporating soft data into site interpretation can combine geologic and geophysical information into a description of hydrologic heterogeneity. The relationship between units defined for the groundwater model and seismic velocity was described with indicator thresholds. This information was used with seismic tomography measurements further to condition multiple-indicator geostatistical realizations at locations between boreholes. Hydrologic information was utilized in unit definition both through air-permeability measurements and through determination of the plausibility of each stochastic unit distribution via parameter estimation using nonlinear regression and application of geohydrologic rules. The change in distribution of the units between the hard-data and hard/soft-data realizations caused large changes in the estimated hydraulic conductivities in units 4 and 5, indicating that conductivities estimated through inverse modeling are very sensitive to fine-scale, geologically based spatial distribution of units. This is fortunate because it allows use of hydraulic data to delineate the small-scale heterogeneity that is critical to migration of contaminants. The character of this detailed heterogeneity is not considered when broad units of homogeneous (or Gaussian distributed) hydraulic conductivity (as typically defined in deterministic conceptual models) are utilized.

DISCUSSION OF USE OF SEDIMENTOLOGICAL INFORMATION TO FACILITATE DEVELOPMENT OF CONCEPTUAL GROUNDWATER MODELS

A higher quality groundwater model is produced more efficiently when the conceptual models are closer to geohydrologic conditions in the field. Examples show that incorporating expert opinion about the proportion of each geohydrologic unit present in an area, together with delineation of unit boundaries, orientations, continuity and relative hydraulic properties, improves estimated parameter values from inverse groundwater modeling, conceptual model configuration and resulting predictions. Valuable expert knowledge includes characterizing the degree of sedimentological preservation (accumulation/erosion) through geologic analysis, correlating geohydrologic units with such soft data as geophysical responses, and characterizing the variability of hydraulic conductivity within units and between units through small-scale permeability measurements and site-specific empirical relationships among sedimentological characteristics. Recent advances, following the work described herein, include development of simulation software that allows definition of "framework" units within which geostatistical simulation is conducted (Wingle and Poeter, 1996). The contacts between these units can be sharp, fuzzy or gradational. Current work by Edington, Cross and Poeter (pers. commun.) includes geologic modeling that can be used to define this framework.

Skeptics may argue that soft information helped in the synthetic case because the expert opinion was true. In field applications, expert opinion is imperfect. The truth was unknown in this field study, and although it appeared, at first glance, that the only way to confirm the improved site simulation was to gather more subsurface data, the improvement was instead demonstrated by the reduction in head residuals after addition of the soft-data reduction of the standard deviations of the estimated parameters in the combined soft/hard realizations versus the hard realizations.

Better conceptual models can be retained and poorer models eliminated by using geologic rules regarding hydraulic conductivity relationships between units. Specifically, such rules may be better developed by characterizing the "pores as well as the grains" (i.e., by assessing conductivity when describing sediment character), through air-permeameter measurements in conjunction with the outcrop description and both vertical and horizontal section measurements. Hydraulic conductivity measurements are best collected at multiple scales if correlation lengths are to be calculated. Alternatively, physical measurements of the dimensions of visually distinguishable geohydrologic units (so confirmed by hydraulic measurements) can be used to establish correlation lengths.

In the second synthetic example and in the field example, we employed multiple-indicator geostatistical simulation to create the spatial distribution of the units. This certainly is not the only simulation technique available to generate a distribution of units for input to inverse parameter estimation. For example, an extension of indicator geostatistical simulation developed by Carle and Fogg (1996) allows information on the probability of juxtaposed units in a given direction (directional transition probability) to be reproduced in the resulting simulation. The transition probabilities allow development of consistent variograms. These transition probabilities are readily calculated in the vertical direction from borehole data, but they must be inferred from geologic information in the lateral directions. Some sedimentary process models (Webb, 1994; Scheibe, 1993) also provide stochastic representation of units in the subsurface, although it may not be possible to condition these models to site-specific data.

Finally, as we discuss the potential contributions of sedimentologists and stratigraphers to improve development of groundwater models, we note that the reverse also is true. Because the groundwater flow field is sensitive to continuity of units, sedimentologists may find that inverse modeling results are a valuable tool for evaluating unit continuity as it applies to sedimentological problems.

CONCLUSIONS

Utilization of soft data to define unit distributions and their relative hydraulic conductivities enhances the value of inverse modeling by reducing nonuniqueness.

The first synthetic example illustrates that accurate distribution of units in a flow model is important to accurate parameter estimation. Sedimentologists can assist by accurately mapping the units, specifically emphasizing the locations of unit boundaries, volumetric proportion of each unit over the site, continuity of units and their connections to each other.

The second synthetic example and the field example illustrates the importance not only of determining accurately the distribution of units but also of collecting measurements that correspond to hydraulic conductivity. In this way, appropriate rules can be defined for use in eliminat-

ing unit distributions that do not reflect field conditions. Of additional value to geohydrologists is the characterization of hydraulic properties or sedimentological measures that serve as surrogates for hydraulic conductivity (e.g., combination of grain size distribution, shape, sorting, packing and degree of cementation correlated with hydraulic conductivity measurements at enough locations to develop rules on relative values).

ACKNOWLEDGMENTS

Portions of this paper were previously published in *Ground Water* (Poeter and McKenna, 1995) and *Water Resources Research* (McKenna and Poeter, 1995). We appreciate the publishers' permission to present the material in this paper. Much of this work was supported by the U.S. Army Waterways Experiment Station.

REFERENCES

ALABERT, F., 1987, Stochastic imaging of spatial distributions using hard and soft information: Unpub. M.S. Thesis, Stanford University, Stanford, Calif., 197 p.

ALLEN, J.R.L., 1965, A review of the origin and characteristics of recent alluvial sediments: Sedimentology, v. 5, p. 89-191.

ANDERMAN, E.R., HILL, M.C., AND POETER, E.P., 1995, Two- and three-dimensional advective transport observations in groundwater flow parameter estimation: Eos, Transactions of the American Geophysical Union, v. 76, no. 17, p. 148.

BRIDGE, J.S., AND LEEDER, M.R., 1979, A simulation model of alluvial stratigraphy: Sedimentology, v. 26, p. 617-644.

CARLE, S.F., AND FOGG, G.E., 1996, Transition probability-based indicator geostatistics: Mathematical Geology, v. 28, no. 4, p. 453-476.

CROSS, T.A., 1991, Field scale reservoir characterization, *in* Lake, L.W., Carroll, H.B., and T.C. Wesson, eds., Reservoir Characterization II: San Diego, Academic Press, p. 493-496.

DAVIS, J.M., LOHMANN, R.C., PHILLIPS, F.M., WILSON, I.L., AND LOVE, D.W., 1993, Architecture of the Sierra Ladrones Formation, central New Mexico—Depositional controls on the permeability correlation structure: Geological Society of America Bulletin, v. 105, p. 998-1007.

FARVOLDEN, R.N., AND CHERRY, J.A., 1991, Are geology departments prepared for the 21st century: Geology, v. 19, no. 5.

FOGG, G.E., 1989, Emergence of geologic and stochastic approaches for characterization of heterogeneous aquifers, *in* Proceedings, New Field Techniques for Quantifying the Physical and Chemical Properties of Heterogeneous Aquifers, Dallas, Texas, March 20-23, 1989: Dublin, Ohio, National Water Well Association, p. 1-18.

GOMEZ-HERNANDEZ, J.J., AND SRIVASTAVA, R.M., 1990, ISIM-3D, an ANSI-C three dimensional multiple indicator conditional simulation program: Computers in Geoscience, v. 16, no. 4, p. 395-114.

HILL, M.C., 1992, A computer program (MODFLOWP) for estimating parameters of a transient, three-dimensional groundwater flow model using nonlinear regression: U.S. Geological Survey Open-File Rept. 91-484, 358 p.

HOWARD, J.D., 1966, Patterns of sediment dispersal in the Fountain Formation of Colorado: The Mountain Geologist, v. 3, no. 4, p. 147-153.

HUBERT, J.F., 1960, Petrology of the Fountain and Lyons Formations, Front Range, Colorado: Colorado School of Mines Quarterly, v. 55, no. 1, p. 1-238.

ISAAKS, E.H., AND SRIVASTAVA, R.M., 1988, Spatial continuity measures for probabilistic and deterministic geostatistics: Mathematical Geology, v. 20, no. 4, p. 313-341.

KIUSALAAS, N.J., 1992, Estimation of groundwater recharge using neutron probe moisture readings near Golden, CO: Unpub. M.E. Thesis ER-1191, Colorado School of Mines, Golden, 186 p.

KONIKOW, L.F., AND PAPADOPULOS, S.S., 1988, Scientific problems, *in* Back, W., Roshenshein, J.S., and Seaber, P.R., eds., Hydrogeology: Geological Society of America, The Geology of North America, v. O-2, p. 503-508.

MCKENNA, S.A., 1994, Utilization of soft data for uncertainty reduction in groundwater flow and transport modeling: Unpub. Ph.D. Dissertation T-1291, Colorado School of Mines, Golden, 325 p.

MCKENNA, S.A., AND POETER, E.P., 1995, Field example of data fusion for site characterization: Water Resources Research., v. 31, no. 12.

POETER, E.P., AND HILL, M.C., 1997, Inverse methods—A necessary next step in groundwater modeling: Ground Water, v. 34, no. 2.

POETER, E.P., AND MCKENNA, S.A., 1995, Reducing uncertainty associated with groundwater flow and transport predictions: Ground Water, v. 33, no. 6.

RIZZO, D.M., AND DOUGHERTY, D.E., 1994, Characterization of aquifer properties using artificial neural networks—Neural kriging: Water Resources Research, v. 30, no. 2, p.483-497.

RUBIN, Y., AND DAGAN, G., 1988, Stochastic analysis of boundaries effects on head spatial variability in heterogeneous aquifers, part 1—Constant head boundary: Water Resources Research, v. 24, no. 10, p. 1689-1697.

SCHEIBE, T., 1993, Characterization of spatial structuring of natural porous media and its impacts on subsurface flow and transport: Unpub. Ph.D. Thesis, Stanford University, Stanford, Calif.

SLATT, R.M., JORDAN, D.W., D'AGOSTINO, A.E., AND GILLESPIE, R.H., 1992, Outcrop gamma-ray logging to improve understanding of subsurface well log correlations, *in* Hurst, A., Griffiths, C.M., and Worthington, P.F., eds., Geological Applications of Wireline Logs II: Geological Society of America Special Publication 65, p. 3-19.

STEPHENSON, D.A., CUTWRIGHT, B.L., AND WOESSNER, W.W., 1991, Hydrogeology—It is: GSA Today, v. 1, no. 5, p. 1, 94-95, 99.

WALKER, R.G., AND CANT, D.J., 1984, Sandy fluvial systems, *in* Walker, R.G., ed., Facies Models (2d ed.): Geological Association of Canada, Geoscience Canada, Reprint Series 1, p. 71-90.

WEBB, E.K., 1994, Simulating the three-dimensional distribution of sediment units in braided-stream deposits: Journal of Sedimentary Research, v. B64, no. 2, p. 219-231.

WEIMER, R.J., AND LAND, C.B., 1972, Lyons Formation (Permian), Jefferson County, Colorado—A fluvial deposit: The Mountain Geologist, v. 9, nos. 2 and 3, p. 289-297.

WEN, X., GOMEZ-HERNANDEZ, J.J., CAPILLA, J.E., AND SAHUQUILLO, A., 1996, Significance of conditioning to piezometric head data for predictions of mass transport in groundwater modeling: Mathematical Geology, v. 28, no. 7, p. 951-968.

WINGLE, W., MCKENNA, S.A., AND POETER, E.P., 1997, UNCERT, a geostatistical tool for evaluating, modeling, and designing remediation facilities: Unpub. computer software, available at http://uncert.mines.edu.

ZHENG, C., 1991, A modular, three-dimensional transport model, users manual: Rockville, Md., S.S. Papadopolous and Associates, Inc.

ZHU, H., AND JOURNEL, A.G., 1993, Formatting and integrating soft data—Stochastic imaging via the Markov-Bayes algorithm, *in* Soares, A., ed., Geostatistics Troia '92, Volume 1—Quantitative Geology and Geostatistics: Kluwer Academic Publishers, p. 1-12.

APPENDIX: BACKGROUND ON GEOSTATISTICAL SIMULATION

Soft data (data with non-negligible uncertainty, such as geophysical data, expert opinion, data used to infer a property of a material) are useful in improving the quality of subsurface simulation of hydrofacies. To illustrate their value, it is important to understand the fundamental aspects of the simulation process. Before undertaking a simulation, a spatial statistical analysis of the available data is conducted, often in the form of fitting a *semivariogram* to the field data. First, an experimental semivariogram is developed by sampling data at various spacings or lags (within a given tolerance), taking the difference of the values, squaring that difference, summing the squares and dividing by the number of pairs (twice the number of data points, n). In short, the semivariogram, $\gamma(h)$, is one-half the mean square error produced by assigning the value at one end of the lag to the other end:

$$\gamma(h) = \frac{\sum[z(x) - z(x + h)]^2}{2n}. \qquad (A1)$$

The function $\gamma(h)$ generally increases with increasing lag (Fig. A1). The *nugget* defines the y intercept, representing variance at essentially zero lag spacing. The *sill* describes the variance of the entire data set, and *range* indicates the spacing at which $\gamma(h)$ reaches the global variance. For random functions that are second-order stationary, the semivariogram and the covariance, Cov, are related by,

$$\gamma(h) = Cov(0) - Cov(h). \qquad (A2)$$

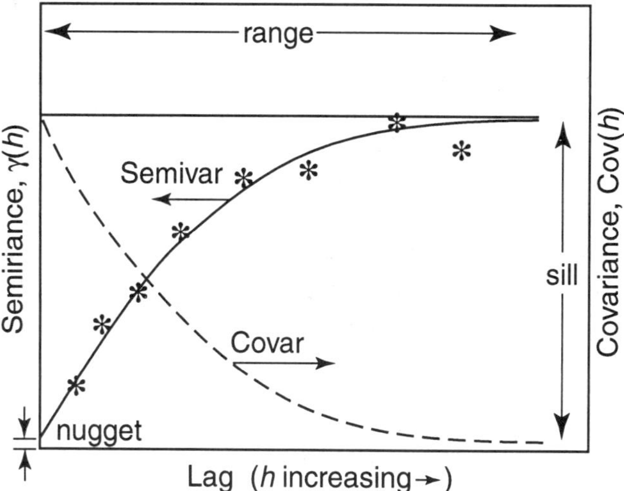

FIG. A1.—Generic example of a semivariogram and its relationship to covariance. Asterisks are the results of applying equation A1. Solid line is the semivariogram model, and the dashed line is the covariance model.

Once the experimental spatial relationship is determined, a mathematical model is fit to the data (Fig. A1). These spatial relationships are used to weight the measured values during the simulation process. When only hard data (typically from boreholes) are available, the experimental semivariogram often exhibits pure nugget effect (i.e., the nugget value equals the sill value even at short lags). In other words, all the lag distances are larger than the correlation range of the feature of interest because the borehole spacings are larger than the correlation range. Soft data are valuable in building the all-important rising limb of the semivariogram (discussed below).

Simulation requires defining a grid over the area of interest. A random path is followed through the grid, and the parameter of interest is simulated for each grid block by generating a random number and selecting the parameter value from a cumulative probability-density function (cdf) for that location. For grid blocks where hard data are available (i.e., the parameter being simulated is known), the block is assigned the known value. At other grid blocks, the cdf is determined using measured values within a search ellipse surrounding the block and the weighting functions discussed above. Generally, the cdf is influenced most by the nearest points and less by the more distant points, although the weighting function may vary with direction. Soft data are incorporated in the development of the cdf as described in the next section. Once a parameter value is simulated for a block, it is treated as hard data and incorporated in the development of the cdf of subsequently simulated grid blocks if it falls within the search ellipse of that block. By starting with different random number seeds, multiple realizations are generated that honor the measured data and the spatial statistics of the site, but individually each is unique.

We base our simulations on indicators (integer values) rather than parameter values. These indicators may be defined as a lithologic type or as a hydrofacies unit determined by statistical analysis of a combination of geophysical responses, cuttings or core observations and hydraulic test results. Alternatively, an indicator may be specified by threshold value for continuously varying parameters. For example, several indicators might be specified that include materials with the following ranges of conductivity (K):

Indicator 1: $K \leq 0.001$ cm/s
Indicator 2: $0.001 < K \leq 0.1$ cm/s
Indicator 3: $0.1 < K \leq 1.0$ cm/s
and so on.

The resulting subsurface descriptions are utilized to define alternative conceptual models (called realizations) for groundwater flow and transport modeling. However, the realizations typically include numerous unreasonable interpretations. Not only might they be visually unreasonable (i.e., the geologic configuration does not have realistic character) but many cannot be calibrated to yield reasonable hydraulic parameters as determined by field tests. These unreasonable realizations are either modified to improve the representation (Wen et al., 1996) or culled to examine only the reasonable alternatives (Poeter and McKenna, 1995; McKenna and Poeter, 1995). The acceptable models are then used to determine the probability of a response function value, for example, the uncertainty distribution of concentration at a point in space and time (Fig. A2).

REMEDIAL ALTERNATIVE 1

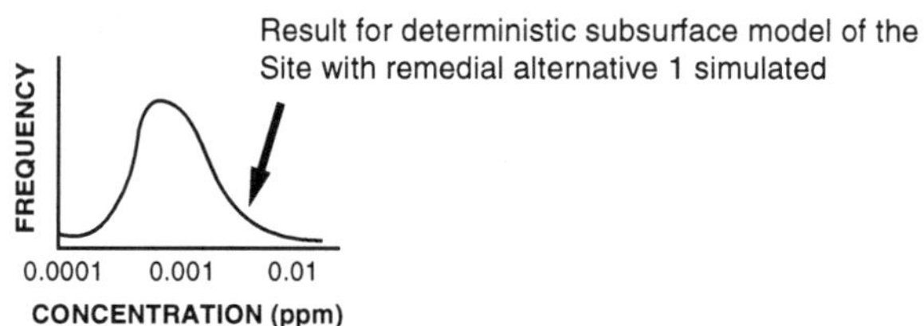

REMEDIAL ALTERNATIVE 2

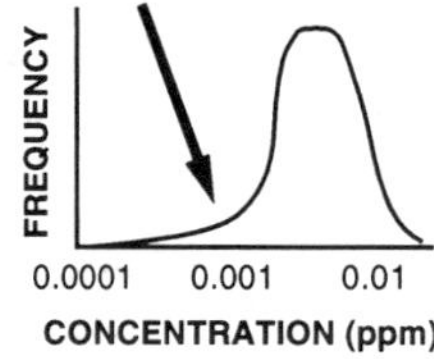

FIG. A2.—Evaluation of alternative remedial actions using a range of interpretations (as opposed to one deterministic interpretation) of the geology may result in selection of a different remedial action. Histograms compare the probability of a given concentration at a point in space and time using two remediation alternatives.

By incorporating soft information (e.g., expert opinion on site conditions; knowledge of the geologic history of the area, geophysical data) into the simulation process (Alabert, 1987; McKenna, 1994), the character of the realizations becomes more geologically realistic, spatial distributions of properties are more accurate and the realizations are better *conditioned*. Consequently, analysis effort is reduced, and results are more certain. Current research in our group involves geologic modeling to establish a framework that further constrains the realizations.

Sedimentologists can contribute to improving ground-water analysis, through acquisition of soft data.

Three types of soft data are useful (Alabert, 1987):

1. imprecise data for which a value is known with uncertainty and which are characterized through calibration at locations where both soft and hard data exist;

2. interval-bound data for which it is known that the value is greater than one value and less than another, but the character of the probability distribution between the bounds is unknown; and

3. prior distributions for which the probability distribution is determined from characterization of similar materials at another location, from knowledge of geologic history or from expert opinion.

Type 1 soft data include such items as lithologic facies determined from seismic velocity; estimates of porosity and permeability from sedimentological data; or concentration of uranium from a scintillometer. Type 2 soft data might include knowledge that a confining unit occurs within 100 to 150 m of the surface from previous exploration; or bracketing of the type of unit underlying an area based on surface observation. Type 3 soft data may be in the form of a distribution type (perhaps log normal) with a mean and variance of hydraulic conductivity based on numerous tests in a unit; or the percentage of each facies from geologic sections measured at an analogous site. The same data may be utilized as any of the soft-data types depending on the application of interest.

Only type 1, imprecise, data require much in the way of exploration because, when building semivariograms or conducting simulations, type 2 interval data are simply treated as hard data outside the interval (i.e., we know it is not that value) and as unknown within the interval. For type 3, cdf, data the prior distribution is honored directly. As mentioned, a calibration must be obtained for imprecise data. This is accomplished with data from a location at which both hard and soft data are available (collocated data). Hard data are plotted on one axis, soft data on the other, and thresholds (z_{c1}, z_{c2} and so on) are selected (Fig. 3A). Then two probabilities, p_1 and p_2, are calculated. The probability that the soft data indicate the location has a value below the threshold when the hard data indicate the value is below the threshold, p_1, $p_1 = A/(A + D)$, where A and D are defined in Figure A3. High-quality soft data have a p_1 near 1.0. The probability that the soft data predict the value is below the threshold when the hard data are above the threshold, p_2, $p_2 = B/(B + C)$. High-quality soft data have a p_2 near zero. These probabilities are not complimentary and, as such, do not sum to one. The difference, $p_1 - p_2$, is a measure of the quality of the soft data, a higher value indicating higher quality soft data. The generic illustration presented as Figure A3 exhibits $p_1 - p_2 = 0.7 - 0.2 = 0.5$, which is representative of low- to moderate-quality soft data. The quality of results from application of this approach depends on having a large amount of good quality soft data.

To build semivariograms with a combination of hard and imprecise data, three estimates of the true semivariogram gamma value are combined: (1) an estimate calculated from the hard data at lags where hard data are present, (2) an estimate from a hard-imprecise data cross-covariance, and (3) an estimate calculated from the imprecise data covariance (Alabert, 1987):

$$C_I^* = \omega_1 [C_I]_{N_H} + \omega_2 [\frac{C_{II'}}{p_1 - p_2}]_{N_H N_S} + \omega_3 [\frac{C_{I'}}{p_1 - p_2}]_{N_S}, \quad (A3)$$

where C_I^* is the estimated true hard-data covariance, C_I is the estimated hard-data covariance from the available sample of hard data, $C_{II'}$ is the hard-soft data covariance, and $C_{I'}$ is the soft-data covariance. The variables N_H and N_S refer to the number of hard and soft conditioning data in the search ellipse (N_{TOTAL} is $N_H + (N_H + N_S) + N_S$). Weights, ω, for the covariance equation are determined to account for the relative numbers of hard and soft data. One simple method for calculating ω values is:

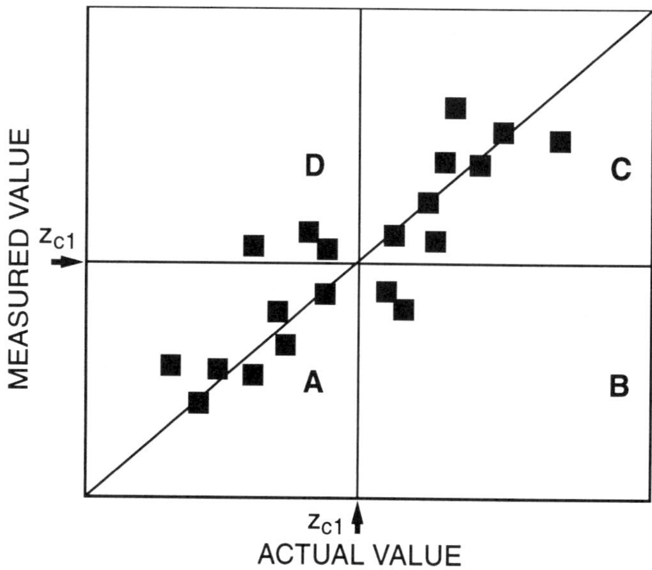

FIG. A3.—Calibration of imprecise soft data. "z" values indicate threshold cutoffs for calculating probabilities that indicate quality of the soft data.

$$\omega_1 = \frac{N_H}{N_{TOTAL}}, \qquad \omega_3 = \frac{N_S}{N_{TOTAL}},$$

$$\omega_2 = 1.0 - (\omega_1 + \omega_3). \tag{A4}$$

The hard-soft cross-covariances and the soft-data estimates are scaled by the quantities $(p_1 - p_2)$ and $(p_1 - p_2)^2$, respectively, to provide an estimate of the true hard-data gamma value at intermediate lag spacings (Alabert, 1987). The final estimate of spatial covariance is then converted to a semivariogram using the nonergodic relationship developed by Isaaks and Srivastava (1988).

We use geostatistical simulation software originally written by Gomez-Hernandez and Srivastava (1990) and modified to utilize soft data (McKenna, 1994) and to provide more flexibility in simulation (Wingle et al., 1997). The software is publicly available at http://uncert.mines.-edu/. In the modified algorithm, the cdf at each threshold is a linear combination of the hard data points within the search ellipse weighted by their respective kriging weights and the imprecise data points weighted by a spatial declustering weight or a kriging weight and scaled by the p_1 and p_2 values (Alabert, 1987):

$$cdf(z_k) = \omega_1 \sum_{\alpha=1}^{N_H} \lambda_\alpha i(x_\alpha, z_K) + \omega_2 \sum_{\beta=1}^{N_S} \upsilon_\beta \frac{i(x_\beta, z_K) - p_2(z_K)}{p_1(z_K) - p_2(z_K)},$$

$$\tag{A5}$$

where ω_1 and ω_2 are weights that account for the relative quantities of hard and imprecise data within the search ellipse ($\omega_1 = N_H/(N_H + N_S)$ and $\omega_2 = N_S/(N_H + N_S)$). The "$i$" terms are the indicator value (0 or 1) of each conditioning point at the current threshold. The kriging weights, λ_α and υ_β, are derived by solving the kriging matrix subject to $\Sigma\lambda_\alpha + \Sigma\upsilon_\beta = 1.0$ (McKenna, 1994). The simulator has several options for controlling the amount of hard and imprecise data found in the search ellipse and the type of kriging process (simple or ordinary) to be used.

Another approach is to assume that hard data alone can be used to sufficiently define the covariance, and the soft data need only be used to better constrain the simulations. Zhu and Journel (1993) accomplished this by assuming that hard data prevail over collocated soft data in conditioning simulations; and calculating calibration parameters after scatter-plotting collocated hard and soft data. The calibration parameters are used to scale the hard covariance, and the noncollocated soft data are used to condition the simulation. This approach is not often useful because the hard data semivariograms are often poor (nearly pure nugget, due to the hard data spacings being larger than the correlation length of the parameter of interest). Our primary reason for seeking soft data is to determine geostatistical parameters at the scale of interest.